Exploring Biology

Pamela S. Camp
Karen Arms

SAUNDERS COLLEGE PUBLISHING

Philadelphia   New York   Chicago
San Francisco   Montreal   Toronto
London   Sydney   Tokyo   Mexico City
Rio de Janeiro   Madrid

Address orders to: 383 Madison Avenue
New York, N.Y. 10017

Address editorial correspondence to:
West Washington Square
Philadelphia, Pa. 19105

This book was set in Melior by Hampton Graphics, Inc.
The art director and cover designer was Nancy E. J. Grossman.
The editors were Michael Brown, M. Lee Walters and Patrice Smith.
The production manager was Tom O'Connor.
The artwork was drawn by Vantage Art.
R. R. Donnelley & Sons Company was printer and binder.
Cover: Fruiting bodies developing on the slime mold *Arcyria ferruginea*. (Biophoto Associates,
    Dept. of Plant Sciences, Leeds University, Leeds LS29JT, U.K.)
Part openings:
    Part One (Penguins), U.S. Navy.
    Parts Two (Cross section of wood), Three (Fungal spores), Four (Chrysanthemums), and Five
    (Snow hare), Biophoto Associates.

**LIBRARY OF CONGRESS
CATALOG CARD NO.: 80-53911**

Camp, Pamela S., & Karen Arms
    Exploring biology.
    Philadelphia, Pa.: Saunders College Publishing
    496 p.
    8101   801010

EXPLORING BIOLOGY   ISBN   0-03-047701-8

1234   39   987654321

CBS COLLEGE PUBLISHING
Saunders College Publishing
The Dryden Press
Holt, Rinehart and Winston

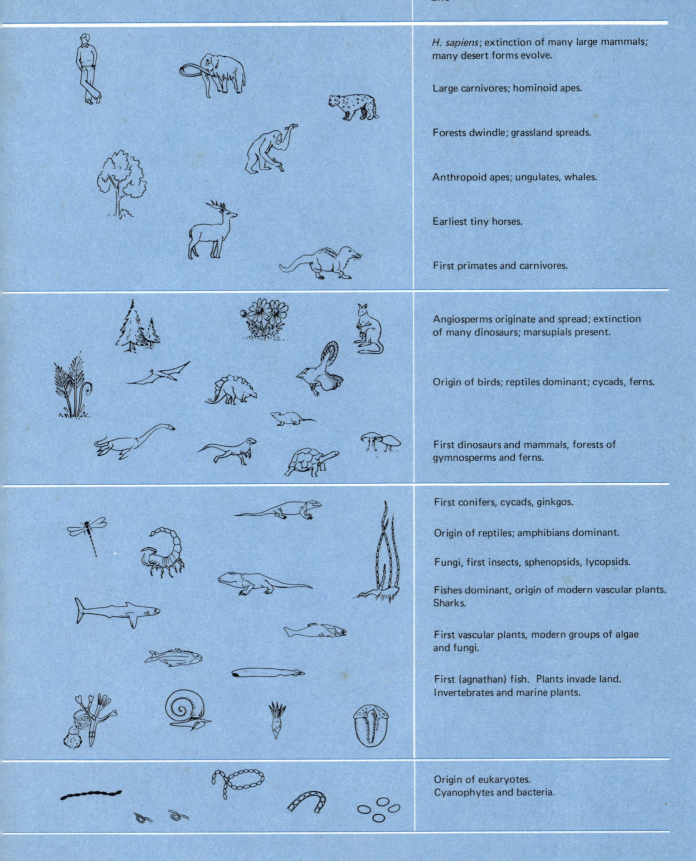

Life

H. sapiens; extinction of many large mammals; many desert forms evolve.

Large carnivores; hominoid apes.

Forests dwindle; grassland spreads.

Anthropoid apes; ungulates, whales.

Earliest tiny horses.

First primates and carnivores.

Angiosperms originate and spread; extinction of many dinosaurs; marsupials present.

Origin of birds; reptiles dominant; cycads, ferns.

First dinosaurs and mammals, forests of gymnosperms and ferns.

First conifers, cycads, ginkgos.

Origin of reptiles; amphibians dominant.

Fungi, first insects, sphenopsids, lycopsids.

Fishes dominant, origin of modern vascular plants. Sharks.

First vascular plants, modern groups of algae and fungi.

First (agnathan) fish. Plants invade land. Invertebrates and marine plants.

Origin of eukaryotes.
Cyanophytes and bacteria.

To Katherine and Nan—
our first audience—
with love

PREFACE

Exploring Biology is our answer to requests for a text suitable for a short course in biology. In writing this book, we realized it had to contain only the rock-bottom basics of biology, and trying to reduce biology to its essentials presented us with many hard questions. What should an educated citizen know about biology? How can biology best be presented to someone with little or no background in the subject? Thinking about such questions, we came to the conclusion that we must emphasize concepts, generalizations, and theories in biology and take a ruthless eraser to excessive jargon and technical terms.

Those familiar with *Biology* will see from the Table of Contents that the order of topics is very different in this book: after an introductory chapter which sums up some important general concepts, we continue with the diversity of organisms, evolution, and ecology. Many experienced biology teachers reviewed the contents and manuscript of this book and they were convinced that starting with chemistry and biochemistry, as we do in *Biology*, was a fast way to "turn-off" students in a short course on biology. So in this book we give students a feeling for some of the broader questions in biology and some acquaintance with the world around them before asking them to bend their minds to the biochemistry that many of them find intimidating at first.

It may be that the order in which the book is written does not in fact matter very much. The content of biology is thoroughly web-like: anything is easier to understand if you already know everything else, so the order in which it is taught is not very important and we, ourselves, have taught it in at least half a dozen different orders. For this reason, we make each chapter as independent as possible of the others. Another feature that allows flexibility is that the chapters are short. Each can be covered in one (or at the most two) lectures, modules, or reading sessions.

Since there may be more material in this book than is needed for some one-semester courses, the Instructor's Manual suggests which chapters should be chosen for shorter courses with a particular emphasis such as plant, animal, or human biology.

Each chapter has specific parts designed to assist the student. Objectives at the beginning of the chapter focus the students' attention on what they should be learning. They point out to the students those topics that are on a reasonably sound scientific basis, and the important general principles that students should grasp in each area. They also point out important words in the working vocabu-

v

lary of students of biology. We hope that instructors using the book will inform students whether they intend to follow the objectives in preparing examinations, and if not, that they will prepare their own list of objectives and distribute those to the class. The objectives can be easily followed in preparing examination questions; many objectives represent possible essay questions.

Introductions tie the chapter to other chapters and give students a glimpse of the material that will be covered in the chapter. Introductions frequently present terms that will be used often throughout the rest of the chapter.

A summary follows each chapter, bringing together the important points and reinforcing the objectives in pointing out the general importance of the material in the chapter.

In some chapters, essays relate interesting anecdotes or outline special topics related to the chapter.

Each chapter but the first has a self-quiz, based on some of the more important objectives of the chapter. The questions are written in an objective format, but the students should be sure to turn back to the list of learning objectives at the beginning of the chapter on successful completion of the self-quiz and use the list of objectives as an essay self-test. Answers to the self-quiz are provided in the back of the book.

Questions for discussion at the end of each chapter help students to extend the knowledge gained to new problems, or ask students to formulate opinions about a controversial topic, providing more ambitious students with extra food for thought.

The marginal notes serve two purposes. First, some of them give pronunciation guides to new words introduced in the text. We have tried to write these guides in such a way that they can be used without the customary symbols and keys found in a dictionary. Wherever possible, we have broken words down into shorter common words or into elements that rhyme with common words; this sometimes means that we had to move breaks between syllables in order to obtain a set of easily understood directions (a good dictionary should be consulted for proper syllabification). The second use of marginal notes is to redefine words introduced in a previous chapter, which the reader may have forgotten, or may not have covered yet if the chapters are assigned out of order. In some cases, we have provided definitions of ordinary words which just may not happen to be part of some students' vocabularies.

We use metric measurements throughout this book, because biology research is universally conducted in metric units. Those who are unfamiliar with certain metric units will find that they soon become familiar, and as when using foreign money, one finally stops mentally translating them into more familiar terms. Metric tables are included in the book's endpapers. At the request of the United States Metric Association, we use the -re spelling of units such as metre and litre, since that form is preferred in the International System.

We hope that this book presents biology as the lively and fascinating subject that we ourselves find it. If, many years from now, a banker or dancer reading of the latest advances in genetic engineering or cancer research finds that reading this book has made the subject more intelligible, we shall have achieved our ambition in writing it.

P.S.C.
K.A.

ACKNOWLEDGMENTS

With this, our second book, we were older and, with any luck, wiser, so we called on biology teachers for advice from the moment we first put together a tentative list of the book's contents. Our reviewers helped us with the order of the book, pointed out mistakes and ambiguities in the manuscript and suggested material that should be included or omitted. Most important, they read the manuscript with teachers' eyes, identifying areas that their students usually find difficult, suggesting ways to make particular subjects more comprehensible and advising us on the hundred and one details that are important to students, and so to teachers. Our thanks to Del Bennett, Florida Junior College; George Bleekman, American River College; Ailene Feldherr, Sante Fe Community College; Richard F. Firenze, Broome Community College; Douglas G. Fratianne, Ohio State University; Michael S. Gaines, University of Kansas; Mathilda L. Girardeau, Florida Junior College at Jacksonville; Robert T. Hersh, University of Kansas; David M. Hoppe, University of Minnesota; David W. Inouye, University of Maryland; Henry A. Levin, Kansas City Kansas Community College; Charles E. Martin, Rutgers University; Richard C. Millien, Grossmont College; J. Thomas Mullins, University of Florida; David C. Newton, Central Connecticut State College; Christina J. Myles, Manchester Community College; Jane Oram, Essex Community College; David Rayle, San Diego State University; Grace Rollason, University of Massachusetts; Robert C. Romans, Bowling Green State University; Donald Scales; Andrew T. Smith, Arizona State University; Andrew J. Snope, Essex Community College; Tom E. Wynn, North Carolina State University.

We are lucky enough to have found ourselves a publisher with whom it is stimulating, infuriating, amusing and a continuing pleasure to work. Our salutations and thanks to all at Saunders College Publishing and Holt, Rinehart and Winston and especially to Kendall Getman, Nancy Grossman, Patrice Smith, and Lee Walters who nursed this book from conception to parturition.

Various friends and colleagues have helped us in an assortment of ways. They read chapters, suggested things we should read, fed, housed, and transported us, translated our illegible scrawl into manuscript and contributed ideas, photographs, and moral support. Thank you Mary Ahl, Eileen Burnett, May Berenbaum, Clive Bransom, William Camp, Neil Campbell, Carolyn Eberhard, Paul Feeny, Jennifer Haarstick, Barbara Higgins, Gordon and Hazel Leedale, Jean McPheeters, Janice and Ian Skidmore, Stephen and Stephanie Sutton. Finally, our deepest gratitude is to each other.

CONTENTS

INTRODUCTION

When you have studied this chapter, you should be able to:

OBJECTIVES

1. Formulate hypotheses that might explain a natural phenomenon scientifically.

2. Design experiments to test a hypothesis.

3. List seven characteristics of living things and state why it is difficult to define life.

Biophoto Associates

1

organism: an individual living thing (e.g., a person, a bacterium, a plant).

Living things first appeared on earth some three billion years ago. Their descendants have diversified into the several million kinds of organisms (plants, animals, bacteria, and fungi) alive today. Throughout the ages people have been interested in the plants and animals that surround them. At first, this interest took the form of observation: people examined and named living things, classified and made lists of what lived where, and collected organisms as some people collect stamps. During this time, the study of living things was generally known as natural history, which is not a science; this was the forerunner of biology, which is a true science. The main thing that distinguishes a science, such as biology, chemistry, or physics, from the "humanities," such as art and literature, is the use of experiments to answer questions. Most of us first study biology because it is a science and because we feel that we should study at least one science if we are to be adequate members of a society in which science and its products are so important.

Science has assumed a position of enormous importance in modern society. Many decisions affecting our future depend on appropriate interpretations of scientific discoveries. Democratic government requires that everyone participate in decisions on such subjects as population control, pollution standards, protection of wildlife, and compulsory immunization. The body of scientific knowledge is already so vast that no one person can understand it all. As responsible citizens, however, we can follow some of the important studies that bear on public issues, and we can apply scientific reasoning to arrive at our own positions on these issues.

There is nothing mysterious about scientific reasoning or experiments. They are merely logical ways of trying to solve problems such as are used by business people, historians and each of us in our everyday lives. We do not need specialized scientific training or knowledge to decide whether conclusions are justified from the data presented. We can request further tests of a theory that does not appear to be well supported by the evidence, and we can agree or disagree with predictions from a theory. We can improve the way we do these things ourselves if we first understand how a scientist arrives at conclusions about natural phenomena through the same kinds of processes.

Fig. 1–1
A collection of sea shells. For thousands of years, people have expressed their interest in nature by collecting and naming living organisms. (Biophoto Associates)

1–A Scientific Method

You may never have thought about how you solve problems, test theories, or decide upon a plan of action. Let us consider how a biologist attacks a problem so that we can examine the main types of thinking involved.

Science usually starts with observations of the natural world and makes a generalization from those observations. For instance, if you are collecting insects, you may notice that many have black and yellow stripes. As you catch them, you probably think they are all bees or wasps and treat them with due caution. However, as you examine them more carefully, you may find that some have features clearly showing that they are flies rather than bees.

Is it merely coincidence that these flies look unlike their drab housefly cousins and resemble unrelated bees? Or do the striped flies gain some advantage from looking like bees? To answer this question, you must think of some ideas, or **hypotheses**, that will account for your observations. You may think of hypotheses such as "a fly's resemblance to bees protects it from predation," or "the flies fool bees into accepting them as members of the hive, and sneak in and steal honey."

hi-POTH-uh-sees (sing.: **hypothesis**)

The next step is to design and perform **experiments** to test the hypothesis. Because hypotheses usually cannot be tested directly, you must first develop a testable prediction from the hypothesis. Some hypotheses are of no use to science because they cannot be tested directly or even indirectly. For instance, the hypothesis "predators think flies are bees" is untestable because you can never know what an animal thinks.

But suppose you use your hypothesis to predict that a predator will not eat a fly that looks like a bee if it has first learned not to eat stinging bees. This prediction *can* be tested.

For this experiment, you need predators that eat insects—toads will do. Toads eat by catching insects flying or crawling nearby. If you put bees into the cage of a naïve toad (one not acquainted with bees), it will catch a few, learn that they sting, and refuse to catch any more. You next put a black-and-yellow striped fly into the cage and see if the toad also refuses the fly. If so, the hypothesis that the fly's resemblance to bees protects it from predation is supported. You should check, though, to be sure the toad is still hungry by offering it a harmless housefly for dessert. But perhaps the bee has nothing to do with it; maybe toads just do not eat striped flies. To test this, you must use a second naïve toad. If this toad cheerfully devours black-and-yellow striped flies, you have gained additional support for the hypothesis that the fly's striped suit is advantageous because it resembles the bee's.

A valid experiment always includes a **control treatment**, such as the second toad above, as well as an **experimental treatment**, in which one (*and only one*) factor is varied—in this case, the opportunity to eat bees. If you had not used a control treatment, the toads might have refused the flies because of some factor that you did not notice, such as not being hungry, and you might have concluded, wrongly, that the resemblance to the bees was responsible.

It is disappointing to realize that a hypothesis can never formally be proved,

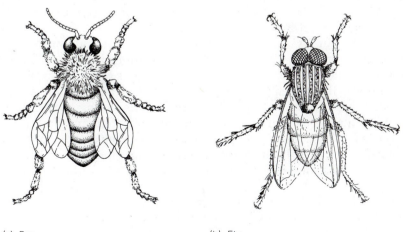

(a) Bee   (b) Fly

Fig. 1–2

Resemblance of a mimic fly to a bee is good enough to fool some of the people some of the time. However, there are differences: the fly has large eyes and stubby, club-shaped antennae. The bee's eyes are smaller and her antennae more slender. A bee also has two pairs of wings to a fly's one pair, but this may be hard to see.

Fig. 1–3

A bee collecting pollen. Bees, and many of their relatives, have distinctive stripes, making it easy for a predator to recognize them and avoid the bee's sting. (Biophoto Associates)

but can only be disproved. You can never "prove" that the flies' stripes protect them from predation; you can only disprove the hypotheses that you think of as alternative explanations for the resemblance and show that the coloration always seems to be effective in discouraging predators of a wide range of species.

A hypothesis supported by many different lines of evidence from repeated experiments is promoted to the status of a **theory** and eventually comes to be regarded as a scientific "fact" or "law."

1–B Fact and Fiction

"It's a scientific fact" is often presented as the clincher to an argument. Most scientists, however, would argue that there is no such thing as a fact. The doubts and uncertainties inherent in the scientific method make a mockery of the belief that, if it is scientific, it must be right.

"Facts" are usually thought of as things or events that are repeated in identical fashion or about which we have unambiguous records. "The sun rises every morning" looks like a fact on first glance, but it is really a prediction about something that will happen in the future, based on what has happened in the past. Another reason that "facts" are less sure than they seem is that they depend on our faith in our senses. A group of people may agree that a particular object is a table. Is that a fact? No, it is a statement resulting from an agreement or convention: all have agreed to call that sort of object a table. Suppose several people look at two photographs, one of a table, the other of an object floating in a lake. Everyone may agree, if the first photograph is a clear one, that the object shown is a table. When they look at the second photograph, however, one of them may say "That is a Loch Ness monster," but the others may legitimately disagree. When technology, in the form of a camera, microscope, oscilloscope, or whatever,

Fig. 1–4

When is a fact not a fact? Nineteenth-century doctors were taught that men and women breathed differently: men used their diaphragms (the sheet of muscle below the rib cage) to expand their chests, whereas women raised the ribs near the top of the chest. Finally, a woman doctor found that women breathed in this way because their clothes were so fashionably tight that the diaphragm could not move far enough to admit air into the lungs. Some fashionable women, like the one in this drawing of 1870 styles, even had their lower ribs removed surgically so that they could lace their waists more tightly.

intervenes between our senses and an object, as it often must in scientific research, the problem of interpreting what we see, hear, or feel becomes even more subject to doubt. Thus, a "fact" is really a piece of information that we, for present purposes, choose to call a fact, probably because we believe in it strongly, because it is useful to us in some way, or because it seems highly likely that it will be repeated without change.

Although scientific facts, laws and truths are much less reliable than is generally believed, most scientists do think that their methods discover useful information about objects and events, and that careful study increases the probability that science's generalizations about nature are a close approximation to "reality." Although scientists are motivated by many different goals and ambi-

Fig. 1–5

We may perceive the same thing differently at different times. If you stare at the star in the middle of the colors for about a minute, cover the colors and then stare at the star in the gray area, you should see colors that are complementary to those in the original. The red, green, yellow and blue patches will be replaced by green, red, blue and yellow patches, respectively.

tions, public support for science rests on the belief that a better understanding of natural phenomena increases our ability to promote human well-being.

Much public support goes to scientific projects on problems of immediate concern such as cancer research or alternative energy sources. However, there is still a great deal of basic or "pure" research to be done to discover the underlying principles of why objects and organisms behave as they do. Although such research may not benefit mankind immediately, it adds to our understanding of the world and will almost inevitably be put to use sooner or later. And even work that does not find an application may be as intellectually satisfying as the painting of a fine picture or the writing of a good play. It is interesting that many people will accept "art for art's sake" but insist that art's stepsister, science, must work for her keep.

1–C What Is Life?

Although biology is defined as the study of living things, it is notoriously difficult to define life. Living organisms have a number of characteristics not found in most nonliving systems, yet none of these is unique to life. With this in mind, we can describe life by listing the main characteristics of living organisms:

1. *Living things are highly ordered.* The chemicals that make up living organisms are much more complex and highly organized than are the chemicals that make up most nonliving systems.

2. *Living things take energy from their environments and use it to maintain and increase their high degree of orderliness.* Most organisms depend, directly or indirectly, on energy from the sun. Green plants use this energy to make food which supports virtually all the organisms on earth. All organisms use energy from their food to maintain their bodies, to grow, and to reproduce.

Fig. 1–6 (below, left)

Living things are highly ordered. This is an organ pipe coral. (Biophoto Associates)

Fig. 1–7 (below, right)

Living things take in energy, like this buffalo feeding in a meadow. The bird on its back takes in energy by eating the parasites on the buffalo's skin.

Fig. 1–8

Living things respond actively to their environments. This swan is chasing away a duck that has come too close.

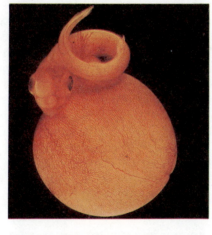

Fig. 1–10

Living things develop. This dogfish embryo, lying on top of a yolk sac from which it draws nourishment, will develop into one of the smaller sharks. (Biophoto Associates)

Fig. 1–9

Living things are adapted to their environments. These orchids do not root in soil but get their water from the very humid air and dripping tree trunks of the forests in which they live. This adaptation makes them frustratingly difficult to grow as house plants. (Biophoto Associates)

3. *Living things respond actively to their environments.* When you push a stone it may move passively. When you push an animal it usually responds actively, by running away, by moving toward you, or by rolling up into a ball. Plants respond more slowly but no less actively: the stem of a plant bends toward light, a root grows downward, and leaves fold or droop during the night. The capacity to respond to outside stimuli is universal among living things.

4. *Living things are adapted to their environments.* Living things and their components are well adapted to their ways of life. Fish, earthworms, and frogs are all constructed in such a way that we can predict roughly how they live merely by examining them.

5. *Living things develop.* Everything changes with time, but living organisms change in particularly complex ways that we call development. A chemical crystal may grow by the addition of identical or similar units, but a plant or animal will develop new branches or organs that may be chemically and structurally different from the chemicals and structures that produced it.

6. *Living things reproduce themselves.* New organisms—bacteria, animals, plants, and fungi—arise only from the reproduction of other, similar, organisms.

7. *The information each organism needs to survive, develop, and reproduce is segregated within the organism and passed from each organism to its offspring.* The only information in a rock is the whole rock. Living things, on the other hand, contain a separate information store, their **genetic material**, which specifies the range of the organism's structure and activity and which is passed from one generation to the next.

1–D The Limitations of Science

Science is only one way of exploring the world around us. Historians try to understand the past (and sometimes to predict the future) by studying what people have done in the past. Religion attempts to explain truths about the human spirit, and philosophy collects information from many sources to draw conclusions about reality and human life.

Because it deals only with things that can be experienced directly or indirectly through the senses (sight and hearing, for example), science is completely excluded from phenomena that cannot be experienced in this way. Thus, by definition, science has nothing to say about the supernatural. As the biologist George Gaylord Simpson put it: "This is not to say that science necessarily denies the existence of immaterial or supernatural relationships, but only that, whether or not they exist, they are not the business of science."

That science does not deal with the supernatural or with the illogical does not mean that scientists are any less emotional, political, or illogical than anyone else. Furthermore, the social environment strongly influences how scientists think and what projects they work on. Not many scientists today study the physics of how to build pyramids to last for a thousand years, because we do not

Fig. 1–11

Living things reproduce themselves. Here the yellow seeds, which will germinate to produce new individuals, are bursting from the fruits of a spindle tree. (Biophoto Associates)

Fig. 1–12

Science often becomes embroiled in political controversy. We have awakened to the fascinating lives and intelligence of whales at a time when many whale species are in danger of extinction because so many of them have been caught and turned into dog-food by modern whaling ships like this one. (Biophoto Associates)

want to build pyramids for the burial of our rulers. Similarly, politics rules science in that funds from government agencies provide most of the financial support for scientific research.

The history of science is replete with scientific dogmas that turned out to be wrong, although for a time they were widely accepted by other scientists. This is one reason why the cautious person—or society—will not place too much faith or invest heavily in a new scientific discovery until it is fairly clear that the theory will stand the test of time. On the other hand, the mistakes of science can be very valuable. Some of the most important discoveries have been produced by obstinate scientists determined to prove a colleague wrong or to back a hunch.

Scientists are fond of saying that science is never good or bad; only society's use of science has moral consequences. From a purist's point of view, this is true. The discovery that the atom could be split was merely a scientific discovery with no moral implications. It was society's decision to use this knowledge to build an atom bomb that produced the moral dilemma of whether or not it was ever right to use such a devastating weapon.

Despite the traditionally ostrich-like approach of scientists to the moral implications of their work, more and more scientists now feel that they must become involved in society's moral decisions about science, if only to make sure that the people responsible for the decisions are basing them on valid information. Some scientists even go so far as to say that certain sorts of research should not be done until society has worked out its moral position on the consequences. For instance, research on producing human babies outside the womb was banned in the United States during the 1970s. This was because it was unclear whether or not a researcher might be held legally responsible for human life produced in this way. Some scientists now feel that they must take, and that society may force them to take, more moral responsibility for the consequences of their research. This attitude, carried to extremes, can destroy science. The

Western world experienced about 500 years in which very few scientific discoveries were made because particular sorts of research or findings violated religious teachings. Few of us want to return to the Dark Ages.

There is no simple solution to this dilemma. The peaceful coexistence of science with society depends on citizens who understand what science is and what it can and cannot do, and do not confuse scientific with moral, economic, or political values.

SUMMARY

Experimentation is science's most characteristic procedure but there is nothing unique about scientific method; it is merely a collection of means for solving problems about the natural world. First, observations are made. Alternative hypotheses that might explain the observations are then formulated, and these hypotheses are tested by experiments designed to disprove one or more of the hypotheses, strengthening the evidence for those that remain. Scientific facts and theories are useful, but they are always open to question.

Biology is the science that studies living things. Life is difficult to define, but living objects have a collection of characteristics which, taken together, are unique. Living things take energy from their environments and use it to maintain a highly complex order; they respond actively to stimuli, are adapted to their environments, and contain all the information that they need to develop, survive, and reproduce in their environments.

QUESTIONS FOR DISCUSSION

1. How would you test the hypothesis that a fly's black and yellow stripes allow it to enter a beehive and steal honey?

2. After every hard rain you find dead earthworms lying on the sidewalk. What experiments would you perform to show the cause of death?

3. Each characteristic of life can be found in some non-living thing. Can you think of examples of these?

4. To what extent do you think scientists should be held responsible for the social and moral consequences of their discoveries?

DIVERSITY OF LIFE  22

When you have studied this chapter you should be able to:

1. Use the following words correctly: cell, photosynthesis, plankton, alga, invertebrate, vertebrate.

2. State some advantages of being one-celled and of being multicellular.

3. List or recognize characteristics of the following groups: sponges, cnidarians, segmented worms, molluscs, arthropods, echinoderms, mosses and liverworts, ferns, conifers, flowering plants.

4. List and explain the difficulties of life on land for an organism that evolved in the water, and give examples of adaptations in plants, insects, and other invertebrates that allow them to live on land.

5. List the two groups of fish and the four groups of land-adapted vertebrates, give or recognize examples of members of each group, and compare members of these groups with respect to body structure and reproduction.

6. Discuss why insects are such a successful group of animals.

7. Explain how flowering plants reflect the fact that they evolved in a world containing abundant land animals.

8. List economic uses of bacteria, protists, algae, fungi, and flowering plants.

Biophoto Associates

11

Everywhere we look around us we see life in incredible variety. The pond teems with plants so small that it takes a microscope to see them; the forest cannot be seen for trees so large that their tops touch the sky. All around us animals crawl, fly, burrow, gallop, wriggle, or "freeze" into motionless statues.

organism: an individual living thing (e.g., a cow, a bacterium, a bush)

How can we make sense of this bewildering array of life? Early naturalists classified organisms by structure, placing forms that resembled one another together—trees with similar leaves and bark, for instance. With the general acceptance of the theory of evolution, however, a different method was adopted. All the organisms on earth must have evolved by descent and modification from one, or at most a few, early forms of life. What relationship among living things could be more natural than kinship? Today, wherever possible, evolutionary relationships are used to classify organisms. Because we have little direct evidence of what life in the past was like, there will always be arguments about evolutionary relationships and about the best way to classify organisms. In this book, we divide organisms into five major kingdoms: animals, plants, **Monera** (bacteria), **Protista** (one-celled organisms), and fungi; we shall consider all five groups in this chapter. Viruses are not considered living organisms but are discussed in Chapter 14.

The first organisms were probably tiny, relatively simple, creatures living in the sea. Organisms of greater size and complexity evolved later. So we start with a visit to the sea.

2–A Straining the Sea

PLANK-ton

If we tow a net of very fine mesh through the sea, especially at night, it will trap **plankton,** floating organisms so small that we can see them only by using a microscope.

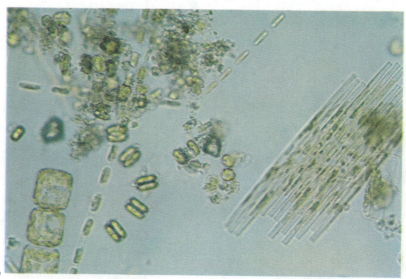

Fig. 2–1

Plankton as seen through a microscope. (Biophoto Associates)

Fig. 2–2

A diatom. This photograph was taken with a scanning electron microscope that magnified the diatom 2,500 times. (Biophoto Associates)

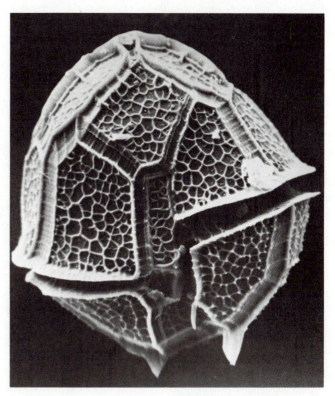

Fig. 2–3

A scanning electron micrograph of a dinoflagellate taken at a magnification of more than 2,000 times. (Biophoto Associates)

Many planktonic organisms are protists, consisting of only one cell each. A **cell** is the basic unit of life—a small, highly organized bit of living substance which can take in food, use it for energy and for growth, and eventually divide into two cells, each essentially identical to the parent cell. **Unicellular,** or one-celled, organisms, such as bacteria and protists, are usually tiny; the large, familiar organisms around us are **multicellular,** consisting of many cells—there are trillions of cells in our own bodies.

As we examine plankton with a microscope, we notice that many of them are green or yellow; these are algae. **Algae** are unicellular protists or simple-bodied multicellular plants which are **photosynthetic,** meaning that they use the energy of sunlight, and carbon dioxide dissolved in the water, to make their own food. These delicate, elegant creatures include the **diatoms,** protists surrounded by cell walls impregnated with silica, the material that makes up sand and glass. Since silica is not biodegradable, the walls of dead diatoms do not decompose, but sink to the sea floor. Huge deposits of diatom walls that were later heaved up by movements of the earth's crust are now mined as diatomaceous earth, used as a fine abrasive in tooth pastes and polishes, and as a packing material in air and water filters.

Almost as abundant as the diatoms are the protistan **dinoflagellates,** so named because each cell has two **flagella,** or whiplike appendages, that whirl about and propel it through the water. Some are covered with "armor plating,"

AL-jee (sing.: **alga;** AL-guh)

DIE-uh-toms

DIE-no-FLAJ-ell-ate
fluh-JELL-uh (sing.: **flagellum**)

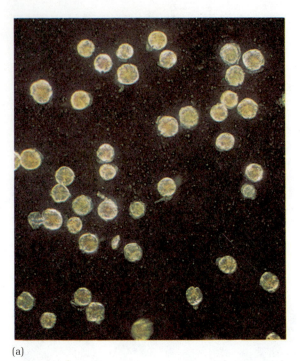

(a)

(a) Gonyaulax. (Biophoto Associates) (b) An oil rig mining petroleum produced from planktonic organisms that lived millions of years ago. (Biophoto Associates)

(b)

while others are naked. Several species are **bioluminescent,** giving off flashes of light. Alister Hardy described an encounter with these organisms in the English Channel:

> I looked over the side to see a small shoal of fish, most likely mackerel, lit up by each individual being covered by a coat of fire; they were being chased this way and that by some much larger fish similarly aflame. On putting over a tow-net, which came up brilliantly illuminated, the sea was seen to be full of very small . . . dinoflagellates of the genus *Goniaulax.*

Gonyaulax (as it is now spelled) has another claim to fame. Some forms produce a nerve poison which may be lethal to humans. During *Gonyaulax* blooms (population explosions), shellfish which have eaten large quantities of this dinoflagellate become unfit for human consumption. Since *Gonyaulax* contains a red pigment, the blooms may color the water red, a fitting warning signal commonly known as "red tide."

pigment: a colored substance.

Both diatoms and dinoflagellates contributed largely to the deposits of petroleum we use at such prodigious rates today. These cells store excess food as oil, and over the years this oil accumulated and underwent chemical changes as the bodies of dead plankton drifted to the sea floor and were later pressed down by the weight of mud and silt settling over them.

Our plankton net has collected a great many protists that are not photosynthetic but live by eating other organisms. Some have one or more flagella; some are covered with shorter projections called **cilia,** which beat like the oars of a galley and move the cell through the water. Still others have no permanent

SILL-ee-uh (sing.: **cilium**)

locomotory structures, but push out part of the cell contents to form temporary **pseudopods** (false feet) (Fig. 2-5). Many make shells of protein, silica, or calcium carbonate (lime). These shells may accumulate in great numbers; the famous White Cliffs of Dover are chalky deposits of plankton shells more than 50 metres thick.

SUE-doe-pods

Some members of the plankton are multicellular. These include the young stages, or **larvae** (sing.: **larva**), of sea animals which are immobile as adults; the larvae, floating in the sea, spread the species by dispersing to wherever the water takes them. The plankton also includes small, actively swimming adult animals of many kinds.

LAR-vee (sing.: larva, LAR-vuh)

2–B The Sunken Forest

When the tide goes out it leaves pools and fingers of water among the coastal rocks. Here live some of the larger multicellular algae, members of the plant kingdom, spreading over the rocks between the tidemarks in a carpet of vivid greens, velvety reds, and inky browns.

These plants need light, but water absorbs light, and the ocean depths are perpetually dark. Thus, most large algae live near the shore, especially where they can attach to sunlit rocks. The forms living between tide lines are left high and dry for part of every day, and many have a slippery gelatinous covering that

Fig. 2–5

(a) Paramecium, a protist covered with cilia, which are visible as a fringe around the body. The pale areas in the cell are vacuoles, sacs collecting water to be excreted. (Biophoto Associates) (b) Amoeba villosa, a protist that uses pseudopods to move and to catch its food. (Biophoto Associates)

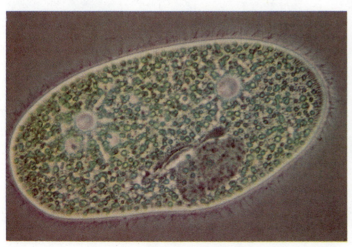

(a)

(b)

slows evaporation and so prevents their drying out. The largest and most complex algae, called kelps, may have air bladders that float the photosynthetic blades in the water and rootlike holdfasts that attach the plant to the rocks (Fig. 2-6).

Multicellular algae provide food for many animals. Algae are nutritious because they accumulate nutrients such as nitrogen, potassium, and iodine in great quantities; they have long been used for both human and livestock food. Alginate, an extract from kelps, finds its way into about half the ice cream produced in the United States, imparting a smooth texture and helping to prevent the formation of ice crystals. The alga known as Irish moss produces carrageenan, an ingredient in puddings, candies, and ice cream. Agar, extracted from another alga, is famous as the basis of nutrient media used to grow bacteria and fungi in the laboratory.

2–C One Cell or Many?

We tend to look upon increases in size and complexity as "progress" and to despise yesteryear's smallness and simplicity. But unicellular organisms do have some advantages over larger creatures. A single cell, for instance, can live in a tiny space and needs only a little food before it is ready to reproduce.

In many situations, however, it is advantageous to be large. When all organisms were small and unicellular, any organism that grew large enough to eat its neighbors could make an excellent living. A single cell cannot just become larger and larger; eventually the center of the cell is too far from the outside to obtain the substances it needs from its environment fast enough. Large organisms are invariably made up of many small cells, and these cells show

Fig. 2–6

Fucus, *a multicellular alga. The rootlike holdfast anchors the body to the rocks, and the flattened blades carry on photosynthesis. (Biophoto Associates)*

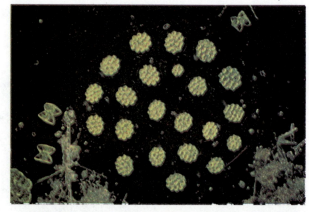

Fig. 2–7

Some protists, like this one, live in colonies with some division of labor between the cells. The black sphere is a colony covered with yellow clumps of reproductive cells which are producing new colonies. The butterfly-like objects are unicellular algae in the water. (Biophoto Associates)

TABLE 2-1 **BASIC FUNCTIONS THAT MUST BE CARRIED OUT BY EVERY ORGANISM**

Feeding or making food	Dispersal (locomotion, scattering seeds or larvae)
Gas exchange	Support and protection
Waste removal	Coordination of all functions (nerves, hormones, etc.)
Internal transport of food, gas, etc.	Reproduction
Sensing environmental stimuli	

Usually each cell can carry out its own basic functions, but in multicellular organisms each cell is also specialized to help carry out one of these functions for the entire body.

division of labor (Table 2-1). Cells of the larger algae, for example, may be specialized as anchoring holdfast cells, as photosynthetic cells producing food, as parts of air bladders, or as conduits for food to pass down to the holdfast; some also specialize in reproduction.

As we turn now to the animals of the sea, we shall find that the simplest are small, with no cell very far from the watery environment that provides food and removes body wastes. As animals evolved, many became larger, and the inner cells were further and further from the environment. Animals that evolved means of circulating food and oxygen to these cells, and of removing wastes to the outside, were able to increase in size, whereas those without such systems had to remain small.

As we examine various animals, we shall point out some of the advances in body structure and complexity that each group showed over its predecessors. The members of any group of organisms show adaptations to a particular way of life. Fish, for instance, are generally adapted to life in the water, birds to life in the air. Some members of every group, however, have undergone **adaptive radiation,** adaptation that suits them to ways of life different from those of their ancestors. Thus the ancestors of unicellular organisms undoubtedly lived in the sea (the evidence for this belief is partly that most of them still live in the sea), but other members of the group have evolved adaptations to life in fresh water and even in damp places on land. We shall find adaptive radiation within every group that we examine.

2–D Sea Creatures Without Backbones

Animals without backbones are called **invertebrates.** The many groups of invertebrates cover a wide range of size and complexity.

Sponges. Sponges, the simplest of the multicellular animals, are so immobile that the ancient Greeks thought they were plants. But, unlike most plants, they have flagella on some of their body cells and produce tiny, mobile larvae. Sponges use their flagella to draw water in through pores in the body wall; food particles in the water, including small organisms, are trapped and digested by cells inside the sponges.

The "natural" sponges sometimes used for bathing are actually sponge

Fig. 2–8

A sponge. Water enters through pores all over the sur-face and leaves the animal via the hole in the center. (Steven Webster)

Fig. 2–9

A cnidarian–a sea anemone. Note the tentacles around the mouth and the soft, muscular body that is anchored to a rock in this particular group. (Bio-photo Associates)

skeletons, with the living cells now dried and decayed. Other sponges have hard, brittle skeletons of lime or silica.

Cnidaria: nye-DARE-ee-uh (the "C" is silent).

Cnidaria. Swimming in the sea, particularly in warmer areas of the world, carries with it the hazard of jellyfish stings. Jellyfish, sea anemones, and corals are common members of the Cnidaria, a group of simple animals with the same basic shape (Fig. 2-10). All cnidarians are carnivores, and the tentacles around the mouth are armed with small organs that sting prey with a paralytic poison or entangle it with sticky threads.

Worms. The world is full of worms. We find flatworms crawling along under stones in a tide pool or swimming in a pond, and roundworms and earthworms in garden soil. The worms differ from cnidarians in having definite left and right sides, top and bottom, and head and tail ends. This permits a streamlined shape and efficient locomotion. In addition, the nerve cells and sense organs (organs that respond to stimuli such as light, chemicals and sound) concentrate in the head end, allowing the animal to sample an area before it moves its body there.

On a muddy seashore you may find a fisherman digging frantically for a lugworm, which burrows through the mud with astonishing speed. It can do this because it is made up of numerous similar segments, each filled with fluid and sealed off from adjoining segments by partitions (Fig. 2-11). The segmented worms also include the land-dwelling earthworms and the leeches, found mainly in fresh water.

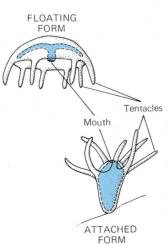

FLOATING
FORM

Tentacles

Mouth

ATTACHED
FORM

Fig. 2–10

The two body forms found in cnidarians are essen-tially the same, except that one floats in the sea and the other lives attached to a rock. Both have tentacles surrounding the mouth. The blind diges-tive cavity (color) doubles as a circulatory system.

The segmented worms are the first animals we have encountered with circulatory systems, with pumping "hearts," which are really the largest of the blood vessels. A circulatory system was made possible by the evolution of the fluid-filled space, the **coelom,** between the digestive tract and the muscular body wall. The coelom allows the internal organs to move separately from each other and from the body wall.

SEE-loam

Molluscs. The seashore abounds with molluscs—chitons, snails, limpets, scallops, clams, mussels, and their relatives. A mollusc has a soft body, with a muscular foot used in locomotion. Most also have shells, which act as protective external skeletons. In clams, oysters, and mussels the shell is double and hinged, while some molluscs—such as the slug and the octopus—have none. Squids are molluscs with an internal rod of cuttlebone instead of an outer shell.

chiton (KITE-on): a mollusc with a flat oval shape and a shell made up of eight cross-wise plates (resembling an armadillo).

The mollusc shell is an important evolutionary advance. Some worms and some cnidarians, notably corals, build hard, protective tubes around themselves, but these tubes are fixed in one place. Some molluscs also cement their shells to one spot, but most of them can move about. Mussels living in the tidal zone, and snails and limpets grazing on the nearby rocks, can close their shells or pull the openings down to the rocks when the tide goes out; this prevents drying out until the tide comes back in.

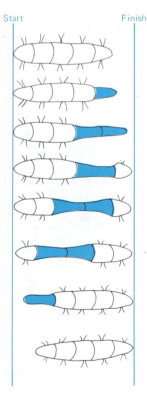

Start   Finish

Fig. 2–11

A segmented worm moving. Each segment acts as a closed sac of fluid surrounded by two sets of muscles that can contract alternately so as to shorten and lengthen the segment. When the segment is short and thick, its bristles push out against the substrate and hold the worm in place. Each segment in turn is squeezed so that it becomes long and thin and extends forward (color). At the end of the sequence, the worm has moved forward.

(a)

Fig. 2–12

Molluscs. (a) A land snail crawls across a leaf. (b) This bivalved mollusc spends its life swimming in the sea, using the two flaps of its mantle and the foot (hanging beneath it) as paddles. (Biophoto Associates)

(b)

marine: of the ocean (salt water).

Arthropods. Crabs, shrimp, and lobsters are familiar marine arthropods—animals with jointed external skeletons (arthro = joint, pod = foot). An arthropod's armor plating, reminiscent of a knight's armor, is fitted together so that it bends at the joints and allows its owner remarkable flexibility.

Originally, biologists speculate, the arthropod body was much like that of a segmented worm, with one pair of jointed appendages on each segment. During the course of evolution, some appendages became claws or walking legs, some became paddles like those of a lobster's tail, some became antennae, and others became mouthparts. Today, arthropods are the most successful of all animal groups, with the greatest number of individuals and of different species.

Echinoderms. Sea urchins look like brittle pincushions, and must be turned over gingerly, for they are well-protected by sharp spines that can pierce soft fingers and then break off, leaving splinters to fester. On the underside is the sea urchin's mouth, with large, tough teeth for scraping algae off the rocks.

Sea urchins, sea stars, brittle stars, and sea cucumbers are all echinoderms (echinos = hedgehog; derma = skin). Most have curious tube feet with suction-cup tips, used for moving around and, in the case of sea stars, for prying open mussels and clams for food. Most are also **radially symmetrical,** with arms and

Fig. 2–13

An arthropod. This lobster shows the typical features of the group, including an external skeleton, and jointed legs. (Biophoto Associates)

Fig. 2–14

A sea urchin, showing the spiny skeleton and delicate tube feet characteristic of echinoderms. (Biophoto Associates)

TABLE 2-2 THE MAIN VERTEBRATE GROUPS

Cartilaginous fish	Fish with cartilaginous skeletons; tail fin usually asymmetrical; gill openings separate. Sharks, skates, and rays.
Bony fish	Fish with bony skeletons; gill openings with a single cover; tail fin usually symmetrical; many have a swimbladder; marine and freshwater. For example, herring, salmon, sturgeon, eels, and sea horses.
Amphibians	Four-legged vertebrates which lay eggs without a shell; respiration via lungs and skin; scales absent. Salamanders, newts, frogs, and toads.
Reptiles	Vertebrates with shelled eggs and scaly skin. Snakes, lizards, turtles, tortoises, crocodiles, and alligators.
Birds	Vertebrates with feathers; high body temperature; most species have more than one mode of locomotion; forelimbs usually modified to form wings. For example, sparrows, penguins, and ostriches.
Mammals	Vertebrates with young nourished by milk from mammary glands of females; usually gestated in female's body before birth; most covered with hair. For example, monkeys, whales, cattle, rodents (see Table 2-3).

other body structures arranged in five parts, or in multiples of five, around a central disc. Despite the fact that echinoderms are sluggish and radially symmetrical, study of their embryonic development shows that they are nearer relatives of **vertebrates** (animals with backbones, including humans) than any other animal group we have yet discussed!

2–E Vertebrates of the Sea

The vertebrates, animals with backbones, are represented in the ocean mainly by fish. The axis of the vertebrate body is the backbone—a long chain of vertebrae with projections to which muscles attach. Their many muscles, and the flexibility of the vertebral column, permit fish to swim by bending the body from

Fig. 2–15

A sea cucumber, an echinoderm. The projections all over the body are tube feet, which echinoderms use to grasp objects and to move around. The mouth and feeding tentacles are at the right. (Biophoto Associates)

Fig. 2–16

The evolutionary family tree of animals according to one widely held theory. The fossil record provides little evidence as to these relationships because there are few early invertebrate fossils. The main evidence for the theory portrayed here comes from the structure and embryonic development of modern animals. Closely related animals generally have similar embryonic development.

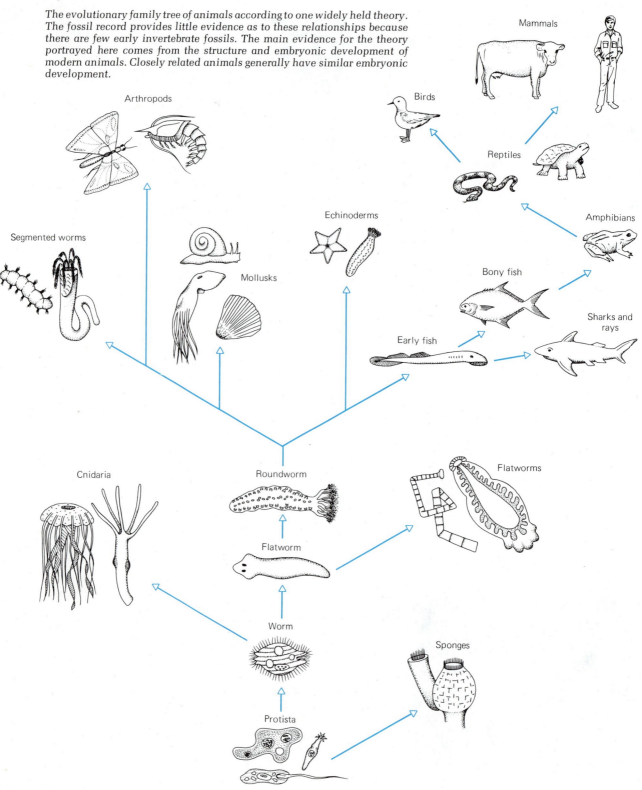

side to side in S-shaped undulations. A fish can move where it will, even against waves and currents which carry most invertebrates along helplessly unless they are firmly anchored.

In coastal waters, we might catch a dogfish, a small shark often dissected in biology classes. As we haul it in, we see the typical row of gill slits behind the head, and feel the rough body surface, covered with scales that are small versions of the formidable pointed teeth in the mouth. Sharks have well-developed senses of smell and pressure, and they can also detect electric currents from the bodies of other animals in the water.

If we dissect a dogfish, we find the typical skeleton of a shark, composed of cartilage, a tough but flexible material like that in the end of your nose, instead of bone, which makes up most of the skeleton of other vertebrates.

By opening the stomach, we can see what the shark has been eating. A large shark killed in the Adriatic Sea had in its stomach two raincoats, part of a horse, an automobile license plate, and a length of rope. The stomach of a small dogfish is more likely to contain small arthropods and fish.

Skates and rays, flattened, bottom-dwelling predators on invertebrates, are the other cartilaginous fish, close relatives of sharks.

Most fish have skeletons of bone. We can recognize bony fish without dissecting them: whereas sharks have a row of gill slits opening separately to the outside, bony fish have a single gill covering with one opening at the back of the head where water leaves the body after giving up its oxygen to blood in the gills. Most bony fish have flat scales, and many use only the tail to swim, instead of throwing the whole body into curves, as a shark does. Many bony fish also have gas-filled **swimbladders,** which allow them to adjust their buoyancy and float higher or lower in the water; sharks have no swimbladders and what buoyancy they have is due to oil stored in their large livers.

Fig. 2–17

A shark, which is a cartilaginous fish. (Paul Feeny)

Fig. 2–18

Bony fish come in many shapes and sizes. (a) A banner fish from a coral reef. (b) This minnow shows the most common body type. (Biophoto Associates, N.H.P.A.)

(a)

(b)

Bony fish show an incredible variety of body shapes and sizes, modifications of fins and tails, and ways of life; their bewildering variety is a result of various evolutionary modifications of an incredibly successful vertebrate body plan for underwater living.

2–F Life on Land

As early organisms evolved in the sea, there came to be more and more competition—among plants for a place in the sun, among animals for food and for refuge from predators. The land was barren, but it had possibilities: for plants, there was abundant light and rocks made of the minerals plants need; for both plants and animals, there was, at first, no competition on land.

However, the land presented problems that kept it uninhabited for a long time. The greatest problem was **desiccation,** or drying out, for water evaporates readily into the air. Land organisms have evolved various waterproof body coverings, but these created a fresh problem: a waterproof coating is also gas-proof, and all organisms must exchange gases with their environment. Both plants and animals need oxygen gas for respiration and must rid their bodies of the gas carbon dioxide, a respiratory waste product. In addition, plants need carbon dioxide for photosynthesis. Most land-dwelling organisms evolved internal gas-exchange surfaces that limit water loss.

respiration: the breakdown of food to release energy needed for life processes.

Another drawback of life on land is the lack of physical support. Water is dense and buoyant, and it helps to support the weight of aquatic organisms. Organisms surrounded by air, on the other hand, need strong skeletons or they will collapse into formless heaps on the ground.

The first land pioneers were probably plants and animals that lived in the intertidal zone of the seashore, or in freshwater ponds and streams that dried up periodically: organisms already adapted to withstanding short dry spells. When an organism developed adaptations allowing it to survive dry spells indefinitely it was truly emancipated from the water.

What adaptations might we have found in these first land-dwellers? A covering of evaporation-retarding slime, like that found in some seaweeds and worms, reduces water loss from the body; so, too, does an external skeleton like the shell of a snail or crab. With a bit of thickening and tightening, so do the scales of fish. The swimbladders of some fish also provide a place inside the body where gases can be exchanged, reducing the exposure of the respiratory surface to the drier outside air and so retarding evaporation; swimbladders have the same origin as lungs—internal surfaces for gas exchange replacing external gills in all true land vertebrates.

2–G Land Animals

Many groups of animals we met before also have representatives living on land: flatworms under logs near streams, the more familiar earthworms in the soil, snails and slugs (molluscs) in low vegetation, and arthropods, such as centipedes, millipedes, and pill bugs, under stones. However, these animals

Fig. 2–19

Adaptations to life on land. (a) The surface of a leaf is waterproof. As a result, the plant loses water less rapidly than it would otherwise. (b) The head of a lizard shows the scales that waterproof the body, reducing water loss. The eardrum (black), eyes, and nostrils of this reptile are also different from those of water-dwelling animals, making the lizard peculiarly sensitive to air-borne sound, smells, and light. (Biophoto Associates)

(a)

(b)

Fig. 2–20

Animals that can survive only in damp places on land. (a) A slug is a mollusc without a shell, active only in cool, damp places. (b) A millipede, an arthropod restricted to life in rotting logs and similar damp places. (Biophoto Associates)

(a)

(b)

usually spend the day in damp places, and they lead restricted lives because they still need a lot of moisture. Only two groups—vertebrates and arthropods—have members able to move about freely on a dry, sunny day. Therefore we see these animals most often.

Land vertebrates. **Amphibians** (amphi = two; bios = life)—frogs, toads, salamanders, and newts—are vertebrates at home both in water and on land, at different stages in their lives. The amphibian egg has a thin covering that cannot withstand desiccation, so amphibians must lay their eggs in a moist place. Usually this is a pond or swamp, but some lay their eggs in a water-filled pouch on one parent's back, or carry the eggs in the mouth or stomach until they hatch.

Fig. 2–21

Amphibians are land vertebrates, but most of them need water to breed. (a) An adult frog in a pond. (b) The tadpole larva of a frog. (Biophoto Associates)

(a)

(b)

The amphibian larva, or tadpole, is fishlike, with a tailfin and no legs. As it grows, it develops legs and, if it is a frog or toad tadpole, absorbs its tail back into the body. Most adult amphibians spend at least part of their time on land, in wet grass or burrows where they do not dry out. An amphibian's lungs are poorly developed, and the moist skin is the surface for most gas exchange. Most adult amphibians are predaceous, eating insects and other small arthropods, worms, and fish.

The first truly terrestrial (land) vertebrates were the **reptiles**—represented today by turtles, lizards, snakes, and the alligator clan. The secret of their early success was evolution of an egg with an almost waterproof shell allowing it to survive on land, well away from egg-eating aquatic predators.

The reptilian body, too, is well adapted to dry air. The body is covered with tight-fitting, water-repellent scales. Along with this barrier to water loss go well-developed lungs, now the only surface for gas exchange. A reptile's skeleton is heavy, able to bear the animal's weight without help from water. The legs are slung under the body, not out to the side as in amphibians, and the toes are clawed; these features enable many reptiles to move quickly.

Reptiles are not really "cold-blooded"; most maintain high body temperatures during the day. A reptile controls its temperature largely by behavior, such as sunbathing. For this reason, reptiles are much more common in the tropics than in colder areas. Ancient reptiles believed to have been "warm-blooded"—able to maintain high body temperatures by retaining body heat—gave rise to birds and mammals.

(a)

(b)

Fig. 2–22

Reptiles. (a) A crocodile. (b) A gecko, an insect-eating lizard whose sucker-like toes permit it to hang upside down. (Biophoto Associates) (c) A snake hatching from its egg. (Biophoto Associates, N.H.P.A.)

(c)

Birds evolved from the dinosaur branch of the ancient reptiles. All healthy adult birds have feathers, which probably evolved from reptilian scales. Birds retain the scaly rear legs and claws of reptiles, but the front limbs have lost their claws and become modified as wings (or sometimes, as in the case of penguins, flippers). The two differently specialized pairs of limbs give most birds at least two different modes of locomotion, such as walking and flying, or swimming and walking, although flightless land birds can only walk.

The bodies of all birds are remarkably similar, due to the demands of flight for a light, streamlined body. The skeleton is reduced to the bare minimum: the dense teeth of a reptile are gone, replaced by a light, horny bill; the hollow, air-filled bones make the body lighter; the bones joining the legs and wings to the body, and the backbone between the legs and wings, no longer move separately, but are fused into a single resilient unit that can absorb the shocks of takeoff and landing. Feathers give the body a streamlined shape, and the tail serves as a rudder.

Bird eggs have harder shells than reptile eggs, and the parents usually sit on the eggs and young, keeping them warm until they develop the insulating layer of fat and feathers needed to retain body heat. The family life of birds may also include elaborate courtship rituals, nest-building and care of eggs and young.

Fig. 2–23

The most characteristic feature of a bird is its feathers. Co-operation between individuals is common and both sexes often devote much energy to raising the young, as shown by these sedge warblers feeding their nestlings. (Biophoto Associates)

Fig. 2–24

Mammals. (a) This ewe and her lamb show the fur and nursing behavior that are characteristics of the group. (b) The dormouse is famed for its sleepiness. Many mammals hibernate, letting their activity and body temperature drop during cold weather. This reduces the amount of food needed to maintain the usually high body temperature at a time of year when little food is available. (c) Otters snoozing in the sun. Many mammals are social, learning hunting behavior and other skills from older members of the family group. (Biophoto Associates) (d) Impala. Impala are ungulates (see Table 2–3), hoofed mammals important to us as cattle, sheep, horses, and other animals we use for food, transport, and leather. Males usually have horns. (Biophoto Associates, N.H.P.A.)

(a)

(c)

(b)

(d)

The **mammals** are another group of land vertebrates that evolved from reptiles. Like birds, mammals are warm-blooded, but they retain body heat with a layer of fur or hair rather than feathers. Like reptiles, early mammals laid eggs, but only two kinds of mammals (the duck-billed platypus and the spiny anteater) lay eggs today. The other mammals gestate the young inside the mother's body, where they are protected from desiccation and from predation, and where the mother's blood supplies food and oxygen and removes wastes. After birth, the mother nourishes the young with milk from her mammary glands. She may also teach them skills such as hunting for food and digging burrows, tasks that the father sometimes shares.

These features of family life are one reason for the success of mammals. Warm-bloodedness is another, for a warm-blooded animal can move around quickly, whatever the weather, and so it has a better chance to catch food or escape predators. Third, early mammals evolved a skeleton built for speed and

TABLE 2-3   **MAJOR GROUPS OF MAMMALS**

Monotremes	Egg-laying mammals: the duck-billed platypus and the echidna (spiny anteater)
Marsupials	Mammals with marsupial pouches, e.g., opossum, kangaroo, koala bear
Insectivores	Small insect-eating mammals, e.g., moles, shrews
Chiropters	Bats
Primates	Lemurs, monkeys, apes, humans
Edentates	Sloths, anteaters, armadillos
Lagomorphs	Rabbits, hares, pikas
Rodents	Mice, rats, voles, beaver, porcupines, guinea pigs, hamsters, jerboas
Cetaceans	Whales, dolphins, porpoises
Carnivores	Dogs, cats, hyenas, bears, raccoons, weasels, pandas, badgers, skunks, mongooses, etc.
Pinnipeds	Seals, sea lions, walruses
Proboscids	Elephants
Odd-toed ungulates	Horses, zebras, asses, tapirs, rhinoceroses
Even-toed ungulates	Pigs, hippopotamuses, camels, deer, giraffe, buffaloes, domestic cattle, gazelles, goats, llama, antelope, sheep, etc.

power, which made them superior predators. The legs are carried right under the body and have strong bones and joints; the muscles that move the back legs are no longer attached to the tail, as in reptiles, but to the hip, so that the tail is lighter and less cumbersome. The jaws are powerful, with fewer bones than in reptiles, and hence fewer joints that might be dislocated while subduing prey. The teeth of mammals have become differentiated and specialized: incisors in front for cutting bits of food, pointed canines for holding and tearing, and molars for grinding mouthfuls before they are swallowed.

Although the earliest mammals were carnivorous (meat-eating), the mammalian body plan has proven very versatile, and many herbivorous (plant-eating) mammals have also evolved. Mammals have colonized habitats ranging from deserts to oceans, from the steaming tropics to the polar circles.

Fig. 2–25

An Atlantic bottle-nosed dolphin, an aquatic mammal. (Paul Feeny)

Fig. 2–26

A useful insect: a lacewing. Lacewings are predators of aphids, destructive garden pests. (Biophoto Associates)

Fig. 2–27

Insects people would rather be without. (a) A locust. Swarms of these insects arise in certain years and may cut a swath across agricultural land, eating every leaf they encounter, particularly in Africa. (Biophoto Associates) (b) A tomato hornworm eating a tomato plant. (Paul Feeny) Billions of tons of crops are lost to insect pests every year.

(a)

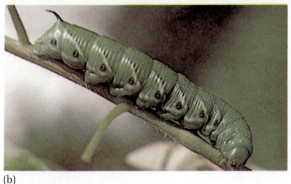

(b)

Insects. The other animal group with members successfully adapted to dry land is the arthropods—notably the insects and, to a lesser extent, the spiders. Of nearly a million different species of insects, almost all are exclusively terrestrial. Although humans have waged war on insects for thousands of years, as far as we know, we have never exterminated even a single insect species. No wonder many people conclude that the insects are destined to inherit the earth.

Many factors contribute to insect success. In insects, the typical arthropod external skeleton is waterproofed, retarding loss of moisture. Insects can also tolerate more water loss than most animals. The small size of insects enables them to live on scanty rations, and to hide from enemies in tiny cracks, on the undersides of leaves, in the hair of an animal, or inside a seed. Most adult insects have wings, and so they can fly to new food sources. And finally, insects reproduce in prodigious numbers; their eggs resist desiccation, and the young develop rapidly to the diminutive adult body size. All in all, insects are a most formidable group!

We depend on insects to pollinate many crops, especially fruits, and we obtain several useful products, such as honey, silk, and shellac, from insects. Nevertheless, humans and insects are usually adversaries. Insects attack human beings directly with bites and stings; blood-sucking insects transmit many diseases. Insect-borne malaria, river blindness, and sleeping sickness affect several million people a year. Insects transmit plant diseases, such as Dutch elm disease and many viral diseases of crop plants. Insects probably do more damage indirectly, however, by competing with people for food and other crops. It is estimated that in 1975 the gypsy moth, tussock moth, southern pine beetle, and spruce budworm destroyed enough trees to build nearly a million houses in the United States alone. In the United States, insects destroy some 10% of all crops grown. Crops in tropical climates suffer from even greater insect damage, since insects grow and reproduce faster in the warm climate of the tropics. In Kenya, officials estimate that insects destroy 75% of the nation's crops. A locust swarm in Africa may be 30 metres deep along a front 1500 metres long, and will consume every fragment of plant material in its path, leaving hundreds of square kilometres of land devastated. Since even modern insecticides have had little impact on this problem, huge losses of crops to insect pests will undoubtedly remain the rule in the foreseeable future.

2–H Land Plants

The plants growing around us are also well adapted to living on land. In land plants, division of labor among cells is more pronounced than in seaweeds. This is because the resources a land plant needs are segregated: sunlight, oxygen, and carbon dioxide are above the surface of the ground, water and minerals below. Photosynthetic cells above ground make the plant's food, while the non-photosynthetic underground parts absorb the plant's water and minerals. In most land plants, the different parts of the body are connected by **vascular tissue,** a transport system that carries water and food. Vascular tissue is reinforced with strengthening material, so that a vascular plant is like a building supported by its plumbing. A waterproof, waxy **cuticle** covers the parts above ground and retards evaporation of water into the air. Many tiny pores in the leaves and stems allow gases to enter and leave the plant with minimal water loss.

CUTE-ick-kull

Fig. 2–28

The leaf surface of a land plant. The slit in the middle is a pore which opens and closes to control how much water vapor and gases enter and leave the leaf. The rest of the surface is covered with a waxy, waterproof cuticle. (Biophoto Associates)

(a)

(b)

Fig. 2–29

Land plants without vascular tissue. (a) A moss covered with the brown stalks on which reproductive cells develop. (Biophoto Associates) (b) A liverwort. The little cups are asexual reproductive organs. (William Camp)

Mosses and liverworts. The earliest land plants lacked vascular tissue, and so had to remain small and live in moist places. They are represented today by the mosses and liverworts. As in the water-dwelling algae believed to be ancestors of land plants, the sperm of these plants must swim to the eggs, and therefore water is necessary for sexual reproduction.

Lower vascular plants. Club mosses and ferns represent some early groups of vascular plants. The fossil record shows that ancient relatives of the club mosses and ferns reached treelike proportions and, indeed, tree-ferns survive today in the tropics (Fig. 2-30).

Fig. 2–30

Ferns. (a) A fern in a northern forest. (b) A tree fern. (Biophoto Associates)

(a)

(b)

Gymnosperms.   Gymnosperms are the conifers, best known as the firs, pines and spruce trees of forest and garden, and their relatives, such as the *Ginkgo* trees planted in smoggy cities because they survive air pollution better than most plants (Fig. 2-31). Two obvious features adapt conifers to the rigors of life on land: strong, woody stems and small leaves. Strong stems enable many forms to grow into very tall trees. Indeed, the tallest living plants are redwoods and Douglas firs, conifers of West Coast forests in the United States. A conifer's needle-like or scale-like leaves have a heavy, waxy cuticle that cuts water loss and enables many conifers to thrive in areas of the southern U.S. where the soil is dry and in the far north, where plants cannot obtain water from the frozen soil in winter.

In reproduction, conifers are also admirably adapted to life with little water. Here for the first time we find plants with air-borne pollen instead of swimming sperm (though the pollen of other gymnosperms releases sperm that swim). Another reproductive advance in the gymnosperms is the **seed,** a small but multicellular plant embryo surrounded by a supply of food and enclosed in a waterproof covering. This reproductive package is admirably suited to drift

Fig. 2–32

A conifer is readily recognized by its needle-like leaves and its cones, the reproductive structures. (Biophoto Associates)

Fig. 2–31

Leaves and seed of a ginkgo. These trees can survive even in very polluted air and so are popular for planting on city streets.

(a)

Fig. 2–33

Flowering plants dominate the world's vegetation and are adapted to life with animals. (a) Marsh marigolds bloom in early spring. Their bright yellow color attracts insect pollinators. (b) Many flowering plants produce fruits like these which must be eaten by animals if the seeds inside are to germinate. The fruits change color as they ripen, a signal to fruit-eaters with color vision (mainly birds). (Biophoto Associates)

(b)

through the air or roll along the ground. The embryo's multicellular body and its stored food supply, provided by the parent, give it a head start in life until its roots can reach water in the soil and its leaves can reach sunlight.

Flowering plants. Most trees, and many low-growing ground plants, are flowering plants. Indeed, the number of different kinds of flowering plants is more than five times the number of kinds of all other many-celled plants put together!

The most obvious feature of a flowering plant is the flower, which comes in a great variety of shapes and sizes. The flower is a reproductive structure; those with large, showy petals and sweet odors attract insects or birds that pollinate them. Many flowering plants, however, produce wind-borne pollen, which, like the pollen of conifers, is broadcast over a large area.

Many flowering plants depend on animals to disperse their seeds. Some seeds hitchhike by fastening on to an animal's feathers, fur, or clothing. Others grow inside tasty fruits which an animal eats. The animal may discard a large,

TABLE 2-4 ECONOMIC IMPORTANCE OF LAND PLANTS	
Mosses	Sphagnum moss: used as fuel (peat) in Ireland and Scotland; used as mulch and planting medium in gardening and nursery industries
Club mosses	Formerly used as Christmas greens, the ground pine is now rare and protected in most areas
Ferns	Foliage may be used in florist industry; plants sold to indoor and outdoor gardeners
Gymnosperms	Lumber: fir, hemlock, spruce, pines, cedar, redwood
	Turpentine: distilled from pine trees
	Pulp for paper: various conifers
	Christmas trees: spruces, pines, firs, eastern red cedar
	Landscape plants: spruces, junipers, yews, cedars, cypress, hemlock, pines, *Ginkgo*
	Gin: sometimes prepared by redistilling spirits with juniper "berries"
Flowering plants	Food: fruits, berries, seeds, nuts, grains, stalks, leaves, roots, tubers; extraction of juices, syrups and fats
	Clothing: cotton, linen
	Lumber: oak, maple, ash, birch, poplar, walnut, cherry, pecan
	Fuel: wood, charcoal
	Landscaping: grass; oak, maple, magnolia, birch, and many other trees; flowering shrubs; annual and perennial herbaceous flowers
	Beverages: coffee, tea; fermentation of many species to make beer, wine, and liquors
	Drugs and medicines: tobacco, aspirin (originally derived from willow bark), morphine, opium, marijuana, atropine (from *Belladonna* plant), digitalis (from foxglove), various tonics (from sassafras, dandelion, coltsfoot, etc.)

Fig. 2–34

Flowering plants of value to humans. (a) A coconut palm, important in many tropical economies. The fruits are edible both ripe and unripe; the fronds and bark supply material for hats, houses, and baskets. (b) Cereals, such as these oats, are the most important human food crops. (Biophoto Associates)

(a)   (b)

hard seed after it eats the fruit, or it may eat small seeds along with the fruit; these seeds usually pass through the digestive system unharmed. The colorful animal-pollinated flower and the equally colorful animal-dispersed fruit reflect the fact that flowering plants evolved when the land was already well populated with animals. Animal life has played a crucial role in the evolution of the spectacular array of flowering plants, from duckweed to dogwood, onions to oak trees.

Human life would be impossible without flowering plants. We may occasionally nibble pine seeds, swill spruce beer, or sauté a spring dish of fern fiddleheads (the opening leaves of ferns), but most of our plant food comes from flowering plants, as does the food for our livestock. Conifers do provide most of our utility lumber, but we turn to flowering trees again for hardwood of greater beauty or strength. Cotton and linen come from flowering plants, and so do many of the dyes that give our clothing variety and the teas and spices that make eating more fun. All in all, the flowering plants are a most important group (Table 2-4).

2-I Bacteria and Fungi

Bacteria and fungi are often lumped together as "microorganisms." The main thing they have in common is that many members of both groups are important decomposers, living on the dead remains of other organisms. Otherwise, however, bacteria and fungi are not alike; in particular, their cells are differently constructed. Fungi have cells similar to those of all the other organisms we have already examined, but bacterial cells have a simpler internal structure (see Chapter 10). For this reason bacteria are thought to have evolved before all other organisms, a theory that is supported by the fact that the oldest known fossils are those of bacteria.

Bacteria. The bacteria all belong to the kingdom Monera, consisting of unicellular organisms with simple cells. Despite their simple structure, however, the chemistry of bacteria is incredibly varied. There are photosynthetic bacteria, bacteria that live as parasites on the living bodies of plants and animals, bacteria that live on dead bodies, bacteria that need oxygen, and bacteria that can survive only in the absence of oxygen. Some live in hot springs at temperatures of

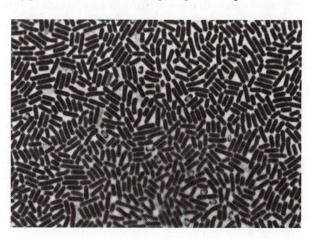

Fig. 2–35

Rod-shaped bacteria, plated on a dish in the laboratory. (Biophoto Associates)

up to 80°C, while others can survive only at the temperature of the human body. Because they reproduce rapidly, bacteria also evolve rapidly, permitting this dramatic adaptive radiation.

Before considering diseases caused by bacteria, let us give the group its due and mention some of its beneficial members. Life on earth would be impossible without bacteria, for they are among the most important decomposers, organisms that break dead bodies down into their chemical components. If all the decomposers were to vanish, the earth would soon be piled knee-deep in corpses, plants would have no soil to grow in, and life would rapidly grind to a halt. Our own bodies are inhabited by bacteria—on the skin, and in the nostrils, mouth, large intestine, and vagina—which keep us healthy by crowding out their disease-causing kin. Although we tend to regard bacteria in food as a bad thing, bacteria are absolutely essential in making various foods and some are even used as preservatives. For instance, bacterial fermentation is used to produce many cheeses, yoghurt, sauerkraut, pickles, and vinegar. Bacterial fermentation produces foods that last indefinitely; this was particularly important in the days before refrigeration.

All the food we eat contains bacteria. Although milk is sterile when it leaves a healthy cow, it contains several types of bacteria by the time it reaches the table. Bacteria in food are often a nuisance since they cause food to decay, and some of them, like the infamous *Salmonella* and *Clostridium botulinum*, cause human disease. *Clostridium* is one of many bacteria that produce **toxins,** poisons that leave the bacterium and survive even if the cell that produced them is dead. (Botulism from *Clostridium* has increased in recent years as home canning has become more popular.) *Clostridium* grows only in oxygen-free, non-acid conditions. The surest protection against it is to boil canned food for 15 minutes, since the toxin is destroyed by heat, or to can only acid foods like pickles and certain fruits, and never to use a can that is pushed outward by the pressure of gas within it. (A tin dented inward is not dangerous as long as the can is intact.) We cannot totally rid food of bacteria; instead, we must constantly try to slow bacterial growth. Both heat and cold slow growth of bacteria in food; this is one reason we cook food and is also the reason for refrigeration. Chemical additives also reduce bacterial growth in food.

Some pathogenic (disease-causing) bacteria destroy the cells of their host, but most cause disease by producing toxins that damage the host. Bacteria that cause diphtheria, tetanus, cholera, dysentery, plague and botulism release toxins that travel through the body. More common are toxins that stay in the outer wall of the bacterium and that produce fever and damage the host's circulatory system. Some pathogenic bacteria live on dead skin cells or on food, and it is their toxins, not the bacteria themselves, that cause disease. The control of many bacterial diseases, by improved hygiene and antibiotics, has created the major health revolution of the past 200 years.

Fungi.   Some fungi are unicellular, like the yeasts that are used to make bread and to ferment wine, while others, like the bracket fungi on trees, are multicellular and often quite large. Whatever their size, all fungi are **saprobes,** which means that they absorb their food across their body surfaces—unlike

Fig. 2–36

Toadstools, the reproductive structures of a fungus. (Biophoto Associates)

plants, which make their food, and animals, which ingest food by way of a mouth.

The feeding part of a fungus usually lies buried in its food and we normally see only the reproductive structures: the black, green, white, or pink fuzz on moldy foods, the brackets on trees, or the mushrooms growing on dead logs or leaves. All of these produce **spores,** single reproductive cells that usually reach new food by traveling through the air. Fungi are important decomposers, but some fungi cause inconvenience when they break down such unlikely substances as leather, hair, wax, cork, and polyvinyl plastics. During their long years of global supremacy, the Spanish and British navies lost more ships to wood-rotting fungi than to enemy action. It was quite common for the bottom of a ship to fall out in mid-ocean.

Many fungi that attack human food are unwelcome though harmless, but some fungi are valuable in food production. English Stilton cheese and French Roquefort, Brie, and Camembert all owe their special flavors to specific fungi introduced as part of the production process. Since prehistoric times, people have preserved and enriched food by fungal fermentation. Consider soy sauce (*shoyu*). Although American-made soy sauce comes from treating soybeans with hydrochloric acid and has little nutritional value, Oriental *shoyu* is made by fermenting boiled soybeans and wheat with a fungus for about a year. This produces a sauce that is rich in vitamins and amino acids, the difference between health and malnutrition for those subsisting mainly on rice. Similarly, wine, beer, cheeses, and fermented sausages all keep longer and are more nutritious than their raw materials.

Although fungi and bacteria often compete for food, the differences between the two groups reduce their competition. For instance, most fungi need high concentrations of oxygen, whereas there are many bacteria that can live at low oxygen levels. Because of the oxygen requirements of fungi, human fungal infections—including ringworm, athlete's foot, and vaginal infections—are nearly all restricted to body surfaces exposed to the air. For the same reason, fungal infections of plants are very common. Leaves are thin and easily penetrated by air, and woody stems and roots have dead centers and living tissue very close to the surface, near the air. Fungi and bacteria also differ in that fungi tolerate acid environments better; thus jelly and vinegar often harbor growths of fungi, but not of bacteria.

SUMMARY

Biologists group organisms according to evolutionary relationships. Each major group of organisms contains members adapted to many habitats and ways of life, often showing remarkable variations from the basic body plan of their group. In this book, organisms are divided into five major groups: Monera (bacteria), Protista (organisms consisting of single, more complex, cells), fungi, plants, and animals.

We believe that life began in the sea, where water surrounds the organisms, provides them with gases and mineral nutrients, carries away wastes, and supports the body. Later, life on land became possible with the evolution of waterproof body coverings, strong skeletons, internal gas exchange surfaces, and waterproof egg shells, seed coats, or other means of protecting the young. The vascular plants, insects and spiders, reptiles, birds, and mammals contain members that can tolerate the dry air and sunshine of the true land environment. Mosses and liverworts, club mosses and ferns, many invertebrates, and amphibians live on land but still require moist, shady habitats or daytime retreats and often depend on at least a film of water for successful reproduction.

Bacteria and fungi are vital as decomposers and are used in the production of many foods. Pathogenic bacteria produce toxins that cause disease. Fungi cause few human diseases because they require air; the chief harm they do is to cause many plant diseases.

SELF-QUIZ

1. Which of the following modes of locomotion would be found among one-celled planktonic organisms?
   a. cilia
   b. tube feet
   c. fins
   d. flippers

2. For each characteristic listed below, select an animal from the list at the right that shows the characteristic.
   ___ a. shell and fleshy, muscular foot
   ___ b. tentacles with stinging organs
   ___ c. tube feet and spiny covering
   ___ d. jointed external skeleton
   ___ e. body divided into fluid-filled segments

   i. arthropod
   ii. cnidarian
   iii. echinoderm
   iv. mollusc
   v. segmented worm
   vi. sponge

3. List two adaptations for reproduction on land that are shared by gymnosperms and flowering plants.

4. Match each of the following animals to the group to which it belongs:

   ___ a. robin   i.   cartilaginous fish
   ___ b. shark   ii. bony fish
   ___ c. codfish   iii. amphibian
   ___ d. snake   iv. reptile
   ___ e. toad   v.   bird
                          vi. mammal

5. List at least two differences in body structure and at least two differences in reproduction between reptiles and amphibians.

6. List three ways that algae contribute to the human economy.

7. Which of the following foods is *not* produced with the aid of bacteria or fungi?
   a. vinegar
   b. marshmallows
   c. cheese
   d. wine
   e. bread

QUESTIONS FOR DISCUSSION

1. Most freshwater invertebrates have lost the mobile larval stage found in their relatives that live in the sea. Can you explain the adaptive advantage of this situation? (Hint: think about the differences between freshwater lakes, rivers, etc., and the sea.)

2. What advantages of social and parental behavior have made it profitable for organisms to spend some of their available energy in these activities? What are the drawbacks of such behavior?

3. Parental care of the young played a major part in the success of birds and mammals. What parallels can you find in the plant kingdom?

4. What energy expenditures must a plant make if it is pollinated by animals? What energy savings does the plant gain by having animal pollination? What is the advantage to a plant of being animal-pollinated rather than wind-pollinated?

SUGGESTIONS FOR FURTHER READING

Banks, H. *Evolution and Plants of the Past.* Belmont, California: Wadsworth, 1975. A lively presentation of paleobotany.

Barnes, R. D. *Invertebrate Zoology,* 4th ed. Philadelphia: Saunders, 1980. An excellent text and reference book on invertebrates except insects.

Buchsbaum, R. *Animals Without Backbones,* 2nd ed. Chicago: University of Chicago Press, 1976. A classic elementary textbook on invertebrates.

Hardy, A. *The Open Sea. Part I, The World of Plankton.* Boston: Houghton Mifflin, 1965. An anecdotal account of plankton.

Essay:

ANIMALS THAT LIVE AS PARASITES

Most familiar animals are herbivores, eating plants, and carnivores, eating other animals. A third interesting group are the parasites, animals that extract food from living host animals.

Parasites can be divided into two main groups. **Endoparasites**—including flukes and tapeworms (flatworms), roundworms, and the one-celled organisms that cause malaria—live inside the host's body, whereas **ectoparasites** live outside the host, sometimes approaching the host only to feed. Ectoparasites include ticks, lice, and fleas (arthropods), and leeches (segmented worms).

Finding food is often difficult for parasites because appropriate hosts may be few and far between. Many tapeworms and flukes compensate by producing hundreds of thousands to millions of eggs, the large numbers ensuring that at least a few find a host; most of the young starve to death, unable to find a suitable host. Many parasitic worm larvae must develop in the body of a secondary host. Snails often serve as intermediary hosts, and efforts at parasite control may aim at eliminating a snail host from the environment. Pinworms have a hand-to-mouth method of transmission: these small, short-lived worms infect mainly young children, who scratch the worms' eggs from the anal area and transfer them to the mouth on the fingers, allowing the next generation of worms to reach the digestive system of the same host. Female fleas have an interesting adaptation: they are most attracted to pregnant females of host mammal species. The flea feeds on the pregnant female's blood and lays eggs which hatch into larvae that live on scraps of skin or dung in the burrow of the mother-to-be. When the young mammals leave their mother, the next generation of fleas hops aboard, provided with new hosts and a means of dispersal.

A parasite's food is pretty much pre-digested by its host. Intestinal parasites are surrounded by digested food, and other parasites need only suck in nutritious blood. There is not much for the parasite's digestive system to do, and most parasites have reduced digestive systems. The energy freed by this savings is

TABLE 2-5 SOME PARASITIC WORMS COMMON IN HUMANS

Name	Symptoms	Means of Infection
Chinese liver fluke	None in mild cases; destruction of liver, bile stones and clogging of liver ducts in severe cases	Eating raw fish
Blood fluke (schistosomes)	Enlargement of liver and spleen Urinary disorders Bloated abdomen, wasted arms and legs	Drinking or wading barefoot in water containing infected person's urine. Infects about 200 million people in 70 nations. Not found where there are modern sewage disposal systems.
Swimmer's itch	Itching after exposure of skin to infested water	Burrowing of fluke larvae that were really looking for other host species
Bladder worm (*Echinococcus*) (immature stage of a worm that lives in a dog when adult)	Cysts up to the size of an orange; symptoms depend on part of body invaded	Infected dogs licking people's hands or faces or contaminating drinking water
Pork, beef and fish tapeworms	Immature worms: cysts Adult worms may cause diarrhea, loss of weight, perforation of intestine	Eating undercooked meat containing worm cysts
Pinworm	Anal itching	Females lay eggs around anal opening: hands may transfer eggs to mouth, maintaining infection in same person. Physical contact may also transfer to other people.
Hookworm	Anemia, lethargy	Young worms burrow through skin (bare feet) from moist soil and grass contaminated by feces of infected humans.

devoted to expansion of the reproductive system. In fact, tapeworms have no digestive system whatever; they absorb all their food across the body wall, and their bodies are little more than "egg factories."

A successful parasite will not cause the death of its host because this destroys its food supply. History is full of examples of parasites, such as the plague or syphilis bacteria, that wiped out huge proportions of new-found host populations, but this also, of course, wiped out most of the parasites. Only those parasites with less destructive effects on their hosts survived to reproduce. Ectoparasites may also cause the death of their hosts, as when bird lice eat the feathers of nestlings as fast as they grow; this prevents the bird from keeping its body warm and makes it unable to fly—either of which may doom a young bird.

A modification of the "don't kill your host" way of life is found among some insects known as parasitoids (= parasite-like). The female lays an egg in a host larva and the egg hatches and uses the host as a food source in its own development. Just as the parasitoid reaches adulthood, it kills its host and cuts its way out of the host's body. So, although the young parasitoid does feed on a living host, killing the host is a programmed part of its life rather than an accident. Female parasitoids can tell whether a host is already parasitized, and lay eggs only on uninhabited hosts, thus ensuring that their own offspring will have enough food to develop.

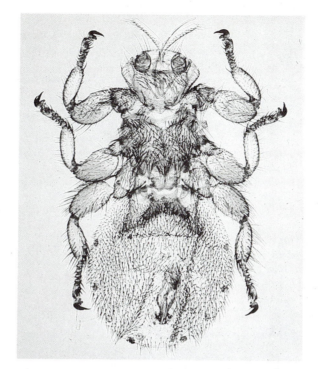

Fig. 2–37

An ectoparasite. This sheep ked lives attached to the sheep's wool, clinging by the hooks on its legs and eating flakes of skin, oil, etc. (Biophoto Associates)

Fig. 2–38

A tapeworm, an endoparasite. The hooks and suckers anchor the worm to the wall of the host's gut. The long string of short, wide sections contains little but reproductive organs (not visible here). (Biophoto Associates)

Essay:
NAMING AND CLASSIFYING ORGANISMS

Living organisms are grouped, or classified, on the basis of kinship. The **species**—a group of organisms closely enough related to be able to interbreed with one another—is the basic unit of classification. Each species is placed in a **genus***, which may contain other, similar species. Each genus is placed in a **family,** each family in an **order,** and so forth; in most cases, each successively higher group includes a larger number of more distantly-related species.

Below are the classifications for one plant and one animal:

	Human Being	Black-Eyed Susan
Kingdom	Animalia	Plantae
Phylum	Chordata	Tracheophyta
Class	Mammalia	Angiospermae
Order	Primates	Asterales
Family	Hominidae	Compositae
Genus	Homo	Rudbeckia
Species	sapiens	hirta

***genus** (JEAN-us) (pl: genera; JENN-er-uh) (adjective: generic; jenn-AIR-ick)

Note that to name a species you have to give the name of both the genus and the species: *Homo sapiens, Rudbeckia hirta*. This is because the species name is often trivial (*hirta* merely means hairy), and many organisms may have the same species names: consider *Hepatica americana* (a spring flower), *Erythronium americanum* (trout lily), *Coccyzus americanus* (cuckoo), and *Veronica americana* (brooklime) (*americana* means American).

We need rules for naming organisms because the same animal or plant is called by so many different names in different countries and even within the same country. The use of common instead of scientific names can be confusing in everyday situations as well as in the laboratory. When you're buying fish, "sea scallops" may sound like a shellfish for which you'd pay a fancy price, and "rock cod" may sound like a relative of the bony codfish popular in New England. In fact, both of them

(a)

Fig. 2–39

Rudbeckia hirta. (*Biophoto Associates*

are common names for the flesh of the shark *Squalus acanthodus*, which is called a dogfish in Europe.

International commissions— one for plants, one for animals and one for bacteria—now regulate the scientific names for members of these groups. The snag? The three commissions never get together, so that *Pieris* is a genus of common garden shrubs and is also a genus of butterflies (including the cabbage white butterfly, *Pieris rapae*).

Scientific names are Latin because, for hundreds of years, Latin was the language of Western scholars. Many of the Latin names are full of information and folklore. A gardener, for instance, finds it useful to know that *Gypsophila* (baby's breath) means "to love chalk," a reference to this plant's preference for lime. Some organisms are named after people. The black-eyed susan, *Rudbeckia,* is named after Olaf Rudbeck, a professor of botany in eighteenth-century Sweden.

Fig. 2–40

(*a*) Pieris japonica, *a shrub.* (*b*) Pieris brassica, *a butterfly.* (*Biophoto Associates*)

(b)

Medieval folklore taught that, as an early herb book puts it, "God . . . maketh grass to grow upon the mountains, and herbs for the use of man, and hath . . . given them particular signatures, whereby a man may read, even in legible character, the use of them." Thus *Hepatica*, a spring flower, was believed to cure diseases of the liver because its leaf is liver-shaped (Fig. 2-41). This practice of naming organisms after parts of the body they resembled was carried to sexually explicit lengths in names like *Phallus impudicus*, a stinkhorn fungus with a fruiting body shaped like a human penis (Fig. 2-42). The great eighteenth-century Swede Linnaeus classified plants by their sexual parts with group names such as *Monandria*, which he described as "one husband in a marriage" and *Polyandria*, "twenty males or more in bed with the same female." This emphasis on sex shocked some of his contemporaries. The Bishop of Carlisle wrote: "To tell you that nothing could equal the gross prurience of Linnaeus's mind is perfectly needless," and Goethe worried about the embarrassment chaste young people might suffer when reading botanical textbooks.

Our system of naming and classifying organisms is probably inadequate to the task. Biologists believe that there are some 10 million different species of organisms in the world, of which only about 15% have been described. (Today we are destroying natural habitats so rapidly, by pollution, our population explosion, and the destruction of forests, rivers and fields, that most of the remaining species will be extinct before they are ever described!) Until better systems of classification become widespread, however, we must make do with what we have. We need scientific names for organisms so that we can communicate with one another and know when we are talking about the same species, and we need some sort of classification scheme to make sense of the vast diversity of organisms all around us.

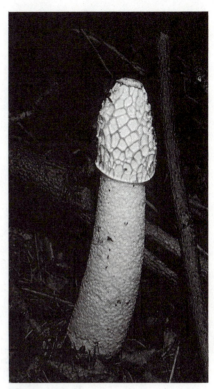

Fig. 2–42

Phallus impudicus. (*Biophoto Associates*)

Fig. 2–41

Hepatica acutiloba.

Part One

EVOLUTION AND ECOLOGY

333 NATURAL SELECTION AND EVOLUTION

The theory of evolution states that today's organisms have arisen by descent and modification from more ancient forms of life. For instance, most biologists believe that human beings evolved from now-extinct animals much like chimpanzees, and that this happened through the accumulation of changes through thousands of generations.

It is hard for us to appreciate today how revolutionary this theory seemed when it was first put forward. Evolution was a shocking idea because we have a tendency, as old as language itself, to think in terms of unchanging types. The ancient Greek philosopher Plato, for instance, believed that God had created eternal, unchangeable prototypes of Horse, Dog, and even of inanimate objects like tables; real horses, dogs, and tables were merely imperfect versions of their ideal types. In essence, the theory of evolution said that this way of thought would not do for living organisms. If the theory is correct, then organisms can evolve and change, and at least some populations of organisms contain the raw materials from which new and different species can evolve. This means that the horses alive today are not deviants from a single "ideal" horse, but are a group of organisms merely displaying some of the variations which the collection of horse **genes** (the particles determining inherited traits) can produce. In other words, there is no such thing as a "normal" horse, or human being, or oak tree; all are variations on a theme. This was a difficult idea.

3–A Evolution by Means of Natural Selection

According to the theory of evolution, organisms arise by modification from ancestral forms of life. We can also define evolution as a change in the genes of a population from one generation to the next. If half of a human population contains the gene for blue eyes in one generation, and in the next generation only a quarter of the population contains the blue-eye gene, that population has evolved, or changed genetically.

Natural selection is one process that produces evolution. **Natural selection** is the phenomenon whereby the bearers of some genes reproduce more successfully than the bearers of other genes. If people with brown-eye genes have more children than anyone else (which would be natural selection for brown eyes), the next generation of the population will have more brown-eye genes than the last generation, and evolution will have occurred.

The theory of evolution is based on three observations: First, as we can see by comparing one cat or human being with another, the members of a species differ from one another. Second, some (though not all) of the differences between organisms are inherited. (Other differences between organisms are not inherited because they are caused by different environments.) Third, more organisms are born than live to grow up and reproduce: many organisms die as embryos or seeds, as saplings, nestlings, or larvae.

larvae (LAR-vee): immature animals that look different from the adults of the species (such as caterpillars).

Fig. 3–1

Variation. The four very different butterflies at the bottom are the offspring of the parents at the top.

The logical conclusion from these three observations is that certain genetic characteristics of an organism will increase its chances of living to grow up and reproduce, over the chances of organisms with other characteristics. To take an extreme example, if you have inherited a severe genetic disease of the liver, you have much less chance of living to grow up and reproduce than someone born without this disease.

Inherited characteristics that improve an organism's chances of living and reproducing will be more common in the next generation than those that decrease its chances of reproducing. Various combinations of genes will be naturally selected for or against, from one generation to the next, depending on how they affect reproductive potential. For natural selection to cause a change in a population from one generation to the next (that is, to cause evolution), it is not necessary that all genes affect survival and reproduction; the same result occurs if there are just some genes that make an individual more likely to grow up and reproduce.

To summarize:

1. Individuals in a population vary in each generation.
2. Some of these variations are genetic.
3. More individuals are produced than live to grow up and reproduce.
4. Individuals with some genes are more likely to survive and reproduce than those with other genes.

Conclusion: From the above four premises it follows that those genetic traits that make their owners more likely to grow up and reproduce will become increasingly common in the population from one generation to the next.

3–B History of the Theory of Evolution

The theory of evolution by natural selection was put forward in a joint presentation of the views of Charles Darwin and Alfred Russel Wallace before the Linnaean Society of London in 1858. Until that time, most biologists accepted the view of such authorities as the ancient Greek philosopher and natural historian Aristotle, and of the Bible's Book of Genesis, and believed that species of organisms never changed but had been created in their present forms.

Starting about 1750, evidence that species changed over the ages accumulated steadily, and many people became increasingly convinced that organisms evolved. Naturalists in France, Britain, and Germany were uncovering fossils of strange animals that no longer existed. These finds captured the popular imagination, and newspapers printed articles and letters arguing about the theological and scientific implications of fossil animals. Some people made the charming suggestion that fossils were not the remains of extinct animals but had been planted in the rocks by God, as fossils, to make life more interesting for everybody. However, various French and British scientists, including Darwin's friend Charles Lyell, became convinced that different organisms had lived at different times.

Darwin and Wallace were not the first to suggest the occurrence of evolution. Their names are linked with the idea of evolution because they proposed the theory of natural selection as the mechanism by which evolution occurs. We are always more likely to believe in a process when people explain how it happens than if they merely assert that it does.

In 1845, Darwin published *The Voyage of the Beagle*, an exciting and readable account of his experiences during a five-year mapping and collecting expedition on a British naval ship. Parts of this book show that Darwin already

(a)

(b)

Fig. 3–2

Fossils. Fossils are the remains of organisms or their imprints preserved in various ways in rocks. (a) A trilobite, member of a now-extinct group of invertebrates. (b) The fossil of a mollusc's shell. (Biophoto Associates)

Fig. 3–3

Position of the Galapagos Islands with part of the Beagle's route (colored). The effects of evolution are exaggerated on these islands because they are isolated from the mainland and from each other. The few plants and animals that have reached the Galapagos have evolved differently on each island.

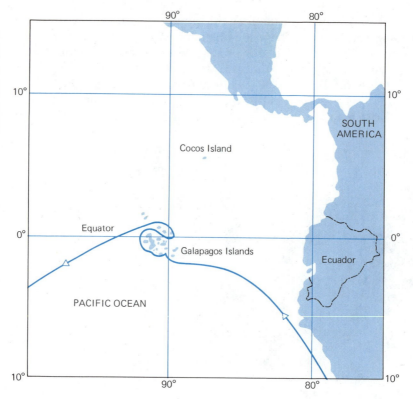

paucity (PAW-city): fewness, smallness in numbers.

archipelago (ARK-ih-PELL-uh-go): a group of islands.

understood how evolution and natural selection work. Of the many species of finches unique to the Galapagos Islands he wrote: "Seeing this gradation and diversity of structure in one small, intimately related group of birds, one might really fancy that from an original paucity of birds in this archipelago, one species had been taken and modified for different ends." In the same book he alluded to natural selection: ". . . some check is constantly preventing the too rapid increase of every organized being left in a state of nature. The supply of food, on average, remains constant; yet the tendency in every animal to increase by propagation is geometrical."

Darwin realized that some force, which he later called "natural selection," constantly checks the potential population explosions of organisms. This idea grew from his reading of an essay on population by Thomas Robert Malthus, a clergyman and economist. Malthus argued that human populations outgrow their food supplies and are eventually reduced by starvation, war and disease. Population size tends to increase geometrically, whereas food supplies, at best, increase only arithmetically. Darwin made famous Malthus's phrase, "the struggle for existence," which they both saw as an inevitable result of the discrepancy between rapid population growth and limited food supply.

Darwin's definitive work, *The Origin of Species by Means of Natural Selection*, was published in 1859 and became an immediate best-seller. Not until the twentieth century, however, did most biologists fully accept the theory that evolution occurs by means of natural selection.

GROUND FINCH

(a)

WOODPECKER FINCH
(A tree finch)

(b)

WARBLER-TYPE FINCH

(c)

Fig. 3–4

Darwin's finches. These are some of the 14 species of finches found in the Galapagos. Darwin was struck by the fact that all the finches were so similar that they seemed to have a common ancestor. But differences in size, beak shape, and feeding habits suggested to him that each isolated population had evolved adaptations to its particular island and way of life.

One widely held objection to the theory of natural selection was the belief that the characters of parents blended in their offspring. If this occurred, the inherited variation in a population would decrease in each generation until all the individuals in the population were alike, leaving natural selection with very little to act upon. We now know that genes do not blend, but Darwin, and most other nineteenth-century biologists, thought that they did. Although Gregor Mendel had published the basic laws of modern genetics (see Chapter 16), his work remained completely unknown until it was rediscovered and publicized in 1901. Mendel's work revealed that genes are inherited independently of one another and do not merge in the offspring; thus, genetic variation is preserved in each generation.

Another major barrier to the acceptance of natural selection was that Darwin illustrated selection with examples of selective breeding of domestic plants and animals. Breeders, for instance, had selected cattle for increased milk production (Holstein), for high-butterfat milk (Jersey), or for meat (Hereford). Darwin never provided a convincing demonstration that selection actually occurs in nature. The example of selection in a wild population described in the next section was not worked out until a hundred years later.

Fig. 3–5

Selection by human beings. These different breeds of cows have been produced by selective breeding for particular characteristics. (Biophoto Associates)

Fig. 3–6

The two forms of the peppered moth. (a) On a lichen-covered tree trunk in an unpolluted area. The gray form is so well-camouflaged that it is almost invisible, below and to the right of the black moth. (b) On a soot-covered tree trunk. (Bernard Kettlewell)

(a)   (b)

3–C The Peppered Moth

lichen (pronounced: liken): a plantlike growth consisting of a fungus that surrounds and protects photosynthetic cells of a simple alga; the fungus and alga are dependent on one another for survival.

In nineteenth-century England, many people collected moths as eagerly as people today collect stamps or Elvis souvenirs, and collectors avidly sought a rare black form of the normally gray, British peppered moth (Fig. 3-6). The moth rests during the day, usually camouflaged on tree trunks covered with gray lichens.

By looking at collections made from about 1850 to 1950, biologists found that the black form of the moth became more and more common during the century, and the gray form scarcer, particularly near industrial cities. Why had this change occurred? Perhaps it resulted from the industrial revolution: the large-scale burning of coal was blackening the tree trunks with soot. Against the darker background the gray form of the peppered moth became more visible to birds, the chief predators on moths, and so the gray moths were easier to catch. Now black moths were better camouflaged from hungry birds.

TABLE 3-1   **NUMBERS OF GRAY AND BLACK PEPPERED MOTHS RECAPTURED AFTER THE RELEASE OF MARKED INDIVIDUALS IN TWO AREAS**

Location		Gray	Black
Dorset (unpolluted) 1953	No. released	469	473
	No. recaptured (%)	62 (13.2)	30 (6.3)
Birmingham (polluted) 1953	No. released	137	447
	No. recaptured (%)	18 (13.1)	123 (27.5)
Birmingham (polluted) 1955	No. released	64	154
	No. recaptured (%)	16 (25.0)	82 (53.3)

Bernard Kettlewell recognized that this was an excellent opportunity to study natural selection experimentally. He raised large numbers of both black and gray forms of the moth in the laboratory, marked them, and released them in two places: one an unpolluted rural area where the black form was more visible to a human observer, the other a polluted industrial area where the gray form was easier to see against the blackened tree trunks. Kettlewell then recaptured as many of the marked moths as he could. The percentage of black moths recovered was twice that of gray moths in the industrial area, but only half that of gray moths in the unpolluted countryside (Table 3-1). This agreed with the prediction that the gray moths were more likely to survive (and so to be recaptured) in the country, and the black moths were more likely to survive near the town.

Next, Kettlewell hid in a blind and watched the moths on tree trunks. On one occasion, equal numbers of gray and black moths were released in an unpolluted area, and birds caught 164 of the black form and only 26 gray ones.

It is clear that, in a polluted area, many more of the black than of the gray moths live long enough to reproduce. Since the color of the moths is inherited, the next generation will contain proportionally more black moths than the last. In other words, the proportion of genes for black color increases in the population with time—and that is evolution.

The natural selective force that brings about this evolution is clear: in polluted areas birds kill a higher percentage of moths with the gene for gray color than of moths with the gene for black color. Natural selection has produced populations of the moth that are well adapted to survive in their environments, populations whose characteristics change as the environment changes.

This theory permits us to predict that if pollution is reduced in industrial areas, black moths will become rarer and gray forms more common in these areas. In fact, the Clean Air Act of 1952 has reduced the air pollution in England. Collections of the peppered moth from industrial Manchester in the years since 1952 reveal a dramatic increase in the ratio of gray to black individuals in the moth population. The ability to predict future events in this way is the most impressive evidence that can be produced for a scientific theory.

3–D Genetic Contribution to Future Generations

The phrase "survival of the fittest," often used in discussions of evolution, suggests that natural selection selects mainly for survival; it does not. Rather, it selects for the contribution of genes to future generations. Survival is important, in that if an individual does not survive for long it cannot reproduce, but even reproduction is not a guarantee of evolutionary success.

Consider Table 3-2, which shows how many young starlings survived for three months after hatching. The female starlings that seemed to be reproducing most efficiently—those laying nine or ten eggs in one brood—could actually be doomed to evolutionary failure and strongly selected against, since hardly any of their young survived. However, females laying four or five eggs per brood had a higher number of offspring surviving for at least three months after they hatched. Apparently the reason more young in the large broods died was that the parents

Fig. 3–7

A Swiss starling landing at its nest in a hole in a tree.

TABLE 3-2   SURVIVAL IN SWISS STARLINGS IN RELATION TO NUMBER OF EGGS LAID*

Brood size (Number of eggs in nest)	Number of young marked	Recoveries per 100 birds marked†
1	65	0
2	328	1.8
3	1278	2.0
4	3956	2.1
5	6175	2.1
6	3156	1.7
7	651	1.5
8	120	0.8
9, 10	28	0

*The number of eggs laid during one nesting period is genetically regulated and, like other genetic variations, is acted upon by natural selection. Lack marked all the nestlings in all the nests he could find, and then recaptured them months later when they had left the nest.

†The only recoveries scored are those for birds over three months old when they were recaptured.
From Lack, *Ecol.* 2, 1948.

could not provide adequate food for more than five or six young birds. It seems reasonable to suppose that in other years, when there is more (or less) food available to the birds, selection would favor birds with larger (or smaller) broods.

The one consistent effect of selection is that it increases a successful gene's representation in future generations.

3–E More Examples of Natural Selection

Several dramatic examples of natural selection in action today are provided by the evolution of resistance to pesticides and antibiotics.

A scale insect feeds on citrus trees in California. In the early 1900s, growers sprayed the trees with cyanide gas and this killed the scale. But in 1914 some of the insects survived the spraying. The cyanide did not kill them because they possessed a single gene, newly apparent in the population, which permitted them to break down cyanide into harmless compounds. As spraying continued, more insects with the new gene than without it survived to reproduce, and they passed on the gene to their offspring. The frequency of the new gene in the population increased until the whole population was resistant to the spray. Since scale insects, like many other insects, have more than one generation a year, they evolve quickly. To combat the evolution of resistance, growers are encouraged to spray only when necessary and to use different chemicals in different months or years.

Precisely the same thing happens with antibiotics used to kill bacteria that cause human disease. When a bacterial population meets a particular drug, bacteria susceptible to that drug are killed. Bacteria with genes conferring resistance to the drug multiply rapidly once competing bacteria have been removed, and they soon become widespread. Some strains of the bacterium that causes the venereal disease gonorrhea can no longer be killed by any known drug. Since antibiotics and disease-causing bacteria frequently meet in hospitals, it is not surpris-

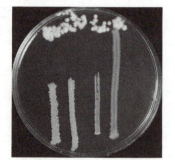

Fig. 3–8

Drug resistance in bacteria. Four varieties of bacteria are growing in lines in this dish. Penicillin, from the fungus at the top, has spread out through the dish. Three of the bacterial varieties have been killed as the antibiotic reached them; the fourth (far right) is antibiotic-resistant and it continues to grow. (Biophoto Associates)

Fig. 3–9

An English robin. (Biophoto Associates)

ing that some hospitals harbor drug-resistant bacteria. In many countries, women are now encouraged to give birth at home whenever possible, because mother and infant are safer from bacterial infection at home than in the hospital.

Most countries have outlawed the use of antibiotics in cattle feed. Cattle fatten faster if fed antibiotics, but they also become a large-scale breeding ground for antibiotic-resistant bacteria. Antibiotics are still added to cattle feed in the United States, and drug-resistant bacteria in cattle are becoming increasingly common.

3–F   Effects of Selection on Variation

If selection acts to produce populations of organisms that are increasingly well adapted to their environments, why aren't all members of a species identical to one another today? In some cases they are, but an organism's environment is always changing, and there are very few genes that are selected for under all conditions. A gene that is advantageous at one time is selected against at another, and thus the population maintains its genetic variability. It is often selectively advantageous for an individual to produce a genetic variety of offspring; this increases the chance that some will survive whatever the environment.

Some English robins migrate south every winter; some stay home. If the winter in England is severe, more of the robins that migrate will survive, but if the winter is mild, the survival rate is higher among the stay-at-homes. The selection pressures for migrating or staying home change from year to year, and so both traits are maintained in the robin population. Similarly, different genetic forms of the sugar maple are maintained in the population by varying selective forces. Many more seedlings of the "southern" than of the "northern" form of sugar maple in Ohio survived the drought of 1954, but the "northern" form survives severe winters better. As a result, genes for both forms are maintained in the Ohio population.

At the opposite extreme, selection has produced nearly identical, almost perfectly adapted organisms in some species. Dandelions reproduce asexually, with the result that they are all similar genetically (see Chapter 8). And dandelions are successful and widespread plants.

asexually: without sexual reproduction.

Selection often reduces the genetic variability in a population. For instance, it has been shown that both duck eggs and human babies are more likely to survive if they are of average size. Selection maintains the size of embryo or newborn which is most likely to survive. In one study, newborn babies that weighed much more or less than 3.6 kilograms (8 pounds) were much less likely to survive the first months of life than were babies that weighed 3.6 kilograms—the norm for that hospital.

Overall, we can say that selection may increase or decrease the genetic variation in a population, but a population with some genetic variety is more likely to survive, in the long run, than one without this variety. When the environment changes, some members of a genetically diverse population will usually survive, whereas the change may wipe out the whole population if its members all have the same genes.

Fig. 3–10

Dandelions. (Biophoto Associates)

Fig. 3–11

Jean Baptiste de Lamarck. (Smithsonian Institution)

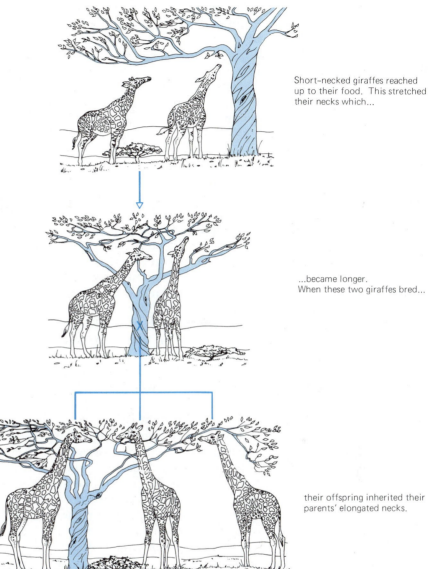

Short-necked giraffes reached up to their food. This stretched their necks which...

...became longer. When these two giraffes bred...

their offspring inherited their parents' elongated necks.

Fig. 3–12

The Lamarckian view of how giraffes evolved long necks. According to Lamarck, giraffes stretched their necks by reaching up and passed this trait to their offspring.

3–G Lamarckism

Natural selection is not the only mechanism that has been proposed to explain how evolution might occur. Before Darwin and Wallace unveiled their theory, the French biologist Jean-Baptiste de Lamarck had suggested that an organism could pass on to its offspring those traits acquired during its lifetime. His most famous example was the long neck of the giraffe: as a giraffe strained to reach leaves growing above its head, Lamarck theorized, its neck stretched, and the added length was passed on to its offspring.

Although this idea seems rather quaint to someone grounded in modern genetics, it dovetailed nicely with pre-1900 theories that different parts of the body contributed to eggs and sperm by sending minute particles through the bloodstream to a collection point in the ovaries and testes. We now know that eggs and sperm contain their own genes, not genes from other parts of the body. This is not to say that nothing an organism does can affect its offspring's genetic heritage—radioactivity and certain drugs may alter the genes in the eggs and sperm and so form new genes that pass on to the offspring.

The political importance of Lamarckism makes an interesting story. The idea that acquired characteristics could be inherited appealed to T. D. Lysenko, a Soviet agriculturalist who gained prominence shortly after the Russian revolution. He claimed he could breed better varieties of grains by giving the parent plants better conditions. Revolutionary ideologists found this theory attractive because it implied that, by improving health and education, they could increase the genetic potential for health and intelligence among people. This was much more acceptable to socialists than the eugenics programs urged by many Western biologists of the 1930s—encouraging healthy, intelligent people to have larger families, and curbing the reproduction of defective members of society. So, for ideological reasons, Lysenko rose to the top in Soviet science, and those who opposed his ideas lost their jobs and even their lives. Not until the 1950s was Lysenko finally discredited. But Lamarckism, as a theory, has had an important impact on society and is undoubtedly quite widely believed to this day.

3–H The Origin of Species

The most dramatic example of evolution is the formation of new species from old. It is difficult to define a species but, for a working definition, we can say that a **species** is a group of organisms that usually breed with one another and not with members of other such groups. Two separate populations of a single species may evolve in different directions until they become two separate species, unable to breed with one another any longer. On St. Kilda, an island off the coast of Scotland, there is a species of bird found nowhere else: the St. Kilda wren. This wren closely resembles the species of wren found on the Scottish mainland and has almost certainly evolved from a population of Scottish wrens which migrated to St. Kilda. The most important factor in this evolution was that the two populations were separated and could not interbreed. If birds from the two populations interbred, they would continually be exchanging genes and would really be part of one population; neither group would evolve adaptations to its own local conditions. Reproductive isolation between two populations is necessary if new species are to form.

Reproductive isolation often exists, as with the St. Kilda wren, when two populations occupy different areas so that their members never meet to breed. If they are forced together, members of some of these species will breed with one another. Lions and tigers, and horses and donkeys, will breed if they are forced together in captivity. But this does not mean that lions and tigers are not members of different species. In fact, the genetic differences between these

(a)

(c)

(b)

Fig. 3–13

(a) The village on St. Kilda, which is now deserted. (b) A St. Kilda wren with an insect in her bill. She is more speckled on head, wings and tail than the mainland wren. (c) A wren from the Scottish mainland bringing insects to her nest. (Biophoto Associates)

species are confirmed by the fact that the offspring of lion-tiger and horse-donkey crosses are infertile.

Populations can form new species even if they are not separated from one another. This usually results from a drastic **mutation** (a change in the genetic makeup of the organism) that produces reproductive isolation in one big step. Many cases are known in flowering plants, and recent research has shown that several animal species undoubtedly arose in a similar manner. Sometimes a mutation which doubles the normal amount of genetic material appears spon-

Fig. 3–14

Flowers of a wild Primula (top) and of a garden polyanthus (bottom), a tetraploid variety bred from the wild plant. (Biophoto Associates)

taneously in a flowering plant. Such plants (called **tetraploids**) cannot breed with normal plants of the same species because the genes of the two are incompatible. However, a tetraploid with both male and female reproductive organs may be able to reproduce all by itself. Its offspring will all be tetraploids and will form a brand new species, reproducing in isolation from the original population. Probably more than a third of all plant species have arisen in this way. Nearly all domesticated species of plants, varieties with larger fruits and flowers than those of their wild ancestors, are tetraploids.

SUMMARY

The theory of evolution asserts that species are not unchangeable, but arise by descent and modification from pre-existing species. Individuals in any population differ from one another, and some of these differences are inherited. Natural selection is any force that makes individuals with some genes more likely to reproduce than individuals with other genes in each generation; it leads to evolution, a change in the proportions of the genes in a population from one generation to the next.

In 1858, Darwin and Wallace put forward the theory that evolution occurred, and that it was caused by natural selection. The evolution of natural populations through natural selection was not convincingly shown until the twentieth century, when Kettlewell found that predation by birds acted as a selective force in the evolution of predominantly black populations of the peppered moth in polluted areas of England.

Selective pressures that shape the evolution of a population often may be contradictory, thus ensuring that genetic variation is maintained in a population. On the other hand, selection may eliminate individuals that do not conform to the population average for a particular characteristic. The traits that survive the process of selection may be thought of as adaptations that fit an organism to reproduce in its particular environment. Adaptations are many and various. The only consistent effect of natural selection is that it makes genes that contribute to an individual's reproductive success more common in the next generation.

A species is a group of interbreeding organisms that do not interbreed with members of other such groups. New species may form after two populations of the same species become so isolated from one another that genes no longer travel from one to the other. Each then evolves under local selective pressures, and they may become so different that they are considered separate species. New species may evolve within a single population when mutations create reproductive isolation between the mutant and other members of the population in one step.

SELF-QUIZ

Choose the letter of the one *best* answer for each question.

1. In the light of the definition of evolution, which of the following is *not* capable of evolving?
   a. the mice in your town
   b. the color of a population of moths
   c. your biology teacher
   d. a herd of sheep
   e. the bacteria living in your large intestine

2. Which of the following did Kettlewell conclude from his studies on the color of moths?
   a. a dark moth lays more eggs than a light moth in industrial areas
   b. dark moths are more resistant to pollution than are light moths
   c. pollution caused some moths to become darker than others
   d. dark moths are more likely to survive in polluted areas than are light moths
   e. birds prefer the taste of dark moths to the taste of light moths

3. Which bird is most successful evolutionarily?
   a. lays 9 eggs, 8 hatch and 2 reproduce
   b. lays 2 eggs, 2 hatch and 2 reproduce
   c. lays 5 eggs, 5 hatch and 3 reproduce
   d. lays 9 eggs, 9 hatch and 3 reproduce
   e. lays 7 eggs, 5 hatch and 4 reproduce

4. Suppose that you have a pack of 50 dogs. You select the largest male and largest female from the group, mate them, and sterilize the other members of the pack. Assuming that food supplies remain adequate, you should expect that, in the next generation of dogs:
   a. the young dogs will be, on the average, larger than their two parents
   b. the young dogs will be, on the average, larger than the older members of the pack
   c. the young dogs will be the same average size as the older dogs in the pack
   d. all of the young dogs will be larger than the older dogs in the pack

5. New species may originate by:
   a. doubling of the genetic material of a flowering plant
   b. gradual accumulation of changes selected for by local conditions
   c. a mutation that prevents reproduction with most other members of the species
   d. all of the above
   e. a and b only

QUESTIONS FOR DISCUSSION

1. Consider Table 3-2. What might be the disadvantage to a starling of laying a very small clutch of eggs?

2. Which of the female starlings in Table 3-2 will leave the most young in the population and hence make the greatest contribution to the genes of the next generation?

3. From what brood size do the greatest number of young survive? Is this also the most frequent family size (assume that the experimenter marked every bird that could be found)?

4. Suppose the environment changed so that only half as much food was available to the starlings. Would you expect a gradual change in the most frequent

brood size? How would this change be brought about?

5. Are all causes of death natural selection? When organisms die in an earthquake, have they been selected against?

6. Huntington's chorea is a human genetic trait in which part of the brain degenerates and the victim experiences uncontrolled, dance-like movements. It usually does not show up until the victim is in his or her 40s, after most people have completed their reproductive life. Can such a trait be selected against?

7. Is human evolution subject to the same pressures as the evolution of other species? Why or why not?

Essay:
CHARLES DARWIN

(Most of this essay is in Darwin's own words.* Our additions are in italics.)

*C**harles Darwin was born in 1809 into the sort of upper-middle-class English society that Jane Austen brought to life in her novels.***

My father sent me to Edinburgh University where I stayed for two years. I became convinced that my father would leave me property enough to subsist on with some comfort; my belief was sufficient to check any strenuous effort to learn medicine. The instruction at Edinburgh was altogether by lectures, and these were intolerably dull. Dr. Duncan's lectures on Materia Medica at 8 o'clock on a winter's morning are something fearful to remember. It has proved one of the greatest evils of my life that I was not urged to practice

*Excerpts are from *The Autobiography of Charles Darwin* (Nora Barlow, Ed.). London: Collins, 1958.

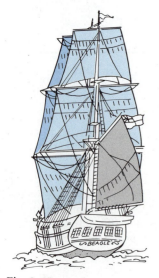

Fig. 3–15

H.M.S. Beagle.

dissection, for I should soon have got over my disgust, and the practice would have been invaluable for all my future work.

Darwin dropped out of medical school and went on to Cambridge to study theology.

From my passion for shooting and hunting, I got into a sporting set. We used often to dine together in the evening and we sometimes drank too much, with jolly singing and playing at cards afterwards. I know that I ought to feel ashamed of days and evenings thus spent, but as some of my friends were very pleasant, and we were all in the highest spirits, I cannot help looking back on these times with much pleasure.

But no pursuit at Cambridge gave me so much pleasure as collecting beetles. No poet ever felt more delight at seeing his first poem published than I did at seeing the magic words "Captured by C. Darwin, Esq." *on the label for an insect.*

In 1831 on returning home I found a letter informing me that Captain Fitz-Roy was willing to give up part of his own cabin to any young man who would go with him, without pay, as naturalist to the voyage of the *Beagle.* Afterwards, I heard that I had run a very narrow risk of being rejected because of the shape of my nose! Fitz-Roy was an ardent disciple of Lavater, and was convinced that he could judge of a man's character by the outline of his features.

The voyage of the *Beagle* has been by far the most important event of my life. As far as I can judge, I worked to the utmost during the voyage from the mere pleasure of investigation. But I was also ambitious to take a fair place among scientific men.

Fig. 3–16

Thomas Malthus. (Biophoto Associates, National Portrait Gallery, London)

Fig. 3–17

Charles Lyell. Darwin read Lyell's treatise on geology during the voyage of the Beagle and was initially more interested in geological than in biological changes. (Smithsonian Institution)

During the voyage, I had been deeply impressed by discovering great fossil animals like existing armadillos. It was evident that such facts as these could only be explained on the supposition that species gradually became modified. It was equally evident that neither the surrounding conditions, nor the will of the organisms could account for the innumerable cases in which organisms of every kind are beautifully adapted to their habitats. I soon perceived that selection was the keystone to man's success in making useful races of animals and plants. But how selection could be applied to organisms living in a state of nature remained for some time a mystery to me. In October 1838, I happened to read for amusement Malthus on population and, being well prepared to appreciate the struggle for existence which everywhere goes on, it at once struck me that under these circumstances favourable variations would tend to be preserved and unfavourable ones destroyed. The result of this would be formation of a new species.

This was in 1838. It was nearly twenty years before Darwin wrote down his theory, although in the meantime he wrote books and articles on a myriad of other biological subjects. Darwin offers no explanation for his long delay except to say: I had at last got a theory by which to work; but I was so anxious to avoid prejudice, that I determined not for some time to write even the briefest sketch of it.

Early in 1856, Lyell [*a geologist*] advised me to write out my views pretty fully and I began at once to do so. But my plans were overthrown, for early in the summer of 1858 Mr. Wallace [*Alfred Russel Wallace, an eminent naturalist*] sent me an essay, and this essay contained exactly the same theory as mine. [*Wallace had written his essay in 3 days!*]

Lyell urged that Wallace's essay and the abstract of Darwin's manuscript be published together. I was at first very unwilling to consent,

Fig. 3–18

Fathers of the modern theory of evolution. (a) Charles Darwin and (b) Alfred Russel Wallace. (Biophoto Associates, Linnaean Society, London)

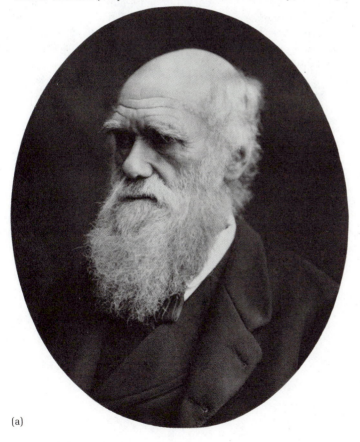

(a)

(b)

as I thought Mr. Wallace might consider my doing so unjustifiable, for I did not then know how generous and noble was his disposition. Nevertheless, our joint production excited very little attention and the only notice I can remember was by Professor Haughton, whose verdict was that all that was new in them was false, and all that was old was true. This shows how necessary it is that any new view should be explained at considerable length in order to arouse public attention.

Finally Darwin described evolution and natural selection in The Origin of Species, *published in 1858. This version of the theory, full of details and examples, immediately attracted attention and the book became a best-seller. It sparked an immense controversy, a fitting tribute to the most original biology book ever written. In particular, it bothered, and continues to offend, some Christians. This is for two reasons. First, the theory of evolution contradicts a literal interpretation of the biblical story that the earth and its organisms were created in seven days. This offends fundamentalists who believe in the literal truth of the Bible. Second, the theory superficially leads to a rather deterministic view of life, since neither human nor divine agency is needed for evolution to occur. Darwin himself became trapped in this second point of view. His mind rejected any reality that could not immediately be tested by observation or experiment.*

Toward the end of his life he wrote: I gradually came to disbelieve in Christianity as a divine revelation. *His wife, Emma Wedgewood, whose mind was more flexible, fought him relentlessly on this point. She wrote to him, "May not the habit in scientific pursuits of believing nothing till it is proved, influence your mind too much in other things which cannot be proved in the same way, and which if true are likely to be above our comprehension?" She pointed out to Darwin that bringing scientific method to bear on religious beliefs is a pointless exercise because the two are philosophically distinct realms. Many people before and after Darwin have, however, confused them.*

By and large, biologists accepted the theory of natural selection with open arms, since it explained so much and made sense of a million unexplained facts. There have always been those who resisted the appeal of the theory of evolution and every now and then declare "Darwin was wrong," in the hope of some profitable publicity, usually revealing that they have never read, and do not understand, the theory. Of course modern evolutionists have refined many aspects of Darwin's work, but there can be no question that the theory of evolution has been the single most fruitful piece of biological thinking ever produced.

THE DISTRIBUTION OF ORGANISMS

4

When you have studied this chapter you should be able to:

1. List the two main factors that determine what type of community of organisms occurs in a particular area on land, and explain how each factor exerts its influence.

2. Describe the distribution of life in the oceans and explain what physical and chemical characteristics determine this distribution.

3. Explain what the term "succession" means in ecology; outline succession from a given starting point, mentioning, in the proper order, the types of vegetation that would be found.

4. Explain why widely separated areas with similar climate usually contain similar species.

5. Define the term "convergent evolution" and give examples.

Biophoto Associates

Fig. 4–1

Organisms new to Western science are found almost every day. This as-yet-unnamed butterfly was discovered in Sulawesi, Indonesia, in March, 1980. (Biophoto Associates)

In the next three chapters we shall study ecology, the branch of biology that examines the interactions between organisms and their environments. Environments include physical factors (sunlight, temperature, water, soil) and biological factors (other organisms present). Although most people had never heard of ecology before the 1960s, ecology was celebrating its centennial by then. The science of ecology grew out of the already-established discipline of natural history, which can be defined as the observation and description of organisms in nature.

The culture of European and American countries is based on the Judeo-Christian religious heritage of most of their people, and the Biblical command, "be fruitful and multiply, and fill the earth and subdue it," is reflected in our civilization. Our traditional attitude has been that humankind is destined for the "conquest of nature." Recently, however, people have become aware that we are not above nature, but a part of it, and our own survival depends on the survival of the plants and animals, bacteria, and fungi around us. In many cases our activities have jeopardized this survival, and we now face the task of making sure that we do not push the system too far out of kilter. But in order to know how we should (or should not) act, we must first understand how organisms interact in nature. This is the aim of ecology, and we begin by studying its findings; in Chapter 7 we shall go on to see how our knowledge of ecology can be applied to the disturbances of nature introduced by the human race.

We begin by considering the question, "Why are organisms where they are?" This question arose out of two general patterns observed by Western naturalists as they explored the world and catalogued its life. First, each newly discovered area contained previously unknown species of organisms, and the list of known species has grown steadily. Second, in spite of the ever-increasing numbers of known species, there are only a few basic types of **communities**, that is, groups of organisms living in the same place. Walking through a tropical forest in South America, we would find tall trees with large leaves and fruits, festooned with immense climbing vines, and we would see colorful butterflies and birds flitting through the gloomy shade. A tropical forest in Africa would look very much the same, but the particular species of trees and vines, butterflies and birds, would be different. Other types of communities—desert, shrubland, grassland, or tundra—look much the same wherever they occur. Why are such similar communities of organisms found in different parts of the world?

4–A Climate and Vegetation

If we look at a map of the world showing the kinds of communities in different places (Fig. 4-2), we find that communities of the same type have similar climates; this is what determines the type of plants that can grow in the area which, in turn, determines the appearance of the community.

Climate depends basically on the sun. Near the equator, the sun's rays strike almost vertically, and this gives tropical plants much more of the sun's energy than is enjoyed by plants outside the tropics, which receive oblique rays (Fig. 4-3). Because of the tilt of the earth on its axis, in nontropical areas the seasons

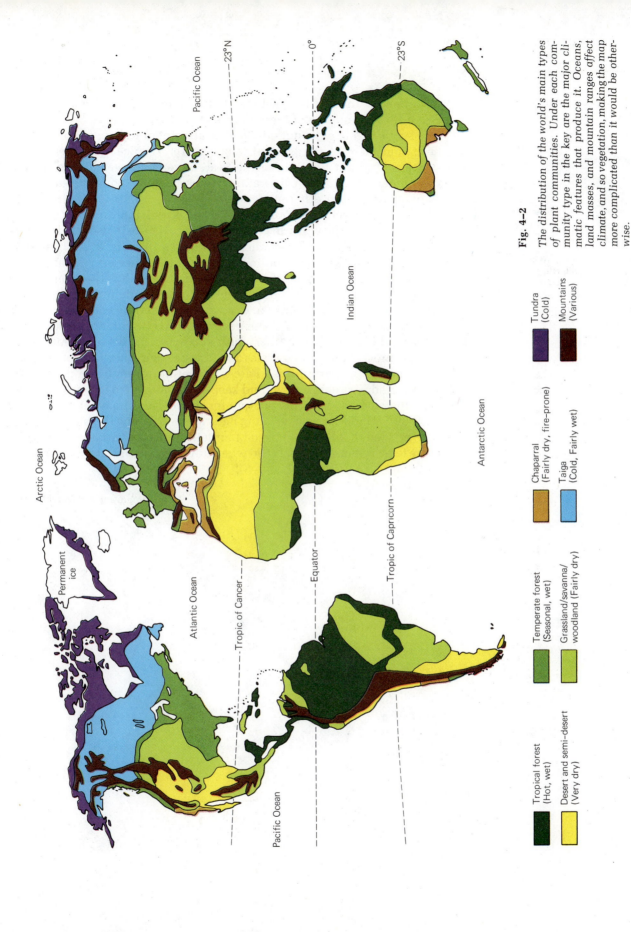

Fig. 4–2

The distribution of the world's main types of plant communities. Under each community type in the key are the major climatic features that produce it. Oceans, land masses, and mountain ranges affect climate, and so vegetation, making the map more complicated than it would be otherwise.

Tropical forest (Hot, wet)

Desert and semi-desert (Very dry)

Temperate forest (Seasonal, wet)

Grassland/savanna/ woodland (Fairly dry)

Chaparral (Fairly dry, fire-prone)

Taiga (Cold, Fairly wet)

Tundra (Cold)

Mountains (Various)

Permanent ice

Pacific Ocean

Arctic Ocean

Atlantic Ocean

Tropic of Cancer

Equator

Tropic of Capricorn

Pacific Ocean

23°N

0°

23°S

Indian Ocean

Antarctic Ocean

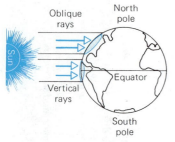

Fig. 4–3

A beam of sunlight striking the earth at high latitudes is spread over a wider area, and is therefore less intense at any one point than a similar beam striking near the equator.

vary at different times of the year, while in the tropics there is little seasonal difference in day length and temperature.

The sun is responsible not only for the amount of light available for photosynthesis, but also for the general temperature. Tropical climates, receiving near-vertical sunlight throughout the year, have fairly steady, high temperatures. In other areas, the temperature varies roughly with the amount and intensity of sunlight at different seasons.

The other important component of climate is moisture, and this depends on sunlight and temperature. Warm air holds more moisture than cool air, and as air cools some of its moisture may condense as rain, snow, or dew. Air heated at the equator rises, expanding and cooling as it mounts higher into the atmosphere, and releasing some of its moisture as it does so; the result is the steamy rains of tropical jungles. The air moves on, at high altitudes, both north and south from the equator, and eventually sinks to earth again, becoming warmer and soaking up more moisture as it does so. The descent of this dry air creates the world's great deserts (Fig. 4-4). Still further north and south, in the temperate latitudes that include most of the United States and Europe, swirling winds pull masses of air, sometimes from warm tropical areas, sometimes from frigid polar regions, giving us varied weather patterns and keeping us faithful to the evening news to see what the weather map has in store for the morrow (maybe).

In 1889 C. Hart Merriam, a young naturalist surveying the biology of part of Arizona, noticed that San Francisco Mountain showed a variation in vegetation from base to peak similar to that seen when traveling further and further north (or

Fig. 4–4

Major air movements over the surface of the earth.

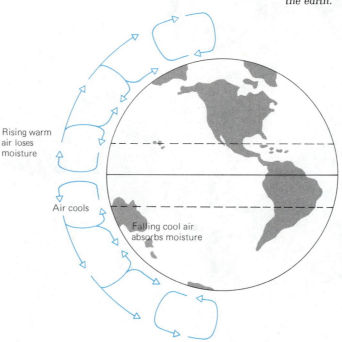

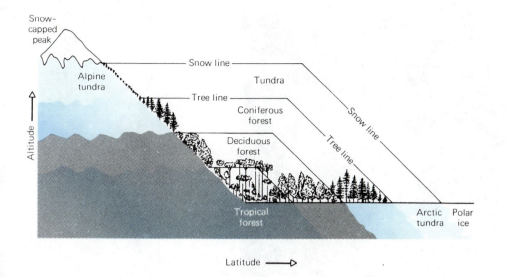

Fig. 4–5

Vegetation changes with latitude and altitude. Temperature, which affects vegetation, falls as one travels up a mountain or away from the equator so that, if there is ample precipitation, vegetation is similar at high latitudes and high altitudes as shown here. (After Colinvaux)

south) from the equator (Fig. 4-5). Since temperature varies with altitude as well as with latitude, Merriam concluded that the type of vegetation in an area is determined by its temperature.

We now know that this conclusion was oversimplified, for moisture plays a role just as important as temperature. Heavy rainfall is needed to support the growth of large trees, while progressively lighter rainfall supports communities dominated by small trees, shrubs, grasses, and finally scattered cacti or other desert plants; in extreme cases, lack of rainfall results in total lack of plants.

4–B   The Tropics

Tropical rain forests occur where rainfall is high throughout the year; the combination of warmth and wetness allows plants to grow year-round. Tropical rain forests are the richest of all communities on land—that is, they contain more different species in a given area than any other. Tall trees, 35 to 50 m high, support climbing vines and smaller plants growing on dust and litter collected in the crotches of tree limbs rather than in the soil of the forest floor; under the tree canopy the shade is too dense for most plants to grow. Dense jungle occurs only along riverbanks and in other light gaps, where there are no tall trees and sunlight reaches the ground.

Many of the animals of the tropical rain forest are brilliantly colored and inhabit the treetops, where most of the food is. On the ground is found only a shallow layer of decaying leaves, feces, and debris. The high temperature and moisture of the forest are ideal for growth of the bacteria and fungi that decompose this material, and there is never much of a backlog of undecayed litter. Minerals released by decomposition are quickly absorbed by the tree roots; the heavy rain would quickly leach minerals out of the soil and carry them away if the trees did not absorb them efficiently.

(a)

(b)

(c)

Fig. 4–6

Tropical rainforest. (a) Clouds shroud this tropical mountaintop almost all the time, providing the moisture that permits rainforest to grow. (b) A rainforest orchid, from New Guinea. (c) A fire-bellied toad. (Biophoto Associates)

If a tropical rain forest is cut down and burned, as it is in the traditional "slash-and-burn" agriculture of many rain forest tribes, rain quickly washes the minerals away and the soil becomes less fertile. The tribe moves on, leaving the surrounding forest to fill in from the edges and return the area to its original condition. Recently, however, developing nations have been clearing their rain forests on a massive scale, in attempts to feed their rapidly growing populations. If this clearing continues at the present rate, there will be no tropical rain forest left by the year 2000.

Areas further away from the equator tend to have distinct wet and dry seasons; with less moisture available trees do not grow as fast. In **tropical seasonal forests**, with pronounced dry seasons, the trees may even lose their leaves during the dry period. In even drier areas we find **savanna**, consisting largely of grasses with occasional trees. Savanna occurs in the interiors of continents where it never gets very cold.

The world's major **hot deserts** occur at about 30° N and S latitudes, about where dry air heated at the equator descends. The sun beats down through the cloudless sky by day, but since heat escapes rapidly through the clear air at night, it may be quite chilly after dark. Desert plants often store water in their tissues or lose their leaves during the dry season. Most desert animals retreat to cool underground burrows during the day, thus avoiding water loss by evaporation; they come out during the cool nights, feed on water-storing plants for their moisture, and further conserve water by excreting concentrated urine and dry feces.

Many of our houseplants come from tropical rain forests. The heat and lack of light in our houses would kill most temperate-zone plants. Rainforest plants, however, are adapted to much more humidity in the air than we can easily give them. Houses resemble deserts more than any other natural community, but many desert plants are rather dull because they grow slowly and seldom flower; plant collectors are continually on the lookout for interesting plants which will grow and flower in the difficult conditions of the average home.

Fig. 4–7

*Savanna with a herd of zebra, Kenya.
(Biophoto Associates)*

Fig. 4–8

Some deserts, like this part of Algeria, are so dry that no plants can survive. (Biophoto Associates)

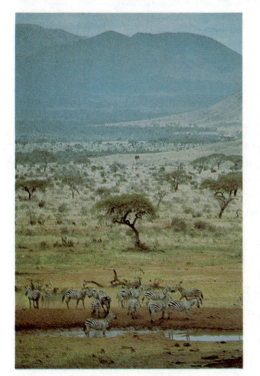

(a)

Fig. 4–9

(a) Chapparal in California. (William Schlesinger) (b) Redwood forest on the Pacific Coast of North America. (Biophoto Associates)

(b)

4–C Temperate Zone

North and south of the tropics and their adjacent deserts lie the temperate regions—so called because their climate seldom reaches the temperature extremes typical of tropical and polar areas. **Temperate forests** occur where there is enough rainfall to support trees. At one extreme are the rain forests, such as those of the North American Pacific coast, where heavy winter rains and summer fog allow growth of the tall redwood forests of northern California and the slightly shorter evergreens of the Olympic Peninsula. **Temperate deciduous forests** are composed mostly of **deciduous** trees, which shed their leaves during the winter. Most of the northeastern United States is (or was) covered by deciduous forest. In the southeast and in parts of the west we find temperate evergreen forests, composed mostly of pines adapted to the poor soil, droughts, and frequent fires of these regions. In somewhat drier areas of central California and parts of the southwestern states is temperate woodland, containing short trees interspersed with shrubs and grasses.

deh-SID-you-us

(a)

(b)

(c)

(d)

(e)

Fig. 4–10

Temperate forest, (a) in spring, (b) in autumn. Animals of the temperate forest include (c) a chaffinch; (d) deer, a fawn is shown here, camouflaged against a background of dead ferns; (e) a wood mouse. (Biophoto Associates)

Drier areas, like coastal California, are populated by **chaparral** or **temperate shrubland** communities, consisting of shrubs that can withstand the droughts and fires typical of the area. Chaparral shrubs may burn to the ground, but their roots send up new growth when the rains come again in the fall.

The vast prairies of North America represent the **temperate grasslands**, also found as the steppes of central Asia, the pampas of South America, and the veldt of South Africa. Temperate grassland occurs in the fairly dry interiors of continents. Savanna occurs in similar, but hotter, areas. The temperate grasslands of the world have largely been turned into rich "breadbaskets," farmed to produce cereals (grains) and soybeans, and very little of the original grassland remains.

Temperate deserts occur where it is too dry even for grasslands, as in the rain shadows of tall mountain ranges. For example, in the North American Great Basin there are large areas containing little but sagebrush and clumps of drought-resistant grasses.

reptiles: lizards, snakes, turtles, and their kin.

amphibians: frogs, toads, salamanders, and newts.

The animal life in temperate regions is less varied and less colorful than that of the tropics. Although birds and mammals are common and varied, there are fewer reptiles and amphibians, which depend on the environment to maintain their body temperatures. Even among the warm-blooded birds there are many species that migrate to the tropics during the winter.

4–D Taiga and Tundra

TIE-guh

BORE-ee-al

The **taiga** (Russian = primeval forest) or **boreal forest,** as it is sometimes called, stretches in almost unending monotony across Canada and Siberia. The monotony is due to the small number of tree species (spruces, pines, and firs) adapted to the extreme cold of the region. There are two seasons: winter and not-winter. Since winter is the longer season, most of the precipitation falls as snow. While the ground is frozen, the trees cannot replenish water lost by evaporation from their needles; the needle-like form and the heavy, waxy covering of the leaves slow evaporation.

Animal life of the taiga includes moose, wolverines, wolves, lynx, gray jays, and many migratory birds, but the dense, cool shade and the small number of plant species severely limit the variety of animal life.

TUN-druh

LIKE-en

North of the boreal forest is the treeless **tundra**, dominated by short grasses, dwarf woody shrubs, and lichens. (A **lichen** is an alga and a fungus growing in intimate association.) The climate is too cold to support the growth of trees, and the soil is shallow, often with a permafrost layer (frozen year-round) below the top few inches thawed by the warmer summers. Because the growing season is so short, tundra vegetation takes a long time to grow back after a disturbance. This was one reason why environmentalists opposed construction of an oil pipeline across Alaska.

Caribou, muskox, and reindeer are among the largest animals on the tundra; lemmings, snowy owls, foxes, and wolverines are also typical. Many migratory birds are summer residents, nesting amid the hordes of biting insects which complete their short life cycles in the puddles and streams of thawed snow during the summer.

(a)

Fig. 4–11

Taiga, or boreal forest, around a mountain lake in Canada.

(b)

(d)

(c)

Fig. 4–12

Tundra and its inhabitants. (a) The treeless landscape; (b) a reindeer; (c) a saxifrage; (d) a dwarf willow, one of the small shrubs that can survive in the thin tundra soil. (Biophoto Associates)

Taiga and tundra are found not only in the far north (the continents do not extend far enough south in the Southern Hemisphere for taiga and tundra to exist there) but also high up on tall mountains.

4–E Communities of the Ocean

The open ocean supports two distinct zones of life. Near the surface is a layer of **plankton**, which are small, floating organisms. Animals in the plankton feed on photosynthetic plankton, and large, actively swimming fish and whales feed on these smaller organisms. As organisms in this top layer die, their bodies fall to the ocean floor, where another community lives on these nutrients falling from above. No plants can exist at these depths because there is not enough sunlight, and animals must survive both scarcity of oxygen and great pressure from the overlying water.

The community at the surface of the sea has abundant light for photosynthesis and oxygen for respiration, but the mineral nutrients needed by plants are scarce, so much of the ocean's top layer contains little life. Regions of upwelling, where water rises from the deep and brings up nutrient-rich sediments, are the only areas with abundant surface life.

Shallow water supports a variety of plant and animal life, for plants can attach to the bottom rocks and still receive the sunlight they need for photosynthesis. These plants may provide both shelter and food for a multitude of animals. Continental shelves, with shallow waters and nutrients carried down to the sea by rivers, are quite rich in life per unit of area. Even richer are coral reefs, in shallow tropical seas, where abundant light and high temperatures encourage rapid plant growth, which in turn supports an amazing variety of animal life.

Fig. 4–13

A coral reef. (Steven Webster)

(a)

(b)

4–F Ecological Succession

We all know that coastal California is predominantly covered, not by chaparral, but by farms, roads, and buildings, and we all know why: human civilization has disturbed the natural communities, clearing the vegetation to make room for human affairs and their adjunct parking lots. So, when we say that climate determines the type of community in an area, we mean the community that would exist if the area were left alone long enough, rather than what may actually exist there. The community that forms if the land is left undisturbed, and that perpetuates itself as long as no disturbances occur, is called the **climax community**.

When a climax community is disturbed, either by human activities or by natural means, such as floods or fires, it begins a slow process of returning to its original state by a process known as succession. **Succession** is a progressive series of changes that ultimately produces a climax community.

A familiar example of succession in the New England area is "old field succession," by which abandoned farms return to the climax deciduous forest. When a farmer stops cultivating the land, grasses and weeds quickly move in and

Fig. 4–14

Periodic fires determine the type of vegetation in many places. (a) A bush fire in Australia. (Biophoto Associates) (b) Fast-growing conifers are replacing those destroyed in a fire several years before. (William Schlesinger)

Fig. 4–15

Succession. (a) Shrubs, goldenrod, and aspens invade an abandoned field which will eventually become forest. (b) Early succession within a forest. Many mosses need little or no soil and can grow on trees and rocks where they will die, forming soil in which larger plants can grow. (Biophoto Associates)

(a)

(b)

(a)

(b)

Fig. 4–16

Fugitive species: (a) fireweed in a forest clearing (Biophoto Associates), (b) a swallowtail butterfly (Biophoto Associates, N.H.P.A.).

perennial (per-EN-ee-al): living for many years.

clothe the earth with a carpet of green: black mustard, wild carrot, and dandelions. The "pioneers" of newly available habitats, these plants grow rapidly and produce seeds adapted to dispersal by wind or animals over a relatively wide area. Soon taller plants, such as goldenrod and perennial grasses, move in. Because these newcomers shade the ground and their long root systems monopolize the soil water, it is difficult for seedlings of the pioneer species to grow. But even as these tall weeds choke out the sun-loving pioneer species, they are in turn shaded and deprived of water by the seedlings of pioneer trees, such as pin cherries and aspens, which take longer to become established but command the lion's share of the resources once they reach a respectable size. Succession is still not complete, for the pioneer trees are not members of the species that make up the mature climax forest; slower-growing oak, maple, beech, and hickory trees will eventually move in and take over, shading out the saplings of the pioneer tree species. After perhaps a century or two, the land returns to the species composition of the original climax forest.

Old field succession occurs relatively rapidly because it builds on the soil left by the original climax forest. If the soil has been severely depleted by poor farming practices, or if there is no soil (as on scoured rock left by a retreating glacier or hardened lava from a volcano), succession will be much slower, for soil must form before most plants can grow. On bare rock, soil may be formed as lichens give off acid that dissolves the rock surface, and as water freezes and thaws in cracks, breaking down the rock. Dead lichens also contribute organic matter to the new soil, and mosses may gain a hold in even a thin layer of lichen remains and rock dust. As the mosses break up the rock more and add their own dead bodies to the pile, the seeds of small rooted plants are able to germinate and grow, beginning a process much like that of old field succession.

In any tract of land, we can always find at least small patches that are undergoing succession following disturbance—a spot where a large tree has fallen, leaving a light gap where pioneer weeds can move in and begin a miniature old field succession; a small pond filled in with dead leaves and gradually developing the species composition of the surrounding forest community; a slope where a landslide has occurred; or a burned forest. The existence of various patches undergoing succession ensures a steady supply of **fugitive** plants, the fast-growing, here-today-and-gone-tomorrow weeds. These species have seeds that can spread over appreciable distances, carried by wind or by animals. In addition, the seeds of many of these fugitive plants are adapted to live for long periods in a dormant state, germinating when a disturbance provides the proper conditions, such as increased light.

Animals, as well as plants, may be fugitive species. Insects that specialize in eating a particular plant species may travel far and use their keen senses to smell out new patches of their food plant some distance away. Some of our agricultural pest problems stem from the fact that most crop plants originated as fugitive species, depending on their sparse distribution and their nomadic habits (never in the same place for many seasons in a row) to protect them from their insect predators. By planting fields exclusively to one crop year after year, farmers create a paradise for such fugitive animals as cabbage worms and cucumber beetles, which no longer have to spend energy to find food and have nothing to do but eat and multiply.

Repeated disturbance will prevent an area from returning to the climax state. Suburban homeowners spend considerable time, energy, and money creating a continuous series of disturbances that maintains a lawn of a few species of short grasses, constantly interrupting the old field succession of tall weeds, shrubs, and light-loving tree seedlings that will take over if vigilance is relaxed. Many areas in the midwest were prevented from reverting to climax forest by native Americans who deliberately set fires after they found that herds of buffalo, which live on the prairie, could be hunted more efficiently than the solitary whitetail deer of the young forests.

Succession can be seen on any city street. Mosses, lichens, and weeds

"John, it's time to go out and interrupt the ecological succession again."

establish themselves in cracks in the sidewalk, quite large plants may grow in a corner where leaf litter and dirt have been deposited by a rain gutter, and moss invades a roof that needs repair. If we stopped cleaning and repairing it, even the center of Manhattan would turn into a rock-filled woodland within our lifetimes.

4–G Why Are Organisms Where They Are?

In our study of succession, we have seen that species may be widespread if they have efficient dispersal mechanisms, such as seeds that float on the wind or are carried by animals. However, there is a limit to how far an organism can travel over territory that is unsuitable for its survival. If we return to the question that opened our discussion, "Why are organisms where they are?" we find that at least two factors determine the answer: climate, and the organisms' ability to disperse to areas where the climate is appropriate.

When we look again at our African and South American rain forests and repeat the question, "Why are there different species in these two areas?" our answer is that, although each area has a climate suitable for growth of the plants and animals of the other area, they are separated by long stretches of ocean, an impassable barrier to dispersal. There are exceptions: for example, small animals called rotifers form cysts that can be blown almost anywhere in the world, but the animals can live only in very restricted types of environments; hence the rotifer species in a marble cemetery urn in Pennsylvania may be the same as that in a marble urn in a South African cemetery, but different from that in the granite urn on the next grave!

Because most species cannot travel between continents, there are different species in the tropical rain forests of the different continents. The similarities of form and color of species in each place result from **convergent evolution**—the evolution of similar characteristics suited to similar environments. For instance, the advantage of being able to conserve water in desert habitats has led to the predominance of plants with thick, water-storing stems in deserts all over the world. Similarly, many different species of plants in different deserts have spiny leaves that deter animals from using the stems as the source of their own water (Fig. 4-18).

Fig. 4–18

Spines on this cactus deter animals that might eat the plant. (Biophoto Associates)

Fig. 4–19

Convergent evolution. (a) cactuses in Arizona; (b) Euphorbs in the Canary Islands. (Biophoto Associates)

(a)

(b)

SUMMARY

The distribution of organisms depends on the availability of suitable places to live and the ability of the organisms to disperse to these places. Suitability depends largely on the climate: the amount of sunlight; the temperature, which depends largely on the amount of sunlight; and the amount of moisture. In the ocean, the availability of nutrients and oxygen also plays a prominent role.

Although the climate of an area determines the composition of its climax community, patches of the area are always in various stages of ecological succession as a result of disturbances of the climax community. Organisms adapted to living in the unstable communities of early successional stages have effective dispersal mechanisms and perpetuate themselves by continuously colonizing the new habitats which arise in the surrounding climax community.

The limits of their dispersal ability prevent most organisms from colonizing all possible habitats, and so in different parts of the world we find similar communities inhabited by similar species, which have arisen by the convergent evolution of similar adaptations to similar environments.

SELF-QUIZ

Choose the *one best* answer for multiple choice questions.

1. One reason that the effect of increasing altitude on vegetation is similar to the effect of increasing latitude is that:
   a. temperature decreases with increases in both latitude and altitude
   b. mountains decrease the angle of sunlight
   c. there are always clouds over mountaintops
   d. it is hard for plants to disperse up the sides of mountains
   e. both are higher up

2. Life near the surface of the open ocean is often limited by:
   a. the temperature
   b. availability of nutrients
   c. lack of oxygen
   d. availability of water
   e. none of the above

3. A pond in a deciduous forest becomes filled in with rock particles and dead leaves, creating soil. List, in order, the types of vegetation that would be seen as this area undergoes ecological succession, and name the climax community that would eventually result.

4. The American prairies and the Asian steppes do not have the same species of grasses because _____. However, both are inhabited primarily by grasses because_____.

5. Which of the following pairs of organisms is an example of convergent evolution?
   a. polar bears and panda bears
   b. oak trees and maple trees
   c. cacti and euphorbs (Fig. 4-19)
   d. American bison and whitetailed deer
   e. skunks and raccoons

QUESTIONS FOR DISCUSSION

1. What is the natural climax community in your area? How have people used the characteristics of the area to suit their own purposes? Are people introducing changes that are not compatible with the natural climate and vegetation? If so, what?

2. Southern Louisiana and northern Florida are the same distance from the equator as desert country in Mexico and Texas. Why is the area in Louisiana and Florida not desert like the area in Mexico and Texas?

3. Why is there less variation in size of vegetation in the tundra than in tropical regions?

4. Old field succession slows down as it proceeds. Explain why this is so.

5. Why does a light gap contain some species of organisms different from those in the surrounding climax forest?

ECOSYSTEMS

When you have studied this chapter, you should be able to:

1. State what an ecosystem is, list its essential components and the nonessential components that are usually present, and give the role of each.

2. Diagram a simple food web and trace the flow of nutrients and energy through it; give the correct trophic level of each component of the ecosytem.

3. State what is meant by the productivity of an ecosystem and list factors that limit productivity in particular types of communities.

4. Explain what is meant by the statement that energy enters and leaves an ecosystem, whereas nutrients cycle within it; recognize approximately how much of the energy available to each trophic level is actually captured, is used in respiration, is used in growth, or is passed on to the next level.

5. Diagram a simple nutrient cycle.

6. State the source of energy for most ecosystems.

7. State the difference between an oligotrophic and a eutrophic lake.

8. Describe what happens when an oligotrophic lake is polluted by nutrients or heat.

9. Describe how far it is possible to "clean up" a polluted body of water.

In Chapter 4 we saw that a tropical rain forest or a prairie may cover thousands of square kilometres. For convenience's sake, ecologists usually select smaller units—for example, a hillside, a lake, or a field. The value of studying such units was recognized in 1887 by Stephen Forbes, biologist for the Illinois Natural History Survey, when he wrote:

> A lake . . . forms a little world within itself—a microcosm within which all the elemental forces are at work and the play of life goes on in full, but on so small a scale as to bring it easily within the mental grasp . . . If one wishes to become acquainted with the black bass, for example, he will learn but little if he limits himself to that species. He must evidently study also the species upon which it depends for its existence, and the various conditions upon which these depend.

Nowadays, we would call Forbes's lake—or any other manageably small unit, with more or less distinct boundaries—an **ecosystem**. Forbes's writing points out the other characteristics of an ecosystem: it consists of all the different organisms living in the area, along with their physical environment. These all interact and change one another, so that the study of an ecosystem is indeed a complex undertaking. For convenience, we usually regard the ecosystem as an isolated unit, but, in fact, things invariably move from one ecosystem to another, as when soil and leaves wash from a forest into a lake, or birds migrate between their summer and winter homes.

flora: the plant life of an area (often includes bacteria and fungi, which used to be classified as plants).

fauna: the animal life of an area.

Not all ecosystems are natural; a space station, an aquarium, and a pot of houseplants are artificial ecosystems. A farm is often considered as an ecosystem because we must recognize the interactions between crop plants, fertilizers, pesticides, soil, climate, and the natural flora and fauna in order to manage the farm effectively.

Fig. 5–1

A lake forms a partly isolated ecosystem enclosed by its banks.

5–A The Basic Components of Ecosystems

If an ecosystem is to exist indefinitely, it must contain water, carbon dioxide, various minerals, oxygen (in most cases), and various kinds of organisms; it must also receive a continuous supply of energy.

The sun is the source of energy for nearly all ecosystems. Solar energy enters the system as green plants carry out photosynthesis, using sunlight to convert water and carbon dioxide into food. In order to do this, a plant must also take in small amounts of simple mineral nutrients, which it builds into more complex chemicals. Green plants are called **producers** because they produce food, which they use to live, to grow, and to reproduce.

green plants: refers to all photosynthetic organisms (including bacteria and unicells as well as plants).

This process cannot go on forever. Eventually all the soil minerals would be locked up in the bodies of plants (many of them dead) and plant growth would

(a)

(b)

(c)

Fig. 5–2

Living components of an ecosystem. (a) Green plants are the producers in most natural ecosystems as well as those produced by humans. (b) Consumers, like this caterpillar, eat the producers (or each other). (c) Decomposers, like these fungi, recycle everything. (Biophoto Associates)

Fig. 5-3

The major components of an ecosystem (colored). The colored arrows show energy flow; the black arrows show movement of nutrients. The solid box encloses the four essential components of a self-sustaining ecosystem. In addition, most ecosystems contain animals, as indicated within the dashed lines. Energy is continuously lost from all ecosystems, mainly in the form of heat, which all organisms produce.

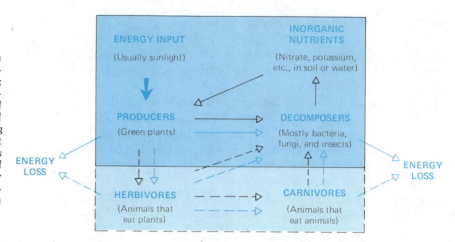

cease. However, every ecosystem also contains **decomposer** organisms—insects, bacteria, and fungi—which digest the food locked up in the dead plants and return their minerals to the simple forms that living plants can absorb and use.

Some ecosystems have few or no green plants. Examples are found in ponds and streams in heavily shaded forests, or on the deep sea floor. These ecosystems depend on green plants growing in sunnier areas to supply their food. Life in a forest stream feeds on leaves and twigs falling from the trees above; likewise, the organisms in the depths of the ocean depend for nourishment on dead bodies sinking from the upper layers of water. Ultimately, though, the source of energy for these ecosystems is still the sun. A few species of bacteria can use energy from chemical reactions, rather than from sunlight, to produce their food. These producers, however, account for an insignificant fraction of the total energy trapped by living organisms on earth; if the sun were snuffed out tomorrow, life as we know it would rapidly grind to a halt.

5-B Food Webs and Energy Flow

An ecosystem is usually much more complicated than the basic chain of producer and decomposer life and death. The energy and nutrients trapped in the bodies of producers are resources that may be exploited by a variety of **consumer** organisms. Consumers include **herbivores** (animals that eat plants) and **carnivores** (animals that eat animals), as well as parasites of both.

All of the living things in an ecosystem can be organized into a **food web** showing what eats what. By analyzing a food web, we can tell how energy and nutrients flow through an ecosystem. Energy and nutrients enter the food web together during photosynthesis, as plants store the sun's energy by changing simple substances (carbon dioxide and water) into the complex form of food. When the plant is eaten, the food it contains is transferred to a herbivore; when the herbivore in its turn is eaten, its body becomes food for a carnivore. Each organism uses most of its food energy to move, respire, and perform other life processes, and very little energy is left to be used by a predator.

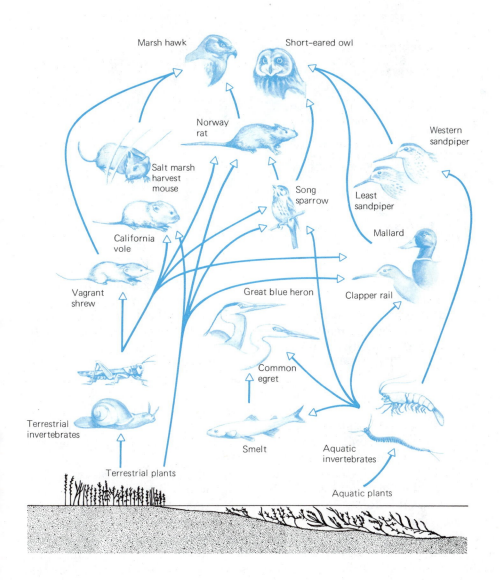

Fig. 5–4

A food web in a salt marsh. This diagram shows, for instance, that voles, rats, sparrows, and rails eat terrestrial plants; rats eat sparrows (mostly eggs and nestlings); and then marsh hawks eat the rats. (After R. F. Johnson. Wilson Bulletin, 68: 91, 1956.)

The flow of energy through the ecosystem, beginning at the point of photosynthesis, can be measured at various points by answering questions such as these: how much solar energy reaches the system? How much energy is present in the food made during photosynthesis? How much of the energy in plant material can be used by a herbivore that eats the plant? How much of the energy in a herbivore can its predator use? Let us consider the answers to some of these questions.

The rate at which the plants in an ecosystem use energy from the sun to build up organic matter (matter in living organisms) is called the **productivity** of the ecosystem. Since energy and nutrients enter living things together, productivity can be measured as the rate of increase in plant body weight. Productivity varies

TABLE 5-1 **PRODUCTIVITY OF SOME MAJOR TYPES OF COMMUNITY* EXPRESSED AS GRAMS OF DRY PLANT MATERIAL PRODUCED PER SQUARE METRE PER YEAR†**

Coral reefs	2500
Tropical rain forest	2200
Temperate forest	1250
Savanna	900
Taiga (boreal forest)	800
Cultivated land	650
Continental shelf	360
Tundra	140
Open ocean	125
Extreme desert	3

*See Chapter 4 for characteristics of the various communities mentioned in this table.
†After R. H. Whittaker, *Communities and Ecosystems*, 2d ed. New York: Macmillan, 1975.

community: the group of all organisms living in an area.

from one type of community to another (Table 5-1). Many communities are less productive than they might be because particular factors limit plant growth; for instance, deserts are short of water and tundra is short of heat. The productivity of oceans depends on nutrients and temperature. Coral reefs, which occur only in warm seas, have plenty of nutrients from the bodies of their dead plant and animal residents. The open ocean, however warm, is short of nutrients. Continental shelves, the shallow areas around continents, are more productive because rivers wash nutrients down onto them.

For those who enjoy figures, the annual productivity of the whole earth is estimated at about 170 billion tons (dry weight) of new plant material—115 billion tons produced on land and 55 billion tons in the oceans (although the oceans occupy about 70% of the earth's surface).

Extraordinary as it seems, the photosynthesis producing these enormous quantities of plant matter traps only about 0.5% of the solar energy that reaches the earth as visible light. Furthermore, only a small fraction of this 0.5% is passed on to consumers and decomposers. Every time an energy conversion takes place—from sunlight to food, from food to new cells or muscle movement—some useful energy is converted into heat energy which escapes into the environment. In addition, energy is lost at every step in a food web: each

Fig. 5–5

In this temperate ecosystem, grasses, trees and other flowering plants are the main producers, creating new plant material by photosynthesis. (Biophoto Associates)

Trophic level

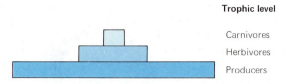

Carnivores

Herbivores

Producers

Fig. 5-6

A pyramid of energy showing how much energy goes into each trophic level in a river in the southern United States. Much less energy exists in the carnivore than in the herbivore trophic level because each carnivore must eat many herbivores to stay alive and up to 90% of the energy is lost at each transfer from one trophic level to another.

organism uses up some energy in its own life and growth, leaving less energy available to the organisms of the next step, and the organisms at each step are startlingly inefficient at intercepting what little energy is available to them. About half the energy trapped by plants is soon released as the plants respire to provide energy for their own life processes, and the remaining half goes into plant growth and reproduction. Herbivores eat only about 10% of the plant material produced each year, and not all of the energy in a herbivore's food becomes herbivore. A caterpillar feeding on leaves absorbs from its gut only about half of the material it eats; the rest is eliminated in the feces. Furthermore, most of the food absorbed from the gut is used in respiration; only about 15% of it finally ends up as more caterpillar.

In general, animals convert roughly 10% of the energy they eat into growth (new animal). The energy loss at each step, or **trophic level**, in a food web explains why food webs seldom have more than about five trophic levels: decomposers, plants, herbivores, and two levels of carnivores, each eating the members of the level below (see Fig. 5-6). A wolf may have to travel 30 kilometres a day to find enough food, and a tiger needs a home range of up to 300 square kilometres. An animal that fed on wolves or tigers would have to cover a vast range to find enough of its widely scattered prey, and it would spend more energy looking for food than it would gain from its meals of wolves or tigers.

5-C Mineral Cycles

Energy enters an ecosystem during photosynthesis and leaves it, mainly as heat, during respiration, muscle contraction, building of new tissues, and so forth. Because it is continually lost, energy must be continually supplied (as sunlight). Water and minerals, by contrast, may travel in a continuous cycle from the water or soil to a plant, to an animal that eats the plant, back to the water or soil via a decomposer, and back into a plant again.

Although the productivity of an ecosystem may be limited by the lack of sunlight, heat, or water, in many cases it is limited instead by a shortage of minerals that are essential plant nutrients. This, in turn, may limit populations of other organisms in the ecosystem.

Living organisms require six nutrients in relatively large quantities: carbon, hydrogen, oxygen, nitrogen, phosphorus and sulfur. These are present as minerals in rocks and are released, by erosion and weathering, into soil, rivers, lakes, and the oceans. Nitrogen and oxygen are also abundant in the atmosphere, and the more modest amount of carbon dioxide in the air is the main source of carbon

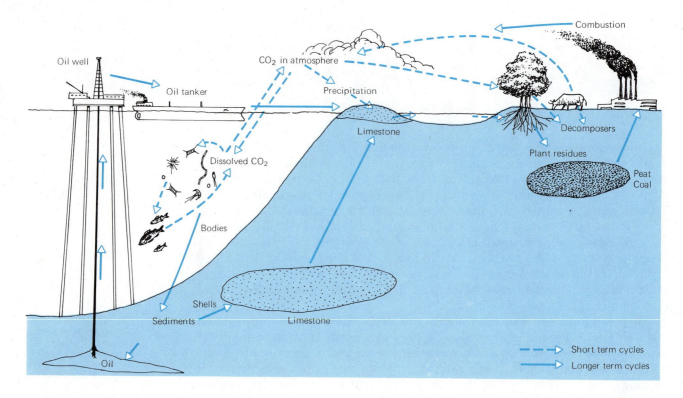

Fig. 5–7

A simplified mineral cycle. This figure shows how one nutrient—carbon—cycles through several ecosystems. Carbon passes through the processes indicated by dashed arrows more rapidly than through those indicated by solid arrows. (CO₂ is carbon dioxide gas.)

available to plants. Substances are so thoroughly recycled that, on the average, every lungful of air you inhale contains several million atoms that were once inhaled by Plato—or by any other person who lived for 65 years.

Mineral nutrients are sometimes recycled rapidly, as in grasslands, where the vegetation dies back each year. Decomposers destroy the dead plants and make many of the nutrients they contain available again the following year. At the opposite end of the time scale, mineral nutrients may spend millions of years locked in the dead bodies of organisms. For example, remains of marine organisms may sink to the bottom of the ocean to form oil deposits or sedimentary rocks. Millions of years may pass before we mine these nutrients as fossil fuel, or before movement of the earth's crust heaves them up as rock that eventually erodes and releases the nutrients.

5–D Lake Ecosystems

In the years since Stephen Forbes wrote the quotation at the beginning of this chapter, we have learned much about lake ecosystems and about how human activities can upset the balance of nature in a lake. Because the poor quality of much of our drinking water is a very real threat to human health today, we will consider this subject in a little more detail.

Sunlight is the lake's source of energy. As light filters down through the waters of a lake it is gradually absorbed; some of it is used in photosynthesis by tiny plants floating near the surface. Every lake has a particular depth, which depends on the clarity of the water, at which the amount of oxygen produced by photosynthesis equals the amount of oxygen in the water used up by the respiration of all the organisms present. Above this depth there is plenty of oxygen; below it there is less and less.

Rooted aquatic plants such as water lilies and rushes grow in the shallow parts of the lake. Fish, tadpoles, insects and other arthropods, snails, and worms live and feed among these plants. In the open surface waters of the lake live floating plants, which need light, and animals that require abundant oxygen, such as fish and small arthropods (Fig. 5-8).

Lower down, where there is less oxygen, the chief source of energy is dead plants and animals that fall from above. Decomposers, fish, and invertebrates that can tolerate low oxygen levels live on this debris, and on one another.

Temperature has a profound influence on life in the lake. A peculiar property of water is that it is most dense at 4°C. As a result, water at 4°C sinks beneath

Fig. 5–8

The structure of a lake ecosystem, showing some organisms found in various parts of the lake.

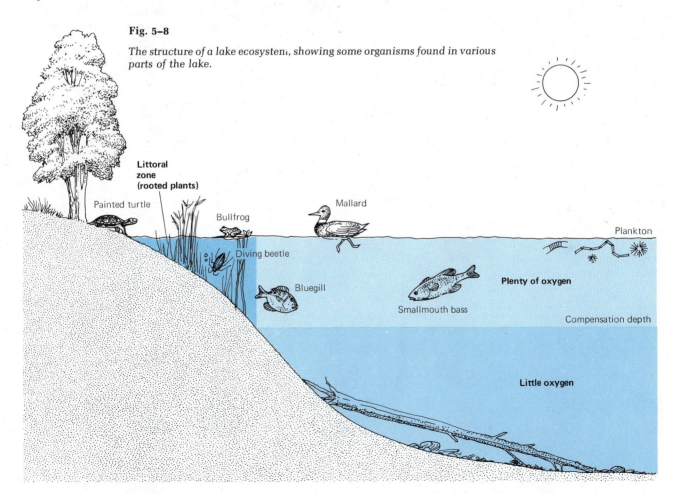

Littoral zone (rooted plants)

Painted turtle

Bullfrog

Mallard

Diving beetle

Bluegill

Smallmouth bass

Plankton

Plenty of oxygen

Compensation depth

Little oxygen

water that is either warmer or colder. In winter, therefore, water cooled below 4°C rises above water at 4°C, and at 0°C the surface water freezes. Plants and animals under the ice survive in water that remains at 0–4°C throughout the winter, protected from the weather above by an insulating layer of colder water or ice. In spring, the ice melts and the sun warms the surface waters, which sink as they approach 4°C, forcing colder water below to rise to the surface. During this **spring overturn**, the waters at different depths mix, helped by wind and waves.

You may have been swimming in the summer in a lake where the surface layer is lovely and warm, but if you let your toes sink they encounter much colder water. This is because the sun-warmed surface waters stay on top and do not mix with the heavier, colder water underneath, which, in a deep lake, stays at 4°C. The boundary between warm and cold water is called the **thermocline** and it lies deeper and deeper in the lake as the summer goes on. In fall the thermocline begins to rise again as the top layer of water loses heat and its temperature falls, eventually leading to another mixing, the **fall overturn**. Trout fishermen know that spring and fall are the times when trout can be found in cold, rising surface waters. Trout need oxygen-rich water, and, since warm water holds less oxygen than cold water, trout spend the summer in the deeper, colder waters.

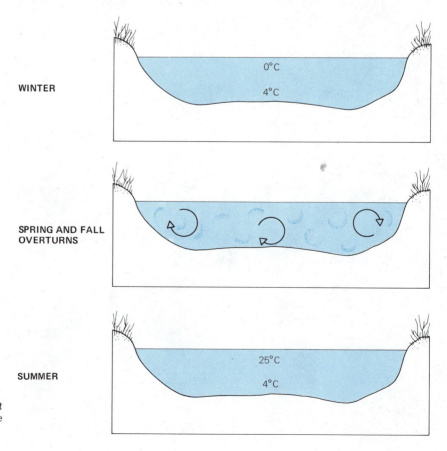

WINTER

SPRING AND FALL OVERTURNS

SUMMER

Fig. 5–9

Temperature layers at different seasons in a lake in a temperate region.

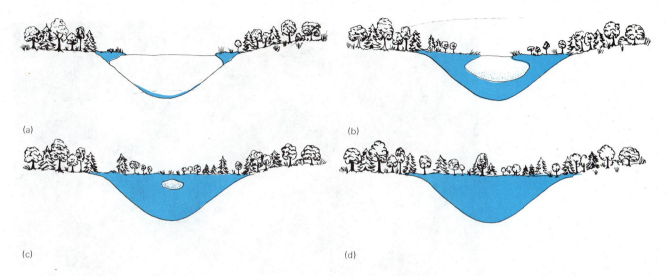

(a)

(b)

(c)

(d)

Fig. 5–10

How an oligotrophic lake may disappear. The lake, (a), first becomes eutrophic, (b), as nutrients and sediment make it more productive. Finally (c) the lake becomes a bog with a layer of plants right across it, and eventually (d) it turns into dry land. This succession of events would normally take thousands of years, even with quite a shallow lake such as Lake Erie. Some oligotrophic lakes, such as Lake Tahoe, are so deep that they might take essentially forever to go through this succession if pollution did not speed up eutrophication.

Lakes can be divided into two categories based on their production of organic matter. **Eutrophic** ("good food") lakes are relatively shallow, rich in nutrients, and short of oxygen. In contrast **oligotrophic** ("few food") lakes are usually deeper, with steeper sides and fewer nutrients; the water is clear and there may be plenty of oxygen in the water right to the bottom of the lake. In the normal course of events, a lake slowly fills with sediment and dead matter and becomes more eutrophic. Eventually the lake turns into a bog or marsh and finally dry land. For a deep oligotrophic lake, this may take millions of years. One effect of human pollution is the speeding up of eutrophication.

Oligotrophic lakes with clear water and game fish are much more appealing and useful to us than eutrophic lakes covered with algae, clogged with weeds, and populated by carp and minnows; this is one reason why our pollution of lakes causes so much concern.

POLLUTION OF LAKES

Broadly defined, pollution is an undesirable change in the characteristics of an ecosystem. Let us consider how human activities like building power plants and dumping wastes pollute a lake and how such pollution can be minimized.

Some forms of pollution are obviously harmful. Chlorinated hydrocarbons, such as DDT and the polychlorinated biphenyls (PCBs), and toxic heavy metals, such as mercury, are just plain poisonous to most organisms—including humans. Many of these substances become concentrated in living tissues because organisms cannot excrete them efficiently. So they accumulate and are passed on, at successively higher concentrations, to predators at higher trophic levels. This is particularly damaging to meat-eating animals such as hawks, humans, and fish-eating sea birds. Pelicans in California and peregrine falcons in the eastern United States were almost wiped out when DDT caused them to lay thin-shelled eggs which broke when the birds sat on them; all ocean swordfish

Fig. 5–11

Water pollution. (Biophoto Associates)

and all fish from Lake Ontario were banned as human food for a time because of the levels of mercury they had accumulated.

Adding nutrients to a lake can also do dreadful damage; the phosphate detergent story is a good example. All the phosphorus in a natural ecosystem comes ultimately from rocks and, because many ecosystems lie on rocks containing little phosphorus, their productivity is limited by the scarcity of this nutrient. Nowhere has this been demonstrated more dramatically than in lakes (Fig. 5-12). Humans add phosphorus to lakes continuously in sewage, detergents, or fertilizer runoff from farms. This increases the productivity of the lake and hastens eutrophication, changing the character of the lake.

Many deep oligotrophic lakes, like Lake Tahoe on the Nevada–California border, have become noticeably more eutrophic since the 1950s because of nutrient pollution. The most obvious symptom is clouding of the water as microscopic plants and animals multiply. In addition, oxygen-loving fish like trout become less common. A nutrient-polluted lake goes through several steps before it reaches the familiar stage of stinking bottom mud and shores littered with dead fish. First, nutrients polluting the lake support a **bloom**, or unusually large population explosion, of algae. As the algae die, bacteria decompose their bodies and use up most of the oxygen. In mud containing no oxygen, certain bacteria produce hydrogen sulfide (a gas that smells like rotten eggs) when they break down dead plants. In a "healthy" lake, bottom-dwelling photosynthetic bacteria, known as sulfur bacteria, use up this hydrogen sulfide as it forms. When there is a lot of fertilizer or sewage in the water, however, there are so many algae and bacteria in the surface water that little light reaches the bottom, and so the photosynthetic sulfur bacteria grow poorly. They cannot use up all the hydrogen sulfide produced and so this gas rises to the surface, giving the nutrient-polluted lake its characteristic stench.

Eutrophication can never be "cured," but it can be arrested. If we stop dumping sewage into a lake, the algae nourished by the sewage will die and sink to the bottom and will not be replaced. The water will become clearer and regain oxygen, and fish populations will return. The 1970s have seen several success stories as people have cleaned up lakes and rivers that had become little more than open sewers. In England, salmon now swim up the River Thames as far as London for the first time in a hundred years. The Thames does not stink any longer, nor does Lake Erie, where towns and industries have reduced the sewage and other waste dumped into the lake. Lake Washington, near Seattle, has also recovered and is now safe for swimming, boating, and fishing.

Nutrient pollution gives trouble precisely because it supplies substances needed by organisms but normally in short supply. Many pollutants, such as toxic chemicals, plastics, and aluminum cans, cause trouble for another reason: they are not biodegradable—that is, decomposers cannot break them down. As a result, they are never destroyed but remain forever to injure and poison people and fish alike. Sailors in small boats report that a scum of bits of plastic floats on the surface of the Atlantic Ocean all the way from the Caribbean to the coast of Europe. Since plastics are chemically similar to some food substances, there is a chance that microorganisms able to break down some kinds of plastic will one day evolve. Already a bacterium that lives on a mixture of aviation fuel and aluminum has caused trouble in the aircraft industry: this is surely a recent adaptation to modern life! Some firms are manufacturing plastics that bacteria *can* destroy—you will sometimes see the word "biodegradable" on an indigestible-looking plastic bottle. Some plastics, however, will probably never be biodegradable, and all we can do is "dispose of them properly" as the label admonishes.

Thermal pollution results from heating water—for example, when lake water is used for cooling in a power station. Since plants grow faster when it is warm, the lake's productivity increases and, as in nutrient pollution, speeds eutrophication. Thermal pollution of lakes can be avoided if water used for cooling is passed through cooling towers or ponds where it can lose its heat to the air before it is returned to a lake or river; in some areas, however, this solution is unsatisfactory because the cooling towers cause fog.

Cooling towers and sewage treatment plants cost money and raise the cost of our water, electricity, and consumer goods. At first it seems as if we must choose between clean lakes and cheap electricity—self-interest tells us to choose cheap electricity. But we have become more far-sighted in recent years and realize

Fig. 5–12.

The effect of phosphorus on the productivity of a lake. This lake in Manitoba was divided in two by plastic sheeting across the narrow neck in the middle of the photograph. Phosphorus was added to the half of the lake in the upper part of the photograph. Several weeks later, the phosphorus-fertilized half of the lake was opaque as a result of a massive plankton "bloom" (population explosion). The lower part of the lake was as clear and oligotrophic as it was before the experiment. (David Schindler)

(a)

(b)

Fig. 5–13

An increasing problem: oil spills. (a) The Argo Merchant founders off the coast of Massachusetts in 1976. (b) A diver after working on the Argo Merchant. (U.S. Navy)

that the choice is not really so simple. For instance, it is cheaper never to pollute a lake than to attempt to clean it up when it becomes a public health hazard. It is cheaper to install a sewage system to protect our local lake, river, or seashore than it is to pipe drinking water hundreds of miles from an unpolluted lake. In many cases prevention, in the long run, proves cheaper than cure, and we can feel a virtuous glow at preserving the ecosystem for our grandchildren into the bargain!

SUMMARY

An ecosystem consists of all the organisms living in an area and depending on each other in various ways, plus their physical and chemical environment. Every ecosystem needs certain chemicals, a source of energy, and producer and decomposer organisms. Most also contain consumer organisms. All of the organisms in an ecosystem can be connected in a complex food web diagram showing what eats what.

Energy and nutrients enter the living world together, as producers make food. Almost all the energy entering the living world comes from sunlight, trapped by the photosynthesis of green plants. This energy is re-released and used to carry out life processes, by plants, consumers, and decomposers. Since some useful energy is lost at each energy transaction, fresh energy must be supplied to an ecosystem continually, and the number of feeding, or trophic, levels is limited.

Although energy flow through the ecosystem is essentially one-way, nutrients may cycle indefinitely. Dead bodies and animal wastes are used as energy sources by decomposers, which release the nutrients in the simple forms that plants can use.

In oligotrophic lakes, productivity is low because of a scarcity of nutrients or heat. Eutrophic lakes are highly productive but contain little oxygen. They are less desirable for recreation and drinking water than are oligotrophic lakes. Pollution with nutrients from sewage and fertilizer runoff, for example, speeds eutrophication, a process that would occur much more slowly without human interference. Thermal pollution speeds eutrophication by raising the activity and productivity of the organisms and by reducing the amount of oxygen in the water. Reducing input of polluting agents can reverse many of the undesirable changes brought about by pollution.

SELF-QUIZ

1. The role of decomposers in an ecosystem is _____.

2. Using the items below, diagram a food web.

   weeds   dung beetle
   rabbit   herbivorous insect
   soil fungi   spider
   berry bush   sparrow
   wolf   hawk

3. Wolves and lions may be said to occupy the same trophic level because they both:
   a. eat herbivores
   b. utilize their food with about 10% efficiency
   c. live on land
   d. are large animals
   e. have a wide range of dietary items

4. Productivity in a coral reef is higher than productivity in most open ocean near the equator because the coral reef has more:
   a. sunlight
   b. nutrients
   c. water
   d. heat

5. Of the total amount of energy that passes from one trophic level to another in a food web, about 10% is:
   a. derived from the sun
   b. used up in respiration
   c. stored as new body growth
   d. converted into the form of useless heat
   e. passed out in feces

6. Nutrient cycles do *not* involve:
   a. movement of some nutrients from organisms to the atmosphere
   b. denser populations of organisms in areas where nutrients are abundant
   c. entry of most nutrients into the food web through animals
   d. limitations of the number of organisms in the ecosystem due to shortage of some nutrients
   e. long-term loss of nutrients from living communities into ocean-bottom sediments

7. Diagram a simple carbon cycle.

8. The ultimate source of energy in almost all ecosystems is _____.

9. Eutrophication:
   a. may be accelerated by excessive phosphorus input
   b. is caused by inhibition of algal blooms
   c. decreases the productivity of a lake
   d. need never happen to a lake if human beings take proper precautions to avoid detergent, sewage, and thermal pollution

10. Phosphate detergents pollute by:
    a. killing producers
    b. killing decomposers
    c. stimulating growth and reproduction of producers and decomposers
    d. killing fish

QUESTIONS FOR DISCUSSION

1. Is there one or more than one food web in any ecosystem?

2. The sparrow in Self-Quiz Question 2 eats berries, herbivorous insects, and spiders. What trophic level would you place it in, and why?

3. How is the flow of energy in an ecosystem linked to the flow of nutrients? How do energy and nutrient flow differ?

4. A "healthy" ecosystem has a balance between life and death, and between the capture and release of energy and nutrients. Use this idea to show how pollution by nutrients, heat, or toxic chemicals disrupts this balance.

6 POPULATIONS

OBJECTIVES

When you have studied this chapter, you should be able to:

1. List three factors that affect the reproductive potential of a population, and explain how the factors act.

2. Give examples of three factors which kill more individuals in a dense than in a less dense population and explain why the factors act in this density-dependent fashion.

3. Explain why most populations remain roughly the same in size from year to year.

4. Draw a graph showing exponential growth of a population and give a reasonable explanation for why that population is growing as it is.

5. Describe why a predator that feeds on only one species is more likely to regulate the size of populations of its prey than is a predator that feeds on many different species.

6. Describe an example of successful biological control of a pest species.

7. Describe the human population explosion, its causes and probable fate, and the role of the demographic transition in the population explosion.

A **population** can be defined as all the members of a species that occupy a particular area at the same time. Each population has its own particular characteristics. Just as the human population of Paris differs from that of Pittsburgh in containing more people who speak French, more business executives, and more children, so the black bear population of Vermont differs in some ways from the black bear population of Wyoming. Different populations of the same species often interbreed to some extent, but the less the interbreeding, the greater the differences between the populations, because each will slowly evolve adaptations to its own particular locale.

Every population has certain characteristics which can be used to describe it or predict its fate. Populations have characteristic densities, patterns of distribution, age structures, and genetic characteristics. In this chapter, we shall concentrate on things that determine the size of a population. Why are there just so many oak trees, lightning bugs, poppies, panthers, and people in the world?

6–A Population Size and Growth

The size of most populations seems to vary little. Mosquitoes, houseflies, June bugs, cockroaches, and weeds in the lawn show up in much the same numbers from year to year. Can this really be the case? Surely these organisms produce so many offspring that their populations could increase greatly each year. A herring may lay a million eggs, a housefly or butterfly several hundred. It is obvious that most of these eggs never live to grow up. Why not? What limits the size of populations?

To answer these questions we must first know how rapidly a population could grow if nothing checked its growth. We find that each population has a characteristic reproductive potential. Experiments show that under ideal condi-

Fig. 6–1

The earth, a planet that is overpopulated with human beings, as seen from space. (U.S. Navy)

Fig. 6–2

Populations of many sea birds, like these gannets, are restricted by limited space to nest. (Biophoto Associates)

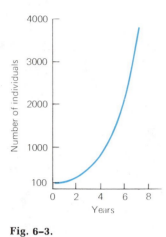

Fig. 6–3.

A graph to show the exponential (= geometric) growth of a population which has access to infinite supplies of space, food, water, and so on.

tions, the number of organisms grows **exponentially**, or **geometrically**, which means that an ever-increasing number of individuals is added to the population in each succeeding unit of time (Fig. 6-3).

A population grows exponentially when its members have a superabundance of the things they need. History records many examples of exponential growth by species imported into hospitable areas where natural enemies or competitors were absent. Dandelions, starlings, and house sparrows introduced into the United States from Europe all underwent dramatic population explosions. Similar population explosions occur when bacteria invade the intestinal tract of a newborn animal or when decomposers invade a newly dead animal or plant. The human population began to grow exponentially in the mid-eighteenth century as better food and hygiene reduced the infant death rate.

An individual's reproductive potential can be increased in any or all of three ways: by producing a larger number of offspring every time it reproduces; by having a longer reproductive life so that it reproduces more times; and by reproducing earlier in life. Of these three factors, the last is by far the most important; a population of bacteria can grow faster than a population of oak trees because a bacterium can reproduce when it is less than an hour old, an oak tree only when it is many years old. A human population in which each woman bore 3.5 children (if such a thing were possible), starting at age 13, would grow just as fast as a population in which each woman bore six children, starting when she was 25. This is why population prophets are so interested in the age at which people marry and have children. It has been calculated that the reproductive potential of a human population in which women lived forever, producing a child every year, starting when they were 20, would be only about twice that of a population in which women bore 5 children altogether, one each year from the age of 20!

(a)

(b)

Fig. 6–4

(a) A cabbage white butterfly. (b) A gypsy moth. (Biophoto Associates, N.H.P.A.)

Cabbage white butterflies, scourge of the home gardener's cabbage, cauliflower, and broccoli plants, were introduced into Quebec from Europe in 1865. In 1869 a New England gardener complained that he had "not been able to raise a respectable cabbage for four or five years, on account of the ravages of this species of voracious rascals." By 1884 the butterflies had spread right across the country to west of the Rockies. In contrast, the gypsy moth, introduced into Massachusetts from Europe in 1869, has spread westward only as far as Minnesota in more than a century. One secret of the cabbage white's faster spread is its greater reproductive potential. Cabbage whites can breed when they are only a few weeks old and may have more than three generations in a single summer, with each female laying hundreds of eggs. The gypsy moth, in contrast, usually has only one generation a year and lays fewer eggs.

In general we can say that the greater the reproductive potential of a species the faster populations of that species can grow if they get the chance, but that most organisms seldom get the chance, and population explosions are the exception rather than the rule.

6–B Regulation of Population Size

No population can grow exponentially for very long. Eventually all the food or space is used up, and then more individuals will die than are born. For instance, when rabbits were introduced into Australia by Europeans an incredible population explosion followed. Eventually, however, rabbits became so numerous that the local predators learned to catch them; some rabbits could not find food; and disease spread rapidly. As a result, the death rate increased and the population ceased to grow.

The number of individuals in a natural population varies with time, sometimes dramatically. If the size of a population declines too drastically, the population may become extinct. Local populations of many species do become

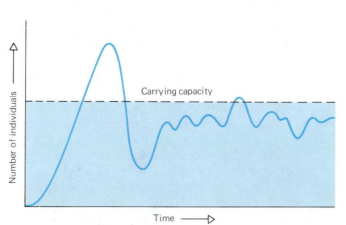

Fig. 6–5

Variation in the number of individuals when a small population is introduced into a new and hospitable environment. At first the number of individuals grows very rapidly until the population runs out of some vital resource such as food or space. This produces a population crash, when more individuals die than are born. Then the population grows back up until the rate of births and deaths is about equal. The population can vary around this size indefinitely without destroying its environment, and this size is known as the carrying capacity of the environment for the population.

Fig. 6–6

A crowded population: penguins in Antarctica. (Biophoto Associates)

extinct quite often, but the population is usually re-established later by immigration from neighboring populations of the same species. A population of dandelions may be exterminated when you spread a herbicide on the lawn, but usually it will be re-established later by immigration.

In spite of fluctuations, the average size of most large populations changes relatively little over the years. This suggests that population size is generally regulated by **density-dependent factors**, factors that kill a higher proportion of a crowded population than of a sparse one. Predation and disease are factors which can cause a higher mortality rate in a large than in a small population. Disease spreads faster through a dense population, and predators catch more victims when prey is abundant. Even more obvious than disease and predation, however, is competition between individuals—for food, space, water, light, or whatever is in short supply. Competition is always more keen and kills more individuals when a population is large than when it is small.

COMPETITION

Competition among members of the same species is almost inevitable in nature. It is the rule in human populations, too, where people compete for money, status, gasoline, space, or anything that is in short supply.

A good example of competition is the fate of maggots of blowflies that infest the carcasses of dead animals. The first flies that find a carcass lay their eggs, and most of the larvae will have enough food to grow to maturity. As the population of flies on the carcass builds up, however, there comes a time when the food runs out and most of the remaining maggots die.

Instead of competing directly for a limited resource, members of many

maggot: the larva (immature stage) of a fly.

species compete indirectly for social dominance or for a territory. A **territory** is an area occupied by one or more members of the same animal species and defended by its occupants against other members of the same, and sometimes of other, species. The territory is of no value in itself, but it contains shelter and enough food for the animal to feed itself and raise its young. In countries with inadequate welfare systems, the same is true of human competition for jobs and the money they bring. The competition is not really for the money but for the food and shelter it buys.

Biologists studying muskrats in an Iowa marsh found that the marsh always contained about 400 adult muskrats, with very little variation from year to year. This was because the marsh contained about 180 territories, each territory providing food and a refuge from predators for a pair of muskrats. Muskrats unsuccessful in the annual competition for territories were forced to live in unfavorable areas at the edge of the marsh, where they and their offspring suffered a high rate of mortality from overcrowding, inadequate food, predation, and interference by other animals. One year bad weather killed many of the muskrats, but their reproductive potential (averaging 11–17 young per female per year) quickly returned the population to the 400 level.

Fig. 6–7

Many organisms spend at least part of their lives in crowded populations. Here a brood of cater-pillars competes for food on a tree. (Biophoto Associates)

Fig. 6–8

Territories. These Adelie penguins in Antarctica nest in crowded groups. Each pair of birds defends a small territory which may include only the nest (a depression lined with stones) and standing room around it. Three nests are visible here: in the left and right foreground and under the bird sitting in the middle. (U.S. Navy)

Adaptations that reduce competition between members of the same species have evolved. In some bird species, males and females have beaks of different lengths, enabling them to feed on different insect prey; in many butterflies, amphibians, and fish with larval stages, the young and adults feed on different foods. Generally, however, members of the same species need the same resources and are bound to compete for them except when the population is very small.

Competition occurs between members of different species as well as between members of the same species. Sometimes the competition is so acute that one species becomes extinct. A unique species of giant tortoise lived on one of the Galapagos Islands off the west coast of South America. In 1962 an expedition reported that the tortoises were extinct, although remains of tortoises that had been dead for only about 2 years were found. The reason for the extinction was not far to seek. The tortoises' food plants had been completely consumed by goats, introduced onto the island by a party of fishermen in 1957.

Extinction is not the inevitable result of competition between species; we coexist, if not happily, with insects such as the corn borers that compete with us for each year's corn crop. Often competing species evolve in such a way as to reduce the competition among them; as a result they can continue to exist together. For example, different species of wood warblers in northern forests forage for insects in different parts of trees, reducing the competition among them (Fig. 6-10).

PREDATION

We have seen that competition for space, food, and other resources can determine the maximum size of a population. When the population size falls

Fig. 6-9

Weeds compete with crop plants for light, space, water, and minerals. Here corn marigolds invade a field of oats. (Biophoto Associates)

BLACKBURNIAN WARBLER   BAY-BREASTED WARBLER   MYRTLE WARBLER

below that level, more juveniles will live to grow up than usual and the population will return to its original size, when it will again be limited by competition.

We often think of **predation**—members of one species eating those of another—as limiting the size of a population. (We can include in predation a parasite eating its host, a herbivore eating a plant, and a carnivore eating another animal.) Predation is a major source of mortality in most populations, but only occasionally is it the main factor that limits population size.

An example of predation limiting the population occurred in the 1880s, when the California citrus industry was nearly destroyed by the cottony cushion scale, a small sap-sucking insect accidentally introduced from Australia. When other methods of control failed, a biologist was sent to Australia to look for the scale insect's natural enemies. He brought back 129 vedalia ladybirds (carnivorous beetles), which were used to start a breeding colony in California. In 1889, 10,000 of the ladybirds were released in California orchards and by October of that year the cottony cushion scale had been virtually eliminated from most of Southern California. The control remained effective for over 50 years, until the advent of the pesticide DDT, which killed the vedalia beetle in some areas and

Fig. 6–10

Different species of wood warblers hunt for insects in different parts of trees in the same New England forests. This means that there is less competition between them for food than there would be if each species fed in all parts of a tree.

Fig. 6–11

A vedalia beetle (the shiny black object to the left of the picture) feeding on cottony cushion scale. This tiny beetle saved the California citrus industry from destruction in the late nineteenth century. (F. E. Skinner)

permitted the scale population to thrive. This remains one of the most striking successes of biological control of an insect pest. Millions of dollars' worth of citrus fruits were saved every year for the $1500 it cost to send someone to Australia to find the beetles.

A predator specializing in one species of prey and reproducing rapidly can control the numbers of its prey. Both eventually survive at low densities. When cottony cushion scale is rare, the vedalia beetle will not find small isolated pockets of the scale. These scale populations will increase until beetles find them and wipe them out, so the presence of the vedalia beetle keeps the scale insect at a very low density. Similarly, if the density of the scale is low, the number of beetles which can survive by eating the scale will be small.

Most well-known carnivores, such as dogs and tigers, are generalized carnivores feeding on many species; they are much less likely to control the populations of their prey than are predators that feed on only one or a few species. This is true because a generalized carnivore feeds on whatever is available; the most it will usually do is to keep a population explosion of one of its prey in check. For example, the survival rate of young sockeye salmon in a lake in British Columbia increased threefold after predatory squawfish and trout were removed.

It was once thought that wolves controlled the population of moose on Isle Royale, an island in Lake Superior. The island has a population of about 600 moose and 20 timber wolves, which feed on the moose and on smaller prey such as mice. Recent studies, however, suggest that the number of moose on Isle Royale is limited by the availability of the essential nutrient sodium. The wolves kill mainly ailing and feeble moose, the very young and the very old. A healthy adult moose can outrun a wolf or stand and fight; one blow of its leg can kill a wolf. It seems that the wolves take mainly moose that would have died anyway and, therefore, do not control the size of the moose population.

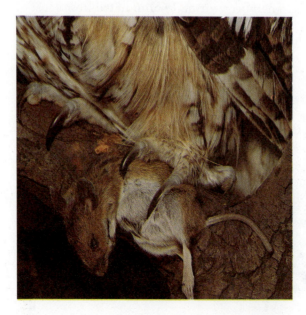

Fig. 6–12

A tawny owl, a generalized predator, with a mouse. (Biophoto Associates)

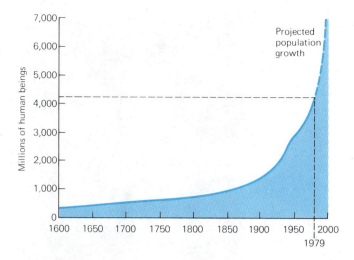

Fig. 6–13

Growth of the human population in the past 400 years.

6–C Decline of the Human Death Rate

The number of people on earth has increased steadily for at least the last 2000 years, but the greatest population growth has taken place in the last 200 years (Fig. 6-13), mainly because the death rate has been dramatically reduced in most countries. Most people assume that this decline in the death rate is a result of modern medical discoveries such as the antibiotics that have reduced mortality in the twentieth century, but this isn't true. Much more important were earlier, less spectacular, improvements in nutrition and hygiene.

antibiotics: substances that combat growth and reproduction of (usually disease-causing) organisms.

In 1665, Samuel Pepys recorded in his diary that bubonic plague, which had devastated Europe, had reached London:

> This day, much against my will, I did in Drury Lane see two or three houses marked with a red cross upon the doors and "Lord have mercy upon us" writ there; which was a sad sight to me, being the first of the kind that, to my remembrance, I ever saw . . . I was forced to buy some roll tobacco to smell and to chaw, which took away my apprehension.

By the end of the century, the plague had claimed some 25 million lives; one fourth of the people in Europe died.

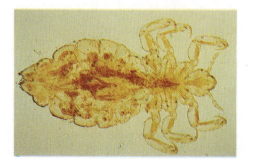

Fig. 6–14

A human body louse, one of the insects that spreads plague. (Biophoto Associates)

Fig. 6–15

Stephen Hales, an eighteenth century botanist (see Section 18–C) wearing a wig. Wigs were popular partly because the wearer could shave his head and so get rid of head lice. (Biophoto Associates, Linnaean Society of London)

Bubonic plague and related diseases such as pneumonic plague, typhus, and trench fever are caused by bacteria transmitted to people by the bites of body and head lice and of rat fleas. In most parts of the world, everybody had lice until the nineteenth century. Body lice live, not on the human body, but in clothes. In Europe the poor never removed all of their clothes; they merely took some of them off in summer and put more on in winter. The wealthy changed their garments but never washed them. The habit of shaving the head and wearing a wig probably originated partly as an attempt to reduce the lousiness of even the most royal head, but since the wigs were made of human hair, they often contained nits (louse egg cases) when purchased. Pepys says:

> Thence to Westminster to my barber's; to have my Periwigg he lately made me cleansed of nits, which vexed me cruelly that he should put such a thing into my hands.

The tutor of a French princess in the seventeenth century wrote:

> One had carefully taught the young princess that it was bad manners to scratch when one did it by habit and not by necessity, and that it was improper to take lice or fleas or other vermin by the neck to kill them in company, except in the most intimate circles.

The new habit of washing clothes and hair, which became more and more common after the eighteenth century, reduced the incidence of typhus and the plague dramatically. Later, it was mainly those living where cleanliness was impossible, such as soldiers on campaign, who caught these diseases. During the Civil War in the United States, 92,000 died in battle or of their wounds, but 190,000 died of disease. World War II was the first modern war in which more people died from enemy action than from disease.

The "clean revolution" that made louse-borne diseases little more than a historic curiosity for most of us began in the eighteenth century. Reasoning purely from circumstantial evidence and trial and error, some people claimed that disease could be reduced if lice and rats, which carried fleas, were eliminated. Physicians started to insist that bedding and clothing in hospitals and homes be washed and fumigated, and buildings cleaned. The efforts of the reformers were hampered by die-hards who clung to the old convictions that disease was borne on the harmful night air, that washing the body endangered the health, and that washing clothes merely ruined the cloth. Typhus was little more than a memory in most civilized communities by 1909—the date when biologists first showed that the disease is transmitted by lice—and yet the antibiotics that can now cure many of these diseases were 30 years in the future.

The effect of modern medical care on mortality in a Navajo reservation in Arizona was studied during the 1960s. Access to the hospital saved only one life in 10 years whereas, in the same period, instructing parents on infant care reduced infant deaths to 6, instead of the 15 which occurred in the previous 10 years. Most deaths in less developed countries are due to respiratory and digestive infections in infants; such deaths can be easily reduced, by improved nutrition and hygiene, without expensive medical care.

Life expectancy is the number of years that a newborn baby can be expected to survive. Not surprisingly, life expectancy is higher in countries whose people are better nourished and educated. Life expectancy in most of Europe and North America, for example, is now more than 70 years; the average for Latin America and East Asia (including Japan and China) is about 63 years; for most of the other Asian countries and nearly all of Africa, it is less than 45 years.

Despite the staggering sums we pay for health care, improved health care will not increase North American life expectancy much in the foreseeable future;

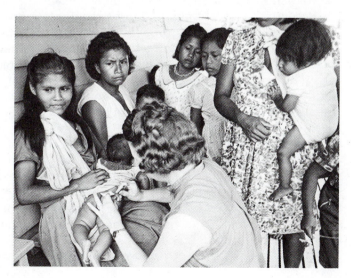

Fig. 6–16

A child in Ecuador receives a polio shot. Most of the decrease in the human death rate in the twentieth century is due to improved nutrition, hygiene and protection from disease which have decreased the infant mortality rate in most parts of the world. (U.S. Agency for International Development)

environmental pollution and our living habits (diets, exercise) are the main things that must be changed if an increase is to occur. The major causes of death in the United States and most of Europe are now cancers (many caused by **carcinogens**—cancer-causing agents—such as tobacco smoke and asbestos), heart attacks, and accidents.

6–D The Demographic Transition

Fig. 6–17

The demographic transition in progress. A doctor counsels a patient at a family planning center in Nepal. A wall poster emphasizes the pleasures of wanted instead of unwanted children. In Nepal, as in many less-developed countries, contraceptives are much more readily available than they are in the West, from village shops and door-to-door salesmen. (U.S. Agency for International Development)

As the death rate has declined since the eighteenth century, a decline in the birth rate has followed. This decline in the birth rate is called the **demographic transition**. The factors most clearly correlated with a lowered birth rate are increasing education and employment of women. The spread of knowledge that lowers the death rate is usually part of a general program to improve education. Educated women learn that they no longer need to bear a large number of children to ensure that a few survive, and they also learn birth control techniques. In addition, women discover that they can contribute to the family's increasing prosperity by holding a job and by spending less time and energy on raising children. This is usually attractive to women, and often an economic necessity, even in countries where religion and tradition dictate large families.

The demographic transition has always taken from one to three generations to spread through the population of a country. While it spreads, death rates are low but birth rates remain high, and the population grows enormously. The population of the world is now growing at the rate of about 70 million people a year; we are gaining the equivalent of the population of New York City every six months. The world population has grown from about 500 million in 1650 to about 4 billion in 1974, and it is expected to approach 8 billion by the year 2000. It took more than a hundred years for the world population to double from 1 to 2 billion; it doubled from 2 to 4 billion in less than 70 years. Table 6-1 shows that the time it takes the population to double varies greatly from one country to another, and that population growth is fastest in those countries which are poorest and can least afford to feed their citizens.

Recently, the world population has begun to grow more slowly. Between 1965 and 1974, birth rates dropped appreciably and, for the first time in more than 50 years, declined faster than death rates. Nevertheless, in 1975 the world population increased by 1.8%, and in Africa the birth rate was still increasing. Even if birth rates are reduced until they equal death rates by the year 2000, the world's population will double to 8 billion people. This is so because of the age structure of the population: more than half of the people in the world are under 25, and even if they have only two children per couple, the world's population will increase greatly.

The rapid growth of the human population is often called the "population explosion." Our exponential population growth has been like the growth of a population of bacteria or rabbits newly introduced to a favorable environment. Like all population explosions, ours must come to an end when there are too many people for each to obtain an adequate amount of some resource that is vital to life but in limited supply.

Food is the vital resource whose inadequacy has already begun to slow the

Fig. 6–18

Combating malnutrition: Milk is distributed to children suffering from protein deficiency in India. (U.S. Agency for International Development)

TABLE 6-1 **DOUBLING TIME AND PER CAPITA GROSS NATIONAL PRODUCT (GNP) IN 1978.**

Country	Per capita GNP* (in U.S. $)	Doubling time† (in years)	Comments
Kuwait	11,510	18	Oil! Immigrants contribute largely to population growth
Switzerland	8,050	173	
U.S.A.	7,060	116	
Japan	4,460	63	
Libya	5,080	18	Highest per capita GNP in Africa
U.S.S.R.	2,620	77	
Argentina	1,590	53	
Albania	600	28	Europe's highest birth rate and lowest per capita GNP
China	350	41	
Bolivia	320	27	Lowest per capita GNP and one of the highest birth rates in Latin America
Pakistan	140	24	One of the highest per capita GNPs in Southeast Asia
Nepal	110	30	One of the lowest per capita GNPs and highest birth rates in Southeast Asia
Yemen	210	24	Lowest per capita GNP in the Middle East
Burundi	100	33	One of the world's highest birth rates

*Per capita gross national product is the gross national product divided by the number of people in the country. (Figures from *World Population Growth and Responses*, Population Reference Bureau, 1978.)

†Doubling time in this table is the time it will take the country to double its population if rate of population increase remains the same as it was in 1978.

growth of human populations. Since our food production has not kept up with the population explosion, there is already widespread starvation. About 12,000 people die of starvation every day and at least 10 million children in the world are so malnourished (poorly fed) that their lives are in danger. In India alone, a million children die of malnourishment every year.

The term "starvation" means death from lack of food. However, most people who are inadequately fed do not die because they take in too little nourishment to sustain life, but because their bodies have little resistance to diseases which would not be fatal to adequately nourished individuals. Most malnourished people eat most of the calories they need, but their food is deficient in protein and vitamins. Protein deficiency diseases, such as kwashiorkor, also lead to mental retardation, particularly in young children.

Malnourishment is, therefore, a more accurate term than starvation for the cause of death in these cases, but this word loses much of its force when we read that half the teenagers in the United States are malnourished. This merely means that American teenagers consume more calories and fewer vitamins than is ideal, not that they are likely to die from a common cold or an intestinal parasite that might kill the truly malnourished person.

Human starvation is not yet necessary. The world's farmers produce enough food calories, proteins, and vitamins to keep more than the world's 4 billion people in good health. The trouble is that food, and the income to buy food, are not equally distributed. Just as with the muskrats, resources are unevenly distributed, so that some individuals live in plenty while others starve. The present problem of food distribution is more economic and political than it is biological. However, only by grasping the biological aspects of the problem can we judge whether or not some of the solutions proposed for our population problems will work.

Before the year 2000, there will almost certainly not be enough food to feed the approximately 8 billion people who will then exist. Extrapolation from animal populations leads to the estimate that the human population must eventually level out at something less than 8 billion people.

SUMMARY

Given ideal conditions, the number of individuals in a population increases exponentially at a rate that varies with the reproductive potential of the species. This reproductive potential is determined mainly by the age of the female parent at first reproduction, but it is also influenced by the number of offspring produced at each reproduction and by the parent's reproductive lifespan. A population seldom, if ever, grows as fast as its reproductive potential would permit. Among the fastest-growing populations are those of organisms introduced into new, favorable environments.

Most populations remain about the same size from year to year. The size of populations is generally limited by density-dependent factors that kill a higher percentage of individuals in a dense than in a less dense population. These factors include predation, disease, and competition for limited resources.

Competition among members of one species for resources is universal. Competition among members of different species is also the rule; it leads either to extinction of one species, the weaker competitor, or to selection for adaptations that reduce the competition. For instance, each species may become specialized so that it uses only part of a limited resource.

Predation is one factor that may cause high mortality in a population. Many specialized predators and parasites are known to keep their prey species at low densities. Predators that feed on many prey species tend to feed on the most abundant species available until its numbers have been considerably reduced, but generalized predators seldom appear to limit the population sizes of their prey species.

Many factors influence the size of a population; the main factor actually controlling population size may vary from population to population, and from time to time within the same population.

Like all other organisms, human beings inhabit an environment with limited resources. The number of people in the world has grown exponentially in the last few years as a result of improved nutrition and hygiene. Historically, the demographic transition (a reduction in the birth rate) follows reduction of the death rate in any country, but only after a gap of 1 to 3 generations; during this gap, the population grows enormously. Many of the developing countries are undergoing a demographic transition in the fourth quarter of the twentieth century.

Experience to date strongly suggests that food is the limited resource halting the human population explosion. Millions of people starve to death (die of causes related to severe malnutrition) every year. At this time, this occurs largely because, for political and economic reasons, food is not always distributed to those who need it. Eventually, there will just not be enough food to go around, and the human population will level out.

SELF-QUIZ ──

1. Which of the following does *not* directly affect reproductive potential?
   a. a female's age at first reproduction
   b. the density of the population
   c. length of time a female is fertile
   d. average number of offspring per brood or litter

2. Which of the following would be *least* likely to act as a density-dependent factor limiting the size of a population of mice?
   a. parasitism
   b. buildup of waste products
   c. predation
   d. a hard winter

3. A population remains roughly the same size from year to year because:
   a. the number of deaths is usually about the same each year
   b. organisms reproduce more when the population is sparse and less when it is dense
   c. various environmental factors counteract the high reproductive potential of the population
   d. organisms stop reproducing when the population size goes above its average level

4. Draw a graph showing the long-term growth of a population of bacteria placed on nutrient medium in a dish.

5. A population can grow exponentially:
   a. when food is the only limiting resource
   b. when first invading a suitable, unoccupied habitat
   c. only if there is no predation
   d. only in the laboratory

6. Is the following statement true or false? Biological control measures work best when a generalized predator which eats many different species of prey is used.

7. The human population explosion is caused mainly by:
   a. the demographic transition
   b. education of women
   c. improved nutrition and hygiene
   d. the industrial revolution
   e. antibiotics

QUESTIONS FOR DISCUSSION ———————————————————

1. Some people think that there are already more people on earth than the earth can support indefinitely. Do you think this is true? Why?

2. People worry about overpopulation of a number of species. What are some of these species? Should we worry? Why?

3. People also worry about the underpopulation of some species, believing that they are on their way to extinction. What are some of these species? Why should we be concerned about their possible extinction?

4. In the early 1970s the birth rate (number of births per 1000 women of reproductive age) of the United States declined; why, then, is the population still growing?

5. What factors have contributed to the increase in the age at which women reproduce for the first time?

Essay:
HOW TO AVOID BEING EATEN

Being eaten is one of the hazards of life for every living thing. Here we will merely hit some of the high spots, the more esoteric of the many and marvelous defenses that organisms have evolved against this particular fate.

Chemical defenses, various types of poisons, are common in both plants and animals. A plant can survive if a herbivore eats a few of its leaves, which sicken the herbivore and warn it away but cause little damage to the plant. Animals are differently made: one good bite by a predator may be a mortal wound. Therefore the chemical defenses of animals are often in the form of sprays, bites, and stings that reach a predator before the predator can inflict damage. Bombardier beetles spray predators with a scalding hot (100° C) secretion; scorpions, bees, and wasps sting; millipedes and many beetles spray poison.

Among the vertebrates, skunks and poisonous snakes are well known for their chemical defenses, but there are many other examples. Colombian Indians tip their poison arrows with batrachotoxin,* found in the mucus on the skin of a tiny, "Day-Glo" green frog; only 3 micrograms of this nerve and muscle poison will kill a human being. A related poison, tetrodotoxin,** was first found in the ovaries and other tissues of puffer fish, an expensive delicacy in Japan. Japanese chefs must be licensed to prepare puffer fish; but again, a couple of micrograms are fatal, and accidents occasionally happen.

Some animals have second-hand chemical defenses. Nudibranchs† are gorgeous sea slugs (see Fig. 6-21) that feed on corals and sea anemones which are well endowed with stinging hairs. In some unknown manner, the nudibranchs move the stinging hairs, complete and undischarged, into their own skins, where they are redeployed against the nudibranchs' potential predators. Similarly, a monarch butterfly is toxic because it contains heart

*buh-TRACK-oh-tox-in
**teh-TROE-doe-tox-in
†NEW-dih-branks

poisons from the milkweed plants that it ate as a caterpillar.

Many animals with effective defenses are strikingly colored—skunks, certain wasps, nudibranchs, poisonous fish, and frogs. This is not chance: the coloration of these animals warns potential predators to stay away.

Warning coloration protects best against predators that have good eyesight and are capable of learning. Predators are apparently not born knowing that they should avoid orange and black butterflies; they have to learn this. Experiments have shown that a bluejay that becomes ill after eating a monarch butterfly will not attempt to eat the similarly-colored viceroy butterfly.

If predators may learn to avoid eating all insects with black and orange wings, then there might be enormous advantage to a tasty, unprotected species if it too had black and orange coloration. And, since nature never seems to waste a bright idea, there are indeed edible animals which protect themselves by copying the warning coloration of a poisonous or otherwise dangerous animal. You yourself have probably reacted to such mimicry by moving carefully away from a harmless fly which wears the black and yellow stripes and tight waist of a bee.

No protection in nature is ever absolute. Since there is always an evolutionary arms race, as it were, between predator and prey, any animal will always have predators that can learn to avoid its sting or that have evolved the body chemistry needed to render its toxic chemicals harmless. Although most birds avoid wasps and bees, some birds such as shrikes and bee-eaters have behavior patterns which allow them to eat these insects without being stung. Evolving the ability to eat a defended species is very advantageous because it opens up a food supply for which there will be little competition.

Fig. 6–19

Thorns on a plant, a deterrent to animals that might eat the plant. (Biophoto Associates)

Fig. 6–20

A Pacific puffer fish. (Biophoto Associates)

Fig. 6–22

The monarch butterfly (left) is protected from birds by the poisons it contains. The viceroy butterfly (right) is edible but birds avoid it because they mistake it for the monarch which it resembles. (Biophoto Associates, N.H.P.A.)

Fig. 6–21

A nudibranch, marine relative of a slug. (The head is to the right.) (Biophoto Associates, N.H.P.A.)

HUMAN EVOLUTION AND ECOLOGY

777

OBJECTIVES When you have studied this chapter, you should be able to:

1. List four resemblances between humans and chimpanzees.

2. Define the following terms and use them in context: hunter-gatherer, agriculture, agricultural revolution, monoculture, pollution.

3. Explain why the agricultural revolution is believed to have been responsible for a large increase in the human population.

4. Contrast the efficiencies of subsistence and modern farms, and of vegetarianism and carnivory.

5. Describe the biological basis of the green revolution and explain its impact on human society, economics, and ecology. Explain why the green revolution has been less successful in increasing food production than its proponents originally expected.

6. Explain how agriculture, soil erosion, use of fossil fuels, and manufacture of plastics contribute to pollution, and discuss possible corrective measures.

7. Explain what is meant by "the tragedy of the commons" and why it makes environmental problems so difficult to solve.

Biophoto Associates.

All the people on earth today belong to the species *Homo sapiens* (= "man, wise"). *Homo sapiens* has a very well-developed brain and flexible hands. These characteristics enable us to manipulate the environment so that we can feed and clothe a huge human population and develop spectacular civilizations. But our cavalier, and often hostile, treatment of all species (including our own) has begun to boomerang on us. We are belatedly awakening to the fact that our planet is a closed ecosystem, with finite space and energy, and that we are consuming and destroying its resources faster than nature can replace them. Some claim that we are merely acting out our inevitable role in nature, altering our environment until it can no longer support us, and paving the way for other species (probably some of the insects) that must one day inherit the earth. Others believe that we shall solve our ecological problems as we have solved so many others.

Human population growth (discussed in Chapter 6) is ultimately responsible for most of our other ecological problems. The population explosion started when early tribes began obtaining food from agriculture instead of from hunting and gathering—a change that had dramatic effects on human history.

In this chapter we shall consider some of the ecological problems that confront us and fill in a little of the evolutionary background that helps to explain how we got ourselves into this mess.

Fig. 7–1

Human environments: (a) country and (b) city. (Biophoto Associates)

(a)

(b)

Fig. 7–2

Some primates. (a) A Barbary rock ape, from Gibraltar, with her young. (b) A baboon. (Biophoto Associates)

(a)

(b)

7–A Human Evolution

Any discussion of human evolution is based on shifting sands. Frequent new discoveries cast doubt on old theories, including the long-accepted belief that humans evolved in Africa. So we give here only a brief sketch of human evolution, and you must bear in mind that many would dispute its accuracy.

Humans are primates and our nearest living relatives are the anthropoid apes: the gibbons, orangutans, gorillas, and chimpanzees. The evidence suggests that humans are more closely related to African chimpanzees and gorillas than to the Asian orangutan. Our resemblance to chimpanzees is almost uncanny, even down to the structure of individual proteins: on the average, human and chimpanzee proteins are 99% alike.

Chimpanzees are not monogamous, but they form strong family bonds, which may last into adult life, and make friends with unrelated animals. They live in groups which defend roughly-defined territories against other groups. They are equally at home on all fours on the ground or swinging from branch to branch in the tree-tops by their arms. At night they sleep in rough nests in trees. Chimps feed mainly on fruits and other vegetation, but they also relish some meat, obtained by hunting cooperatively for small rodents or monkeys, by digging ants out of anthills (sometimes using tools), and occasionally by cannibalism.

anthropoid (AN-throw-poyd): man-like.

monogamous (mon-OG-uh-muss): mating with only one member of the opposite sex in each breeding season.

Fig. 7–3

A chimpanzee, member of the modern-day species most closely related to humans. (Biophoto Associates)

Chimpanzees communicate with one another by sounds, by facial expressions, and by body language. Chimpanzees in captivity have been taught sign language, and some people think they have an essentially human ability to use language. Once they have learned sign language, these chimps will form short but novel sentences to communicate with their human handlers and with each other. They even abstract ideas to some extent: on one occasion when two chimps were communicating in sign language, one of them angrily made the sign for "feces." It had apparently invented a swearword—one that is also common in human languages!

Fig. 7–4

Fig. 7–5

The descent of our ancestors from the trees was strongly influenced by the force of gravity.

The search for the common ancestor of apes and humans has centered on African fossils about 10 million years old, which appear to be the remains of animals with a tendency to **bipedalism** (walking on two legs). From this point on, humans and apes diverged as the apes became increasingly specialized for swinging through trees using their long, strong arms, and human ancestors abandoned life in the trees for a bipedal life on the open plains. Just why humans came down to earth is hotly debated, but the move was important as it freed the hands for catching animals, throwing stones, and other activities.

By 2 million years ago, human ancestors in Africa were fully bipedal tool-users, living in family groups. Since the most dramatic difference between modern humans and apes is the remarkable size of the human brain, it is surprising to find that these tool-using human ancestors had brains no bigger than those of today's chimpanzees. The large human brain did not evolve until about 200,000 years ago. As yet we have no way of knowing when human traits, such as speech and particular behavior patterns, which leave no evidence in fossils, first appeared.

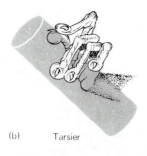

(a) Tree shrew

(b) Tarsier

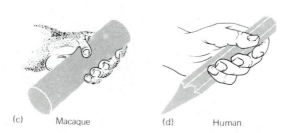

(c) Macaque

(d) Human

Fig. 7–6

Primate hands, from the relatively immobile fingers of a tree shrew (with claws) to the human hand with nails and an opposable thumb.

Fig. 7–7

Australopithecus, *tool-using, bipedal human ancestors, as the artist imagines them, in their home on the African savanna about two and a half million years ago.*

7–B Hunter-Gatherers

For most of our evolutionary history, human beings were nomadic **hunter-gatherers**, killing wild animals and collecting plants for food. Relatively late in our development, permanent settlements appeared in Eurasia.

Many game animals in Europe and Asia migrate during the spring and autumn. Cooperative effort permitted pre-human groups to trap migrating animals which supplied food for the long winter. During summer, smaller mammals were probably trapped, but the main source of food would have been plant materials (seeds, fruits, nuts, roots, and berries) gathered as they were used. The use of fire opened up a new range of plant foods. Cooking can remove many volatile poisons from plants and also makes them softer and more digestible.

A settled hunter-gatherer society of this type must have been a precursor to the development of agriculture. The social organization and communication between individuals demanded by such a society would create strong selective pressures for the development of language, social rites, laws, and customs. We find these reflected in the decorated tools, pots, and dwellings that began to appear in Europe and Asia about 20,000 years ago.

By the time of Columbus, farmers had driven hunter-gatherer groups out of much of the world. Hunting groups survived mainly in parts of Africa, the Americas, and Australia. Today, very few hunting populations remain.

Fig. 7–8

Changes in the skull from great ape to human. Size of the brain case increased; the place where the neck joins the head (arrow) moved as hominids became more upright; and the size of the teeth and jaws was reduced with the change from a herbivorous to an omnivorous diet. (cc = cubic centimetre)

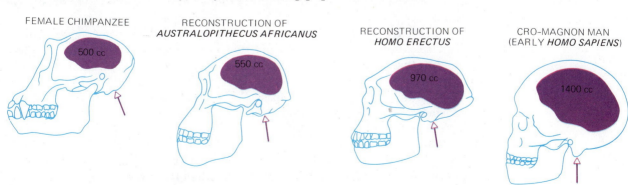

FEMALE CHIMPANZEE
500 cc

RECONSTRUCTION OF
AUSTRALOPITHECUS AFRICANUS
550 cc

RECONSTRUCTION OF
HOMO ERECTUS
970 cc

CRO-MAGNON MAN
(EARLY *HOMO SAPIENS*)
1400 cc

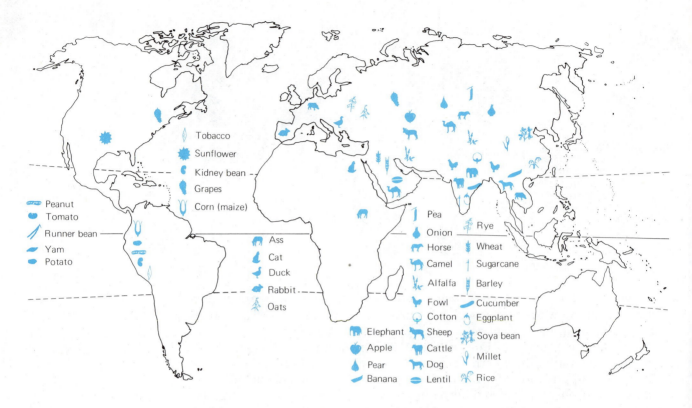

Fig. 7–9

Areas where particular plants and animals were probably first domesticated.

Tobacco
Sunflower
Kidney bean
Grapes
Corn (maize)

Peanut
Tomato
Runner bean
Yam
Potato

Ass
Cat
Duck
Rabbit
Oats

Pea
Onion
Horse
Camel
Alfalfa
Fowl
Cotton
Sheep
Cattle
Dog
Lentil

Rye
Wheat
Sugarcane
Barley
Cucumber
Eggplant
Soya bean
Millet
Rice

Elephant
Apple
Pear
Banana

7–C Origins of Agriculture

Agriculture is the process of breeding and caring for animals and plants that are used for food and clothing. It originated in many different parts of the world at about the same time, some 10,000 years ago.

Once people begin to plant seeds, cultivated strains of plants soon come to differ from their wild counterparts. For instance, cereal (annual grass) seed heads which do not burst and scatter their seeds are easier to collect for planting than those that burst. So the human planter selects for plants that do not shed their seeds easily, and this trait rapidly increases in any population of cultivated plants.

Similarly, docile animals with small horns and woolly coats would have been those most likely to become the domesticated breeding stock. We can imagine that an early agricultural society would have pushed a vicious ram with big horns back into the wild again or, more likely, would have made him the *pièce de résistance* at a tribal feast. So by a combination of conscious and unconscious selection, early farmers rapidly produced animals and plants which bore little resemblance to their wild forebears.

The change from hunting and gathering to agriculture had such a dramatic impact on human societies that it is often known as the **agricultural revolution**. But why did people switch to agriculture when there seem so many advantages to living as hunter-gatherers? Hunter-gatherers do not face the constant battles with pests, droughts, and famines that beset agricultural communities. During a recent drought, South African farmers starved while the hunter-gatherer Bushmen in the Kalahari desert remained well fed. This ability to survive in times of scarcity is possible because most hunter-gatherer groups keep their numbers well below the population size which their territories could support. Their population-control practices include abortion, infanticide, late marriage, abstention from sexual intercourse, and late weaning from the breast. (Women seldom become pregnant while nursing.) Furthermore, these hunter-gatherer populations have a more balanced diet than most farmers, and their incidence of chronic and disabling diseases is no higher.

Some people assert that the Bushman diet contains too few calories for perfect health, but this may well not be so because these people spend few of their waking hours finding food; they could find more if they felt the need. Even though they live in inhospitable deserts, Kalahari Bushmen and Australian aboriginals devote only about 15 hours a week to collecting and preparing food; children do not have to work until they are married, and the aged are cared for and revered. In contrast, most people in agricultural societies work at least 60 hours a week, and spend about 70% (42 hours' worth) of their pay on food. Even in the affluent West, with the world's best agricultural land, we devote about a third of our incomes to buying food. In many ways hunter-gatherers have an easier life.

Fig. 7-11

African Masai. Originally hunter-gatherers, the Masai now also make part of their living as nomadic cattle herders. (Biophoto Associates)

Fig. 7-10

Two varieties each of ancient and modern wheat. Macaroni wheat was bred from a variety of emmer, bread wheat from a cross between emmer and goat grass.

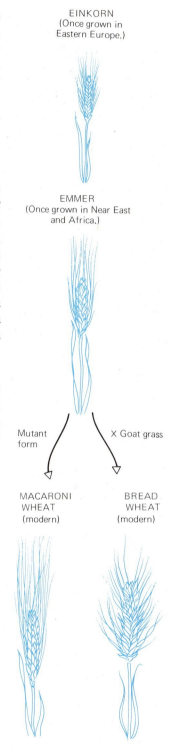

EINKORN
(Once grown in
Eastern Europe.)

EMMER
(Once grown in Near East
and Africa.)

Mutant
form   X Goat grass

MACARONI
WHEAT
(modern)

BREAD
WHEAT
(modern)

Fig. 7–12

A small mixed farm producing several different animal and plant crops, some for home use, some for sale. (Biophoto Associates)

It was probably population pressure, not easy living, that induced early human populations to exchange their hunter-gatherer existences for the hard life of a primitive farm. Agriculture can usually feed more people in a given area than can hunting and gathering. In an area where some disruption of the social structure caused people to abandon their methods of population control, agriculture might have become the preferred way of life. Presumably, at first farming was combined with hunting and gathering; complete dependence on agriculture for food came later.

The agricultural revolution had profound effects on human evolution. One of the most important was that it permitted the accumulation of material goods. Nomadic hunters travel with few possessions; farmers, living in one place, can accumulate as much as they can afford. Land itself can be owned and passed on by inheritance.

The population explosion, which worries us so much today, is a result of agriculture. Population control is usually abandoned by agricultural communities. This is partly because children, who are not important food collectors in hunter-gatherer societies, are useful as labor on a farm. In addition, the inheritance of land and goods becomes more important. The desire to have children who will inherit the family land and care for their aged parents is a recurrent theme in mythology and literature. Largely because they have reproduced faster, agricultural peoples have pushed hunter-gatherers out of most areas of the world today.

A striking consequence of agriculture is that a division of labor immediately springs up within the group. Because a few people can produce food for everyone, the rest are free to become builders, bakers, and merchants. Eventually, the population may even be able to afford the luxury of full-time poets, scholars, and artists who contribute nothing to the group's immediate physical well-being but are the basis of its cultural life.

7–D Agricultural Pests

Since agriculture began, it has been plagued by "pests"—insects, weeds, disease-causing organisms, parasites, birds, and rodents—that destroy livestock, crops, or stored food and so compete with us for food.

In most parts of the world, domesticated plants are much more subject to wholesale destruction by pests than are domesticated animals. The ecological reason for this is not far to seek. First, because we do not want thorns or large quantities of toxic chemicals in our food plants, we have steadily bred out of plants the physical and chemical defenses that permitted their wild ancestors to avoid being eaten by herbivores. A second reason is that plants like the ancestors of most of our vegetables live scattered in the wild, and being hard to find is one of their defenses against herbivores. When they are planted in **monocultures** (large areas of a single species), it is very easy for pests to find them. The contention of gardeners that interplanting one type of vegetable with another will reduce pest damage is correct: experiments have shown that fewer pests specializing in one plant species will find the plants they prefer to eat when these plants are interspersed with other species of different appearance and chemistry. Thus, on small, "mixed" farms (which used to be the only kind there were), where many different plants are cultivated, the problem of insect damage is not nearly so acute as it is on larger, more modern farms.

(a)

(b)

Fig. 7–13

Pests that destroy agricultural crops include (a) a slug (Milton Love), (b) a Colorado potato beetle (Paul Feeny), (c) the carrot-eating caterpillar of a swallowtail butterfly (May Berenbaum).

(c)

On large modern farms, which may have thousands of acres planted with a single crop, the solution has usually been to apply ever-increasing doses of **pesticides**, chemicals that kill pests (mainly insects). The doses are ever-increasing because insects evolve resistance to pesticides, just as they have evolved resistance to the natural insecticide chemicals produced by plants. An additional disadvantage is that pesticides run off into waterways, polluting our water supply. Pesticides are invariably more or less toxic to human beings, and they contribute to a growing number of human deaths. As a consequence, agriculturists are studying other methods of defeating crop pests. Some are experimenting once again with mixed plantings; others are trying biological methods of control.

Biological control usually consists of introducing a predator or parasite that will reduce the population of the pest, or it may interfere with the pest's reproduction by such techniques as trapping males using synthetic sex-attractant chemicals as bait.

In California, experiments with **integrated control**—combining biological and chemical control—have shown that we can save money and reduce pollution by applying small quantities of pesticides only as needed to supplement biological control. In most states, extension agents employed by the state now keep track of breeding times and outbreaks of important pests and advise farmers about spraying their crops. This eliminates the hit-or-miss mass sprayings that used to be the rule and reduces costs and pollution. Neither biological nor chemical control can completely eliminate pest damage to crops. For instance, more pesticide is used on cotton than on any other crop in North America, yet more than $500 million worth of cotton is lost to insect pests every year.

7–E The Efficiency of Agriculture

Food production depends ultimately on photosynthesis, in which plants use sunlight as an energy source to produce food. Crops seldom suffer from lack of sunlight; lack of moisture, nutrients (especially nitrogen and phosphorus) and heat nearly always limits the growth of crops to a fraction of their photosynthetic potential. In most parts of the world, the most productive agriculture occurs on artificially irrigated land. Very few areas have both good soil and enough rain because adequate rainfall quickly washes nutrients out of the soil.

Plants can use nitrogen only in certain forms, mainly nitrates and ammonium. In nature, these forms are produced by bacteria when they decompose plant and animal matter or when they take in nitrogen gas from the atmosphere and "fix" it in solid form. Before about 1940, farmers increased the nitrogen supply to their crops by spreading manure and by planting legumes (clover, alfalfa, peas, and their relatives), whose roots contain nitrogen-fixing bacteria. Recently, modern farms have switched almost completely to nitrogenous fertilizers made by combining natural gas with atmospheric nitrogen. This is expensive but can supply much more nitrogen per acre than can traditional practices. In Indonesia, rice yields doubled when nitrogen was supplied in inorganic fertilizers.

Fig. 7–14

A modern farm uses gas-guzzling equipment and very little human labor to produce a high yield of crops. (Biophoto Associates)

Subsistence vs. modern farms.   The world's farms can be loosely divided into subsistence farms, which provide the family's food and usually a small crop sold for cash, and "modern" farms, where all crops are sold. Modern farms are much less self-sufficient since they need to obtain seed, fertilizer, equipment, credit, and marketing facilities from the rest of society. Modern farms produce more per farm worker per year, but less per unit area, partly because plants must be far enough apart to allow machinery to pass between the rows. A subsistence farmer, or a vegetable gardener, can produce much more per year in an area of the same size by planting one crop between the rows of another and by planting more than one crop per year.

However, the crops of labor-intensive, small farms are more expensive to produce since, in most places, it costs more to cultivate or harvest crops by hand than by machinery, especially in countries like the United States where gasoline for machinery is still relatively cheap and where a flood of government rules and regulations swamps anyone who employs labor. As a result, produce from small farms cannot compete on the open market, and the owners of small farms struggle for financial survival.

Eating lower down the food web.   People feed themselves most efficiently when they eat plants or parts of plants. Usually, from 30 to 40% of a plant's productivity can be harvested, and about 70% of such a harvest can be digested

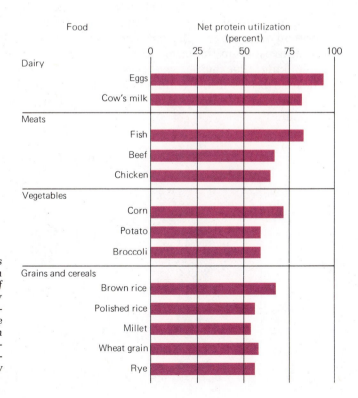

Fig. 7–15

Protein "quality" of various foods. Net protein utilization is a measure of how much of the food's protein is used by the body. The higher the percent utilization, the closer the food comes to meeting human protein needs. Foods deficient in one or more essential amino acids have low utilization values.

by humans. The protein content of plants varies from about 6 to 20%. A plant must expend more energy to produce protein than to produce carbohydrate (sugars, starches). Thus, protein synthesis reduces the amount of energy a plant stores, and it is more efficient to grow and eat plants with a lower protein content, so long as they meet human protein requirements. Cereals produce around 10% protein, which is nutritionally adequate for an adult human. This is why world agriculture is now so dependent on cereals.

Eating animals is much less efficient than eating plants. The animals must eat plants first, and animals convert less than 15% of their food calories into more animal, available as human food. In no case is the efficiency of conversion higher than the 25% attained in milk and egg production.

As societies become wealthier, their consumption of animal products increases. This makes it even more difficult for food production to surpass population growth as the number (if not the proportion) of people with comfortable incomes increases.

On the other hand, some grazing land is unsuitable for cultivation, and growing animals in these areas is the most effective means of increasing food production. Similarly, we shall continue to eat meat from the sea for some time to come. Fish are large enough to catch, whereas most of the sea's plants are tiny floating organisms so small that they are too expensive to harvest directly.

Food processing.   In both traditional and more developed societies, the energy used in processing, distributing, and cooking food is greater than that used to produce the food in the first place. It takes about twice as much energy to cook a kilogram of rice in rural India as to grow it. In the United States, each calorie on our dinner tables costs an estimated nine calories to put there: half a calorie invested on the farm, and the rest in processing, packaging, shipping, and cooking. (Packaging was the single largest contributor to the increase in retail food prices in the United States between 1967 and 1977, partly because packaging uses a lot of plastics, which are produced from increasingly scarce petroleum.)

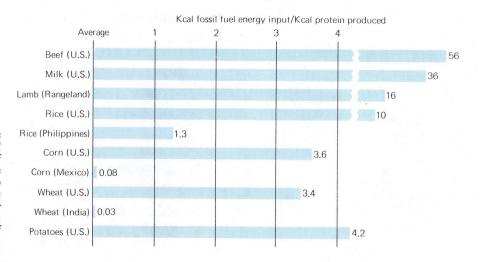

Fig. 7–16

The number of kilocalories (kcal) of fossil fuel energy used to produce each kilocalorie of edible protein, for various crops in various parts of the world. The more machinery is used in cultivation, the higher the number of calories used in production. (Data courtesy of David Pimentel)

Reducing world hunger.   From this discussion, what is the most efficient way to attack the world's food problems? It seems that hungry nations should increase their production of crops with the lowest adequate protein content, and that they should encourage more efficient agriculture among their small farmers, who make up about half the population and occupy more than half the land in many countries.

In the past, underdeveloped countries have usually preferred guns and heavy industry to agriculture as the route to modern living. This is not entirely chance. If heavy industry grows at the expense of agriculture and rural industry, the demand for food grows slowly. Industry yields its wealth to only a small fraction of the population, and the income of the rural poor does not increase. Rural development, on the other hand, puts heavy demands on the food supply; the malnourished spend the bulk of any income increase on food.

Toleration of an economic situation in which the poor get poorer and the rich get richer is one reason that many countries have invested little in agricultural development, so that food production has increased only slowly while population growth has soared. Despite this difficult situation, in the 1950s food production began to increase dramatically in some countries, including India, Mexico, and China, although most of Africa has not yet experienced any agricultural gains.

7–F The Green Revolution

Wheat production in Mexico increased more than eightfold from 1950 to 1970; twice as much land was brought into production and the yield quadrupled. In the same period, India doubled its production of grain. These spectacular increases, hailed as the "green revolution," resulted from intensive efforts to breed new strains of plants, to educate farmers in new growing techniques, and to change the economic structure of agriculture in less developed countries.

The story of wheat in Mexico illustrates the achievements of the green revolution, and also its shortcomings. First, researchers developed disease-

Fig. 7–17

Irrigation systems, like this one in a South African valley, increase the production of many crops. (Biophoto Associates, Geoslide Photo Library)

resistant varieties of wheat that responded to irrigation and fertilization by producing very large crops. Next, an education program was launched to bring this information to the farmer. In addition, economic conditions had to favor the gamble of changing from a proven crop to a novel one. In this case, the market price of wheat was high and the cost of fertilizer low, and so there were strong incentives for farmers to make the change. Finally, Mexico's efficient transportation network permitted farmers to obtain the fertilizer they needed and to ship their crops to market.

By the late 1960s, food production in Mexico began to level off. Although the green revolution had reached the larger farms, it had not affected the less prosperous farmers, who work 80% of Mexico's agricultural land but produce only 55% of its crops. As a rule, education, transportation, irrigation, and money are all more readily available to large than to small farmers. Large farmers can take more risks; they can borrow money to tide them over a crop failure, whereas the subsistence farmer usually cannot borrow enough even to pay for this year's fertilizer, much less survive a crop failure. As a result, the green revolution has often worsened the plight of small farmers, whose crops cannot compete with those produced by modern practices.

7–G Soil Erosion

Soil erosion is both an agricultural and an ecological problem. It occurs when wind or water moves soil from one place, which is often agricultural land, to another place, usually a river or non-agricultural land.

In the United States, up to 20 metric tons (1 metric ton = 2200 lb) of soil are lost from each hectare of agricultural land every year as a result of erosion. The Great Plains, which once were covered by more than 2 metres of topsoil, have lost about one third of their topsoil to erosion in a hundred years. Many areas in the Eastern United States and in California have lost virtually all their topsoil to erosion and have become impossible to farm. Because soil is formed very slowly,

hectare: a unit of area, equal to 2.5 acres.

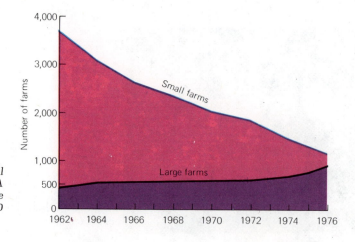

Fig. 7–18

Decline in the number of small farms in the United States. (A small farm is defined as one that sells less than $20,000 worth of produce in a year.)

Fig. 7–19

(a) Soil erosion: every year, water washes 18 tons of soil per hectare off this farm in Wisconsin. (U.S.D.A. Soil Conservation Service)
(b) The better practice: terraced fields on this hillside in Nepal prevent soil erosion. (U.S. Agency for International Development)

by the weathering of underlying rock and the decomposition of organic matter, it takes a long time to replace.

Soil erosion is largely a problem of newly developed farmland. It is rare in Europe and Asia, where the same land has often been farmed for 2000 years or more. There farmers must prevent erosion, and put back into the soil what they take out of it, or their children may starve. In newly developed agricultural areas, on the other hand, people often move on when they have exhausted the soil in one area, and incentives to preserve the soil are fewer.

Soil erosion in North America is gradually being reduced as farmers adopt simple preventive measures. First, land should not be left unplanted for long;

Fig. 7–20

Soil erosion by wind. Much of the dry soil disturbed by this harrow will be lost to agriculture forever. (Biophoto Associates)

plant roots hold the soil in place. Terracing and contour plowing prevent soil from washing down slopes, and planting windbreaks checks wind erosion. Also, since organic matter can bind a great deal of water, soil with a high content of organic matter is less likely to wash away; fertilizing with manure instead of chemical fertilizer is a considerable deterrent to erosion.

7–H The Fossil Fuel Shortage

Fossil fuel, such as oil and coal, is formed from animals and plants that lived millions of years ago. Because it takes so long to form, it is essentially non-renewable; when we have used all the oil and coal in the earth's crust, there will be no more.

Most countries today are completely dependent on fossil fuel, not just for the niceties of civilization but to feed their populations. At this time, the end of fossil fuel would mean the end of civilization as we know it and the deaths of billions of people.

Since the world is so dependent on fossil fuel, and particularly on oil, the oil-exporting countries can, in theory, ask any price for oil and hold the rest of the world at ransom, causing economic chaos. Because of this, the oil-importing nations are trying, belatedly, to turn to other sorts of energy. This has the additional merit that the world will not be completely helpless when the supply of oil, whatever its price, eventually runs out.

Nuclear fission was once hailed as the energy supply of the future, but it is now clear that, even if nuclear power plants and their radioactive wastes can be made acceptably safe, the supply of uranium fuel (another non-renewable resource) is much more limited than once thought.

Both the sun and hydrogen bombs produce heat and light through nuclear fusion. Harnessing nuclear fusion for useful energy on earth is not yet technically possible although some researchers believe that we shall have energy produced by nuclear fusion by the end of the twenty-first century.

Wind, sunlight, and wood—the traditional sources of energy—are renewable. Most people doubt that these sources can be developed to supply enough energy in time to make much difference as the fossil fuel supply dwindles. Every little bit helps, however, and in the long run we are undoubtedly going to have to use not one main source of energy, as we have in the past, but many different ones.

Synthetic fuels, which are not really synthetic, are liquid fuels produced from fossil fuel sources such as oil shale and coal. They are becoming economical to produce in the 1980s because of the increasing price of oil. The main advantage of synthetic fuels is that they can be substituted for liquid fuels, such as gasoline and fuel oil (which are derived from oil), without modifying existing automobile engines or heating systems; also, they produce less air pollution than most types of coal.

The use of alcohol as an additive or alternative to fossil fuel was pioneered in Brazil in the 1970s. The city of São Paulo has a fleet of cars that run on pure alcohol (which is sometimes used as an additive to gasoline in the United States). The alcohol is produced inexpensively from sugar cane, manioc, and other crops; the necessary energy is supplied largely by the sun (to grow the plants) and by manual labor, which is inexpensive in Brazil. Alcohol may prove to be a useful alternative to gasoline in countries where labor is less costly than oil—this will eventually be true everywhere.

Unless we reduce our use of fossil fuels and develop new energy sources with unprecedented speed, energy shortages alone will reduce the standard of living in developed countries and cause widespread starvation in less developed countries, since food production depends on fossil fuel for fertilizers and transportation.

Fig. 7–21

Alternatives to oil for energy. (a) This electricity-generating plant in New Zealand is powered by hot springs (the clouds are steam, not smoke). (b) Windmills were once used to grind flour; nowadays, many generate electricity. (c) Horses for transport or farming. (d) These banks are cakes of cow dung being dried for fuel in Madras, India. (Biophoto Associates)

(a)

(b)

(c)

(d)

7–I Pollution

All organisms expel their waste products into the environment around them. These wastes include carbon dioxide, feces, and inedible parts of plants and animals. Although waste products make the environment less favorable for the organism that produced them, in a balanced ecosystem one organism's wastes are another's food and drink, and so waste does not accumulate. **Pollution** results when wastes are not destroyed as fast as they are produced and therefore accumulate, making the environment less hospitable to humans and to many other organisms as well.

Modern pollution dates from the industrial revolution, when the use of fossil fuels in densely populated areas became widespread. Wordsworth, in his poem "On Westminster Bridge," celebrated a rare morning in nineteenth-century London when he could actually see the skyline of the city instead of the smog that usually obscured it. London's "pea-souper" fogs (because of their yellow-green color) were a mixture of fog with the sulfur and particulate matter produced by burning coal, and they caused thousands of deaths; happily they have disappeared since passage of the Clean Air Act in 1952. Modern "smogs" are caused mainly by the combustion of gasoline in cars, although wood-burning stoves are again contributing their share to air pollution in some areas. In cities such as Los Angeles and Tokyo, automobile smog frequently renders the air unfit to breathe. Calculations suggest that living in New York City, or in a house with a wood stove, causes lung damage equivalent to that produced by smoking one pack of cigarettes a day.

Since about 1950, pollution from solid wastes has become an acute problem. The use of fossil fuel, this time as the raw material for the manufacture of plastics, has contributed to solid-waste pollution in our affluent, "throw-away" society. Before plastics were invented, biodegradable substances—such as rubber, cot-

Fig. 7–22

Air pollution, Los Angeles.
(Biophoto Associates)

ton, silk, leather, paper, and wood—produced by animals and plants were used for clothing and packaging. These natural products have been largely replaced by plastics—such as nylon, acetate, polyester, and vinyls—which cannot be digested by decomposer organisms.

Another source of pollution from the plastics industry is plasticizers, substances that make plastics more flexible. (Plasticizers tend to evaporate with age, allowing the remaining plastic to become brittle and crack.) Among the plasticizers are very toxic compounds such as the polychlorinated biphenyls (PCBs), now largely banned in the United States. Since they are not biodegradable, these substances still surround us everywhere, from the middle of the Pacific Ocean to the African deserts. It is obvious that we do not yet have the technology or the organization to dispose safely of solid wastes, including plastics, dangerous chemicals, and radioactive waste from nuclear power stations.

7–J   Private Interest and Public Welfare

It is easy to become angry and frustrated at the shortsightedness of a town that puts off building a sewage treatment plant until only much more expensive measures will save the local drinking water, but it is worth understanding why we tend to bury our heads in the sand in this way. The procrastination results from a conflict between the short-term welfare of individuals and the long-term welfare of us all. Garrett Hardin has called this phenomenon "the tragedy of the commons." He illustrates it with the case of the commons in medieval Europe. A common was grazing land that belonged to a whole village; any member of the community could graze sheep and cows there. It was in the interest of each individual to put as many animals on the common as possible, to take advantage of the free animal feed. However, if too many animals grazed there, they eventually destroyed the grass; then everyone suffered because no one could raise cattle. For this reason, common land was eventually replaced by individually owned, enclosed fields. When the land is privately owned, the owner is careful not to put too many cows on one patch of grass, because over-grazing this year means that fewer cows can be supported next year.

In the same way, if we, as individuals or as corporations, were made to pay directly for our contributions to overpopulation and pollution, we would do something about it. An example is the effect of laws making employers liable for on-the-job injuries. When a firm knows that injuries will cost it money, either directly or in higher insurance premiums, it takes steps to ensure greater job safety. In the case of ecological problems, however, it is not usually possible to assign responsibility directly to the people who cause the problems.

If I decide not to use my car, I can make little difference to the overall level of air and water pollution. Furthermore, I shall be at a disadvantage compared with my neighbors because I cannot take a job that would require me to use my car. Because of the conflict between private welfare and public interest, most environmental problems can be solved only by effective action by governmental agencies which, at least in theory, can look beyond the immediate interests of individuals and plan for the long-term good of society.

A government can act in three main ways. First, since most people can understand the "tragedy of the commons" and do not really want to make the earth uninhabitable for future generations, programs to educate the public about the problems are useful. Second, probably the most effective way for government to reduce pollution is to offer incentives (usually in the form of tax reductions) to those who pollute less and to levy pollution taxes on individual and corporate polluters. Third, when all else fails, government can resort to regulation, banning particularly dangerous practices and products altogether. Regulation becomes necessary in two kinds of cases: first, when a practice is so damaging that no financial value can be placed upon it (for example, it is not enough to permit firms to market thalidomide or exterminate the blue whale and tax them for it afterward); second, when it is impossible to assign responsibility with any degree of justice (for instance, how should the cost of reducing the amount of sulfur discharged by a coal-fired power station be distributed between the utility that owns the station and the consumers who use its electricity?).

To solve our ecological problems, we must change our individual and collective behavior, led—or driven—by responsible government. In a democratic society, this can be done only with the support and consent of well-informed citizens who realize that an intricate web connects us to all other living things.

SUMMARY

Humans probably evolved from tree-dwelling, ape-like ancestors. The descent to the ground seems to have spurred evolution to a bipedal posture and so freed the hands to manipulate the environment.

Like all other organisms, human beings inhabit an environment with limited resources. We have recently realized that we are causing changes in the ecology of the earth that are deleterious to us now, and that threaten the survival of the human species in the future.

The invention of agriculture permitted people to live in populations that are more dense than is possible with a hunter-gatherer economy. Today, the populations of many countries are growing faster than their food supply. Food production can be increased dramatically by the techniques of the "green revolution." Among the disadvantages of switching from traditional to modern agriculture, however, are increased dependence on dwindling supplies of fossil fuel and wider gaps between rich and poor.

The population explosion and extravagant use of fossil fuel are the main factors responsible for pollution that threatens human life and health. Because we seldom have to pay directly for our contributions to ecological problems, these are some of the most difficult problems for societies to solve.

SELF-QUIZ

1. Monoculture planting:
   a. decreases pest populations
   b. is utilized on modern farms
   c. lessens dependency on pesticides
   d. all of the above

2. The agricultural revolution is believed to have caused a dramatic increase in the human population. Which of the following was a factor in this increase?
   a. methods of population control were abandoned
   b. the same area of land could feed more people
   c. possessions accumulated, and children to inherit them became more important
   d. all of the above

3. True or False: Subsistence farms produce more food per unit area than do modern farms.

4. The main factor which produced the "green revolution" was the:
   a. intensive irrigation of desert areas
   b. development of new forms of nitrate fertilizers
   c. development of new strains of plants
   d. massive application of inorganic fertilizers on marginal lands

5. Soil erosion can be lessened by:
   a. planting windbreaks
   b. fertilizing with manure
   c. continued planting of land
   d. all of the above

6. The most effective action an individual can take to improve the ecology of the human race is to:
   a. recycle glass bottles
   b. join a powerful citizens' lobby for environmental legislation with clout
   c. ride a bicycle instead of driving a car
   d. join a commune that "lives off the land"
   e. become a vegetarian

QUESTIONS FOR DISCUSSION

1. Apply the "tragedy of the commons" principle to the problem of limiting human family size.

2. Some nations have embarked on government-sponsored propaganda programs to encourage limiting family size; others have passed mandatory sterilization measures. Will it eventually become necessary for all governments to take such steps? Why?

3. If you were minister of agriculture in a developing nation, would you try to bring the green revolution to small farmers as well as large ones? If so, how? If not, do you have your escape route worked out in case of revolution?

888 EVOLUTION AND SEX

OBJECTIVES

When you have studied this chapter, you should be able to:

1. State the advantages and disadvantages of sexual reproduction contrasted with asexual reproduction.

2. Discuss the advantages enjoyed by organisms that combine sexual and asexual reproduction in their life history.

3. Give some reasons why selective pressures acting on a female may be different from those acting on a male, and describe some resulting differences between the two sexes in a species.

4. Give evidence for the theory that female choice almost invariably exists in mating systems.

5. Define polygyny, polyandry and polygamy.

6. Discuss possible reasons for the evolution of monogamy in humans.

At first glance it seems that sex *is* necessary. Most animals, and many higher plants, rely on sexual reproduction to perpetuate their genes. On the other hand, sexual reproduction is much less common among lower plants, bacteria, fungi and unicellular (one-celled) organisms like amoebas.

Although it is widespread, sexual reproduction wastes more energy than does asexual reproduction. If sperm, pollen, or eggs are released into the water or air to find each other by chance, millions of them are lost and so the energy spent to make them is wasted. In animals with internal fertilization, fewer eggs and sperm are lost, but the animal has to spend a lot of time and energy finding and courting a mate. Further energy is wasted because sexually reproduced offspring are so small at first that many of them are killed by predation or starvation.

By contrast, when an organism reproduces asexually, only one or a few offspring usually form at a time, and the offspring often remain attached to the parent until they reach a much larger size. Many unicellular organisms reproduce asexually simply by dividing into two identical, smaller cells. **Budding** of smaller individuals, which eventually detach from the parent, and **parthenogenesis**—development of an unfertilized egg into a new individual—are common forms of asexual reproduction in animals. Many plants also produce new individuals which remain attached to the parent until they are quite large. These forms of asexual reproduction require more energy per offspring than does sexual reproduction, but each asexually produced offspring receives a better start in life than a sexually produced organism and has a good chance of surviving. Thus, asexual reproduction is less apt to waste energy than is sexual reproduction.

As a result, an asexually reproducing organism like a strawberry plant can, on the average, produce many more surviving offspring per season than can a similar plant that reproduces sexually. Why, then, do so many organisms use so much time and energy and so many wasted cells reproducing sexually? This seemingly wasteful method of reproduction must have tremendous adaptive value, or it could not have survived and become so widespread.

Beyond the question "why sex?" lies the question of why an organism has one type of sexual system rather than another—the variations are almost endless.

genes: units of hereditary information that determine the possible range of an organism's form, growth, chemistry, and behavior.

Fig. 8–2

Budding in Hydra, a cnidarian. The bud is a precise replica of its parent; when it is large enough, it breaks off and starts an independent life. (Biophoto Associates)

Fig. 8–1

Butterflies mating. (Biophoto Associates, N.H.P.A.)

Fig. 8-3

A strawberry plant with stolons, each of which may produce several new plants. (Biophoto Associates)

We shall discuss some of the probable reasons for the evolution of particular kinds of sexual systems and end with some speculation on the origin of human sexual systems.

8-A When Is Sex a Good Thing?

The main biological difference between sexual and asexual reproduction is that sexual reproduction produces more genetic variability in a population. An organism that reproduces asexually passes on all its own genes to all its offspring, and therefore they are all alike genetically—that is, the original organism plus its offspring form a **clone.** The only genetic variation that can arise in such a population comes from **mutations**—occasional changes in the genetic material.

Fig. 8-4

Dandelions, which reproduce asexually by developing from unfertilized eggs. So all dandelions contain identical genes except for the variation introduced by mutation. (Biophoto Associates)

Fig. 8–5

Different varieties of domestic fowl show the enormous genetic variability that is possible within a species that reproduces sexually. (Biophoto Associates)

Sexual organisms can also undergo mutation, but their main—often enormous—variability arises during formation of every sperm, pollen grain, or egg and during assembly of new sets of genes when an egg is fertilized. Each of us inherits only half our mother's genes and half our father's genes, and these genes were shuffled in sperm and egg formation. This is why we show many genetic differences from our parents and our brothers and sisters, and why we are even more different from people who are not close relatives.

Some organisms can reproduce both with and without sex. When we study such organisms, we find that there are times when the genetic variation produced by sexual reproduction is advantageous and other times when it is not. Many bacteria, unicellular organisms, and arthropods (insects and their relatives) reproduce asexually during the summer and then reproduce sexually when the temperature declines and days become shorter in the fall. Laboratory experiments have shown that many of these organisms change from asexual to sexual reproduction because of changing environmental conditions. Depriving them of food, heat, light, or oxygen will often effect the switch.

It makes sense that environmental conditions should cause the switch from no sex to sex in this way. Under this system, an organism that is well adapted to its environment can reproduce numerous, equally well-adapted copies of itself while conditions remain constant during the summer. When conditions then change in the fall, the organism reproduces sexually, creating many genetically different offspring. Because they are different, there is a good chance that some of them will survive the variable conditions.

Many species of bacteria and other "lower" organisms occur all over the world in a variety of habitats, and some of these species have existed virtually unchanged for more than 500 million years; it is hard to imagine an environmental change so catastrophic and widespread as to threaten them with extinction. Most higher plants and animals, on the other hand, are specialized and can live only in the few places that supply their particular needs. Individual species survive for only a few million years (which is not long in evolutionary terms).

A species with only asexual reproduction (dandelions are a good example) may do very well for a while, but is ultimately doomed to extinction much more surely than is a species that reproduces sexually. A change in the environment that kills dandelions will kill *all* dandelions in the area because they are all very similar genetically. Organisms need the genetic variability produced by sexual reproduction if they are to form new species adapted to new environments. If they do not form new species, the whole group may become extinct. We think of the ruling reptiles (including dinosaurs) as a large group of higher animals which became extinct. But in fact the ruling reptiles' descendants, the birds, are alive and well and living all over the world. Without sexual reproduction, the reptiles could not have left these very different, successful descendants.

Sexual reproduction, then, is inefficient for a well-adapted organism that lives all over the world and is tolerant of environmental change. More specialized species live in constant danger of dying out, leaving no descendants if they do not have the genetic diversity that sexual reproduction provides.

8–B Evolutionary Roles of Male and Female

In the rest of this chapter we will discuss some aspects of sexual systems in higher animals and their effects upon social structure. We begin by considering differences in the role of male and female in sexual reproduction and go on to consider some of the evolutionary consequences of these roles in different animals.

Fig. 8–6

A shag (cormorant) sitting on her eggs. Females of nearly all species devote more energy and time directly to reproduction than do males. (Biophoto Associates)

The common cormorant or shag
Lays eggs inside a paper bag.
The reason, you will see no doubt,
It is to keep the lightning out.
But what these unobservant birds
Have never noticed is that herds
Of wandering bears may come with buns
And steal the bags to hold the crumbs.
(W. H. Auden and John Garrett)

Fig. 8-7

Human egg surrounded by sperm. The human egg is one of the smaller eggs in the animal kingdom; sperm of different species are all about the same size and much smaller than the egg. (Biophoto Associates)

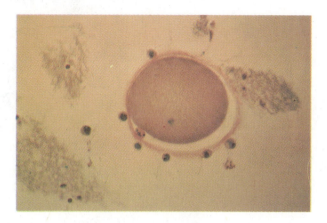

The female is, in a sense, a limiting resource for the sexual reproduction of a species. At the simplest level, because eggs are bigger than sperm, they take more energy to produce. A male's evolutionary success is usually limited, not by the number of sperm he can produce, but by his ability to deliver his sperm to as many eggs as possible. The female's evolutionary success is limited by the number of her eggs that survive to become part of the breeding population. For some females this means laying as many eggs as her energy budget will permit and letting them raise themselves, as in the case of a herring (a fish), which may lay a million eggs a year. At the opposite extreme, raising offspring to maturity may mean gestating them in her own body and caring for and feeding them for years afterwards, as in the case of many mammals. Which strategy is favored by evolution depends on the ecology of the species.

The fact that a female's parental investment in future offspring is usually greater than the male's has the fascinating consequence that the selective pressures acting on a female may conflict with those acting on a male. While it may be advantageous for a male to copulate with as many females as possible in order to raise his chances of fathering surviving offspring, it is apt to be advantageous for a female to be much more choosy. She produces fewer eggs and so has fewer second chances if her first mate is genetically unfit.

Under this selective pressure, it is not surprising that females of all species of animals studied show discrimination in their choice of mates. A simple example comes from a laboratory experiment with fruit flies *(Drosophila).* Of the females, only 4% failed to reproduce, whereas 24% of the males failed to copulate even once. These celibate males courted females just as vigorously as did successful males, but no female ever accepted them. In other words, females do discriminate between different potential mates. The female who discriminates and copulates only with genetically fit males will be at a selective advantage. On the other hand, it may be to a male's advantage to appear genetically fit even when he is not, because females may then be deceived into mating with him. This has been envisioned as an evolutionary battle of the sexes, with skilled salesmanship among the males and an equally well-developed sales resistance and discrimination among the females.

Fig. 8–8

Drosophila *courtship. Here the male (right) vibrates his wings to produce his genetically-determined "song" to which the female (left) may or may not respond. (Biophoto Associates by courtesy of Dr. W. L. Burnet)*

Fig. 8–9

A male skua defending his territory. (Biophoto Associates)

8–C Sexual Differences

A female's reproductive success is not usually limited by her ability to find mates. It is more likely to be limited by her inability to rear her young. A male who demonstrates that he can contribute to raising offspring will be attractive to females. For instance, among birds, the male of choice is often the holder and defender of a territory which provides food and shelter needed by the female and her young. Defending a territory is also of direct selective advantage to the male, for the territory he holds promotes the welfare of his offspring. Once a female finds a good territory, she need not bother about preliminaries but may mate with the resident male without further courtship. In such a situation, a male's ability to hold and defend a territory will determine his breeding success. Males of other species may compete for control over valued resources other than territories, because possession makes them irresistibly attractive to females. For instance, we know that a man with wealth and status attracts some women no matter how unattractive he is physically.

The different sexual roles of male and female may lead to their having different appearances, a phenomenon known as **sexual dimorphism** (di = two, morph = form). For example, female birds are much more likely than males to have drab colors. Because females are vulnerable as they sit on their eggs, it is advantageous for them to be inconspicuous. This camouflage works; when the male is the more conspicuous sex, mortality is invariably higher among males than among females. When defending a territory, a male may flaunt vivid coloration or unusual, exaggerated postures, making him more visible not only to other males who might think of invading but also to females, who may notice what a nice territory he has. An interesting variation on males' use of color to advertise their valuable property to females is found among some bowerbirds and weaverbirds. The male African village weaverbird, for instance, is dull-colored, but he builds a colorful nest and jumps up and down beside it saying, in effect, not "look at me" but "look at the gorgeous nest I have built for you." If no females are attracted, and if the color of the nest starts to fade, he will tear it to pieces and build another one.

Another type of sexual dimorphism is found in the possession of weapons by the male. Large antlers or horns in many hoofed animals, long tusks in boars, and the enormous size of male seals all give a male an edge in combat against other males for mates or breeding territories. Hence there is selective pressure for the males, but not the females, to possess these traits.

Strong sexual dimorphism of this type is rare in **monogamous** species (species whose members take just one mate for a whole breeding season). If the sex ratio is more or less equal, a male is almost bound to find a mate if he stays around long enough. Selective pressure on the monogamous male is mainly to stay alive; he is not involved in much competition with other males and tries to stay out of the way of predators. So he is not under selective pressure to evolve

Fig. 8–11

A bowerbird with his nest. (Biophoto Associates, N.H.P.A.)

Fig. 8–10

Male and female Barrow's goldeneye, members of a sexually dimorphic species. (William Camp)

(a)

(b)

Fig. 8–12

Male decorations and weapons. (a) A peacock displays his tail, used in courtship displays, and (b) the antlers of a red deer, used in fights with other males and to identify him to females. (Biophoto Associates)

Fig. 8–13

Swans are monogamous. Male and female look alike and cooperate to protect and feed their offspring.

the huge horns, elongated canines and other offensive weapons which are the hallmarks of a male destined either for glory or, much more likely, for early death. In a monogamous species such as beavers or geese, the male shows as much discrimination in his choice of mate as the female. Like the female, he has only one chance to choose a mate, and it will be just as advantageous for him as for the female to choose a mate who will produce healthy, successful offspring.

canines: pointed teeth at the sides of the mouth; enlarged as tusks or fangs in some species.

8-D Mating Systems

The courtship behavior of any species ensures accurate recognition so that there is little chance of a female's copulating with a member of the wrong species. In **polygamous** species, those in which each animal may mate with more

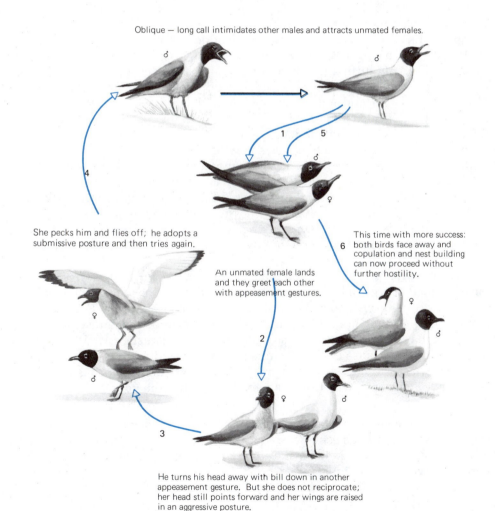

Oblique — long call intimidates other males and attracts unmated females.

She pecks him and flies off; he adopts a submissive posture and then tries again.

An unmated female lands and they greet each other with appeasement gestures.

This time with more success: both birds face away and copulation and nest building can now proceed without further hostility.

He turns his head away with bill down in another appeasement gesture. But she does not reciprocate; her head still points forward and her wings are raised in an aggressive posture.

Fig. 8–14

A courtship sequence in the black-headed gull. Follow the arrows around from the top.

than one other, males frequently have vast sex drive and little discrimination. They will court almost anything vaguely appropriate and the females must recognize and pick out a male. The male's appearance, physique, and courtship behavior assist in this. As a corollary, in monogamous animals, which mate with only one other individual, the sex drive in both sexes is about equal, the sexes are often indistinguishable in appearance and behavior, and courtship is mutual.

Polygamy. Polygamy may be divided into **polyandry**, in which one female mates with more than one male, and the much more commonly found **polygyny**, in which one male mates with many females. Polygyny may evolve where a female gets a better share of some limited resource by joining a mated pair or a male and his harem than by mating with an unmated male. The resource in question might be limited food or better protection from predators in a group than in a twosome, or it might even be peace and quiet and a helping hand to raise the young. An example of such a situation is known from studies of Japanese hamadryas baboons. A particular male was quite capable of defending his territory against all comers, but he was so belligerent that there was constant turmoil in his troop, and females always abandoned him after a short stay. They left to settle down with a troop where there might be less food but where they had a chance of giving birth and raising their young in peace and quiet.

Another selective force for harems with only one male may be that females do not have to share food and other resources with males whom they do not need for copulation. In such situations, the females help the resident male to defend their territory. On the other hand, where it's "all hands to the barricades" to fight off predators, non-reproductive males may be welcome members of the troop. Both kinds of social organization occur among various monkeys and apes.

All other factors being equal, polygyny is a more favorable system for the male of a species (or at least for the successful males who will be the only ones that reproduce). However, polygyny is possible only where the female does not need the male's full-time help in bringing up the young. Its presence or absence, therefore, will usually depend on the advantages or disadvantages to the female.

Monogamy. Under what circumstances does monogamy evolve? When the combined energy of both parents is needed to raise the young, there will be selective pressure on males to exhibit monogamous behavior. For example, monogamy (more or less) is the most common form of human sexuality (although both types of polygamy arise under various circumstances). We can infer that this occurs because the human infant is so demanding to raise: males who stayed and helped their mates to raise the children left more offspring in the next generation than those whose mates were left to raise the young alone.

Ecology and mating systems. An important factor in determining the mating system of a species is its ecology, since such factors as the distribution of food, water, nesting sites, and shelter in the environment determine the distribution and social behavior of individuals.

Where food is scarce and found in small, isolated pockets, individuals tend to be solitary and come together only for a short time, in the breeding season. (Bears, badgers and moose behave like this.) Couples may come together only to mate, or they may form short-lived pairs or colonies while they raise the young. On the other hand, it seems that most monkeys and apes can live in troops because their diet is mainly plentiful plant food; the all-important (in evolutionary terms) females and young can eat well and still enjoy the protection of living in groups. As we would expect from this, in situations where food is occasionally inadequate, groups containing only one male are the rule and there will be considerable competition among males to enter a troop, since this is the only way they can breed.

8–E Human Sexual Systems: Some Speculations

Studies on mate selection in modern human societies show that people do not choose their mates at random from the population; rather, they tend to marry according to certain characteristics, choosing people of similar educational background, race, religion, and social status. Marriage customs are governed by all sorts of rules and traditions. The most widespread of these are the taboos and laws against incest—mating with a genetically related individual.

A tendency against incest occurs in most species. For instance, female pigeons discriminate against males of the same genetic type as themselves. In fruit flies, females who do not discriminate against their male relatives leave only a quarter as many offspring as those that do. Incest taboos are more extensive in human than in most animal societies. One reason may be that, as studies have shown, human mother–son, father–daughter, and sister–brother matings have a very high chance of producing abnormal offspring. Humans seem to carry a relatively large proportion of the harmful genes that invariably show up in inbred stock.

Another example of nonrandom mate choice in humans comes from a nineteenth-century study on marriages of members of the British peerage. The study was undertaken to find out why the families of great men died out much more frequently than did families of the British population in general. The study showed that a considerable proportion of peers married heiresses. An heiress, almost by definition, comes from a small family, since she would not be much of an heiress if her parents' wealth were divided among 10 children. Heiresses who married peers were found to perform reproductively almost exactly as had their mothers before them. If an heiress were one of two children, there was a very high chance that she would also have two children. (In this case, mate selection by the peer and the heiress had the result of coupling relative infertility with high achievement of money and status.)

Modern *Homo sapiens* has a tendency toward monogamy but has presumably evolved from a polygynous primate ancestor. The evolution of human monogamy was probably made almost inevitable by the evolution of an infant that takes so much energy and time to raise. Humans mature more slowly than any other animals, and it seems likely that, throughout much of human history, both parents have had to do their share to raise human offspring to sexual maturity with consistent success. It is as much in the interest of the man as of the woman that his offspring reach maturity, and so the man, too, would be better off monogamous unless he could provide for the children of more than one wife.

There was probably another selective pressure, unique to humans, tending to make early humans monogamous: the possession of material goods, which was of vast importance to human evolutionary success. It would be strongly advantageous to bequeath objects like clothes or a cave to your genetic children. It is possible that before people acquired material possessions, selection favored polygyny for male and monogamy for female strategy, so that the two sexes were at evolutionary war with one another. From the moment the first man invented his first tool, however, woman had him in an impossible bind. (This is not to suggest that men rather than women invented tools, merely that only the man's tools matter to this argument.) Man's bind at this point was that woman could obscure a man's paternity, so that he wouldn't know which were his children or who should inherit his axes. What happened, in essence, was that woman promised man sexual fidelity and, therefore, identifiable offspring to leave his axes to, in exchange for a permanent commitment on his part to helping her raise the children. It did not matter to primitive woman whether he mated with other women, except where doing so led to injury at the hands of jealous husbands or to neglect of his obligations to her and their children—hence the origin of the double standard.

This highly speculative theory of the origin of human monogamy would account for another interesting fact. In societies which condemn adultery among their members, a woman's adultery is considered a much greater offense than a man's. This may well be because female adultery threatens a man's ability to identify his own children. If he cannot identify her children as his own, a man is wasting his time, in evolutionary terms, in helping a woman raise the children.

SUMMARY

All species have a certain amount of genetic variation among their members as a result of mutation. The amount of such variation is slight and members of species which reproduce asexually are, therefore, genetically very similar to one another. Members of sexual species are much more variable because genes are reshuffled during the formation of sperm (or pollen) and eggs and by fertilization. Asexual reproduction wastes less energy than sexual reproduction and is common in species that occur in many places in the world and are adaptable to changing conditions. For more localized and specialized species, the energy wasted in sexual reproduction is worthwhile; at least some of the genetically different individuals in a sexually reproducing species can usually survive and evolve in changed conditions.

Because eggs take more energy to produce than do sperm, females are usually more selective in their choice of mate than are males. Sexual dimorphism may arise when the selective pressures on the two sexes conflict, or when males and females have different roles in reproduction. Sexual dimorphism tends to be greater in polygamous species. In monogamous species, the members of both sexes need to choose their mates with discrimination, and the roles and behavior of the two sexes are more similar.

The sexual system of a species is determined by the amount of energy each sex puts into producing and rearing offspring and by ecological factors such as the distribution of food, prevalence of predators, etc.

Human monogamy probably originated for two main reasons. First, the human infant required the energy of both sexes to raise to sexual maturity. Second, it was important to the male's reproductive success that he passed on his knowledge and his material possessions to his genetic offspring. He could do this only if his mate were sexually faithful. Early woman, in a sense, traded her sexual fidelity to the male for his assistance in feeding and rearing their children; the traces of this system are prevalent in modern human societies.

SELF-QUIZ

1. An advantage of sexual reproduction is that it:
   a. increases the mutation rate
   b. increases genetic variability in a population
   c. produces larger offspring than does asexual reproduction
   d. reduces the risk of death during development for the offspring
   e. gives organisms something to do on Saturday night

2. Some organisms reproduce asexually when environmental conditions are (favorable, unfavorable) and sexually when conditions are (favorable, unfavorable) for growth.

3. It is generally true that males are an abundant resource, and that to be evolutionarily successful they should try to mate with as many females as possible. One possible exception to this generalization might occur when:
   a. there are many more females than males
   b. there are about equal numbers of males and females
   c. the father's care is required to raise the offspring
   d. there are many predators
   e. the male holds a territory against other males

4. Studies on female choice of mates in *Drosophila* showed that:
   a. more males than females reproduce
   b. females who reproduce leave more offspring per individual than do males who reproduce
   c. males who do not reproduce fail because they never court females
   d. males who do not reproduce fail because females will not accept them
   e. females who do not reproduce fail because males do not court them

5. When food is distributed in such a way that an animal must spend a large part of its day wandering from one place to another to find enough to eat, what type of mating system would you expect it to have?
   a. monogamy
   b. polyandry
   c. polygamy
   d. polygyny

6. Monogamy in humans probably evolved because:
   a. our food is sparsely distributed
   b. our ancestors had to hunt wild animals for food
   c. our ancestors were so aggressive that polygamy would have resulted in much bloodshed
   d. our large brain takes so long to develop and train before we become self-reliant

QUESTIONS FOR DISCUSSION

1. Animal-pollinated plants have less "pollen wastage" than do wind-pollinated plants. However, animal pollination demands energy expenditure in other ways. What are some of these? Why might they be "worthwhile"?

2. Does the desirability of genetic variations imply that monogamy in humans is a counter-evolutionary tendency? Why? (Hint: compare the average human family size with the number of possible genetically different eggs or sperm that a person could produce; for humans, with 23 pairs of gene-bearing chromosomes, each person could produce 2^{23} different kinds of reproductive cells.)

3. What are the selective advantages of internal over external fertilization?

4. List as many possible selective advantages as you can think of for courtship rituals.

5. Do humans have courtship rituals? Is courtship in humans mutual or is it carried out predominately by one sex? Why?

6. What criteria do *you* use in mate selection? Will the characteristics that you seek in a mate confer a selective advantage on your offspring?

7. What selective pressures might be responsible for the evolution of polyandry?

8. Female walruses must bear their young on land, but suitable stretches of beach are scarce. Similarly, nest sites for gulls are scarce. Both have crowded breeding grounds. Walruses eat molluscs (mussels, clams, etc.) whereas gulls will eat almost anything— including the egg next door. Explain what mating system you would expect each of these animals to show.

9. In many species, individuals are hermaphroditic, which means they possess both male and female sexual reproductive organs and cells. What is the advantage of hermaphrodism over separation of the sexes? What selective pressures might have led to evolution of species with separate sexes? Why are there never more than two different sexes in a species?

SUGGESTIONS FOR FURTHER READING

Bates, M., and P. S. Humphrey, eds. *The Darwin Reader.* New York: Charles Scribner's Sons, 1956.

Bishop, J. A., and L. M. Cook. "Moths, melanism and clean air." *Scientific American,* January 1975.

Carson, R. L. *The Sea Around Us.* New York: Oxford University Press, 1951. Contains a delightful discussion of the importance of algal photosynthesis.

Commoner, B. *The Closing Circle: Nature, Man and Technology.* New York: Bantam, 1972. An excellent paperback tracing the effect of human activities on parts of the world's ecosystems.

Corliss, J. B., and R. D. Ballard. "Oases of life in the cold abyss." *National Geographic,* October 1977. Describes a sunless ecosystem in the ocean depths.

Darwin, C. *On the Origin of Species.* New York: Cambridge University Press, 1975, and Cambridge, MA: Harvard University Press, 1964. These are recent reprintings.

Ehrlich, P. R. *The Population Bomb.* New York: Ballantine Books, 1968. An ecologist's gripping story of the human population explosion with some frightening science fiction about our lives in the future if we don't do something about population growth.

Eisely, L. C. "Charles Darwin." *Scientific American,* February 1956. An abbreviated biography of Darwin and discussion of his work.

Eisely, L. C. "Alfred Russel Wallace." *Scientific American,* February 1959. A fascinating account of the life of the energetic Wallace, including how he came to believe in the evolution of species, and his view of human evolution.

Hardin, G. "The tragedy of the commons." *Science* 162:1243, 1968. Hardin's thought-provoking argument that ecologically ethical individuals are evolutionarily doomed.

Hardin, G. *Exploring New Ethics for Survival: The Voyage of the Spaceship Beagle.* New York: Viking Press, 1972. Science-fiction treatment of the resource–population–pollution crunch from an evolutionary viewpoint.

Moorehead, A. *Darwin and the Beagle.* New York:

Harper and Row, 1969. A short, beautifully illustrated account of Darwin's travels, based on his diaries.

Morris, D. *The Naked Ape.* A male chauvinist's approach to the evolution of human sexual systems. Highly readable.

Morgan, E. *The Descent of Woman.* New York: Stein and Day, 1972. A feminist's reply to Morris.

Schumacher, E. F. *Small is Beautiful: Economics as if People Mattered.* New York: Harper & Row, 1975. The fascinating and influential book on "intermediate technology" which has led people and governments to rethink traditional ideas about the technology appropriate for our overcrowded planet.

Scientific American. "Food and Agriculture." September 1976 issue. A consideration of the problems and achievements of agriculture in many countries.

Scientific American. "The Biosphere." September 1970 issue. A collection of articles relating to energy and nutrient flows, including the effects of human activities.

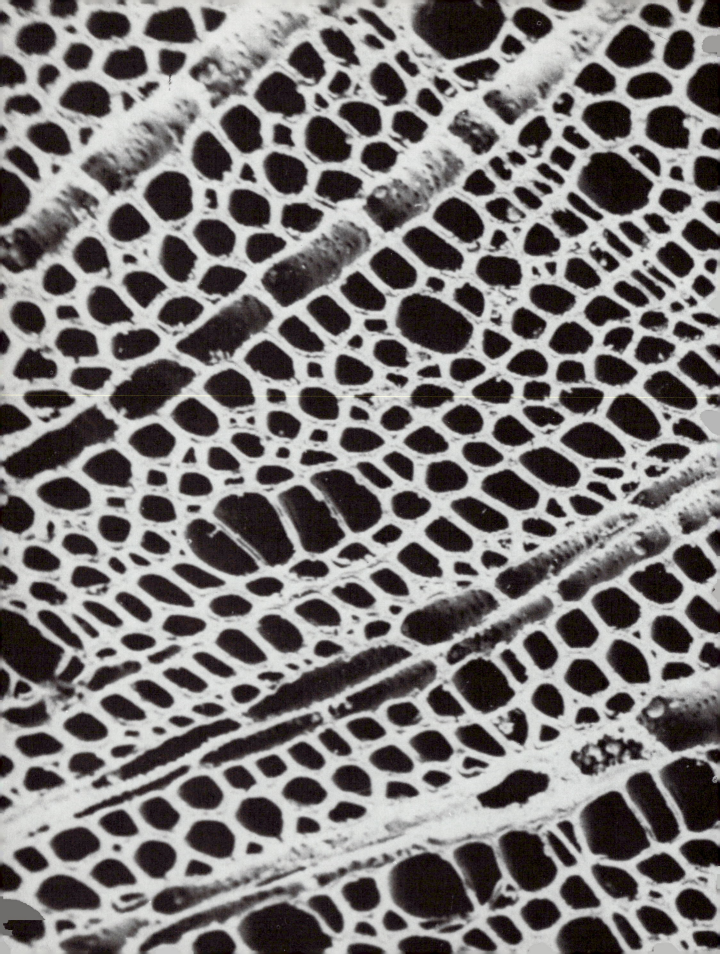

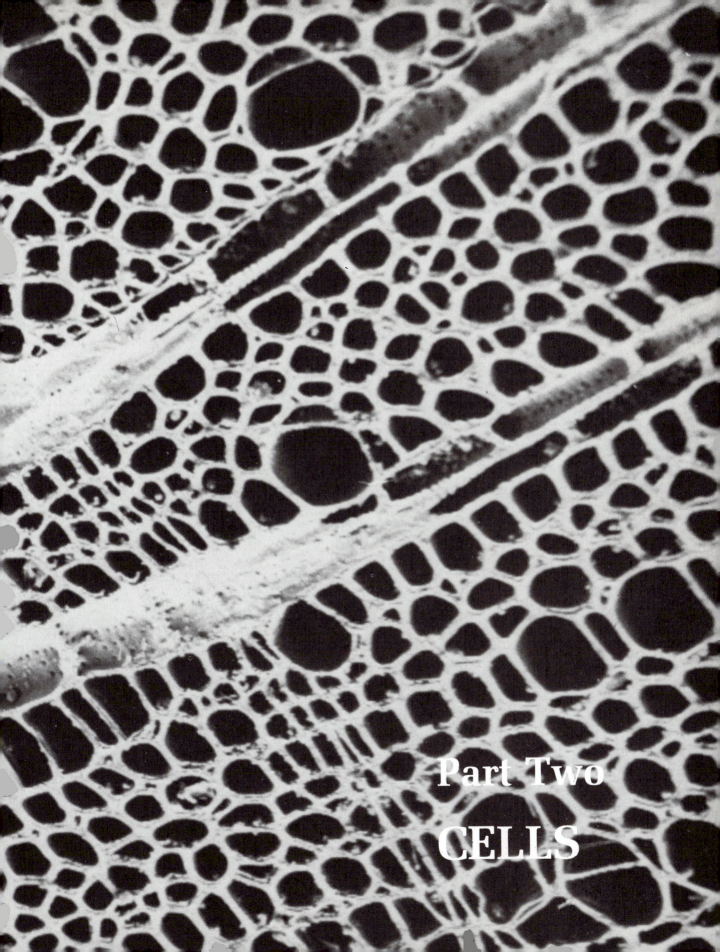

Part Two

CELLS

THE CHEMISTRY OF LIFE

9

OBJECTIVES

When you have studied this chapter, you should be able to:

1. Recognize and use the characteristics of the following: atom, proton, neutron, electron, ion, molecule, single bond, double bond, acid, base, polarity, dissociation, pH scale, organic compound, carboxyl group, amino group, polymer, monomer, carbohydrate, monosaccharide, disaccharide, polysaccharide, lipid, triglyceride, fatty acid, amino acid, polypeptide, catalyst, enzyme, nucleic acid, nucleotide.

2. Recognize examples of ionic, covalent, and hydrogen bonds, and explain how they differ from each other.

3. Write the correct molecular formulas for water, oxygen gas, carbon dioxide, and table salt.

4. List and discuss six reasons why water plays an important role in living systems.

5. Classify the following in the appropriate classes of molecules listed in Objective 1, and state the function of each in a living organism: glucose, cellulose, starch, glycogen, enzyme.

6. List four main classes of organic molecules found in living things; state the roles of each.

7. Explain why temperature, pH, and cofactors affect enzyme-catalyzed reactions.

A human being has been defined as "twenty gallons of water and $5 worth of assorted chemicals." However, it is not just these raw ingredients that count, but how they are put together. Not only the kinds of chemicals but also their arrangement determine the fascinating properties that we call life. The chemistry of animals, plants, bacteria, and fungi is impressive in two ways: first, living things are composed mainly of water; and second, most of the other chemicals in living things have "skeletons" of carbon. In this chapter we shall look at the basic structure of matter, concentrating on water and the carbon-containing chemicals of life.

9–A Elements and Atoms

A **chemical element** is a substance that cannot be broken down into other kinds of substances. If you take a piece of the element silver and divide it into ever-smaller pieces, you will ultimately have a single atom—still of silver. An

TABLE 9-1   CHEMICAL ELEMENTS FOUND IN ANIMALS AND THEIR APPROXIMATE ABUNDANCE BY WEIGHT

Element	Symbol*	Weight (%)
Oxygen	O	62
Carbon	C	20
Hydrogen	H	10
Nitrogen	N	3
Calcium	Ca	2.5
Phosphorus	P	1.1
Chlorine	Cl	0.2
Sulfur	S	0.1
Potassium	K	0.1
Sodium	Na	0.1
Magnesium	Mg	0.07
Iodine	I	0.01
Iron	Fe	0.01
Trace element†		
Copper	Cu	
Manganese	Mn	
Molybdenum	Mo	
Cobalt	Co	
Boron	B	
Zinc	Zn	
Fluorine	F	
Selenium	Se	
Chromium	Cr	

*Each element has been assigned a one- or two-letter symbol used in writing chemical formulas and equations.
†Needed in very small amounts.

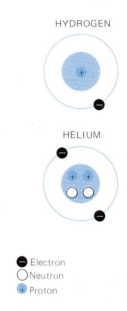

Electron
Neutron
Proton

Fig. 9–1

Models of hydrogen and helium atoms as imagined by Niels Bohr in 1913. These are not drawn to scale; the size of the nucleus (protons + neutrons) with respect to the whole atom is equivalent to that of a grape in a football field.

atom is the smallest unit that shows all the properties of the element. The names of many elements are familiar, everyday words—for example, oxygen, nitrogen, gold, helium. The element carbon is familiar in two of its pure forms—graphite (the "lead" in a pencil) and diamond. Carbon is very important to life because it is extremely versatile; it combines with other elements to make thousands of different substances, many of them vital to living organisms.

Chemists have discovered more than 100 elements, but living things contain only about 20 (Table 9-1). The chemicals that organisms need are not always the most common substances in the environment. For instance, silicon is about 300 times more abundant than carbon in the earth's crust; yet carbon is vital to every living thing, whereas silicon is found in very few.

One of the most dramatic achievements of twentieth-century science was the splitting of atoms into their component parts to understand their structures. The simplest atoms are those of the element hydrogen (Fig. 9-1). A hydrogen atom is made up of one positively charged particle, called a **proton**, and one negatively charged particle, called an **electron**. The electron whizzes around the proton in a relatively large space. In addition, a small percentage of hydrogen atoms have particles called **neutrons**, which bear no electrical charge. Atoms of every element other than hydrogen contain all three types of particles.

All atoms with the same number of protons belong to the same element: for example, a hydrogen atom has one proton, carbon has six, and oxygen eight (Fig.

PRO-tonn (rhymes with "on")
ee-LECK-tron

NEW-tron

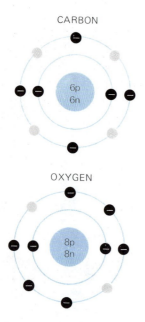

Fig. 9–2

The Bohr models for carbon and oxygen (p = proton; n = neutron). Carbon has room for four more electrons in its outer shell (Section 9–B); oxygen can take two more before its outer shell has a full complement of eight electrons.

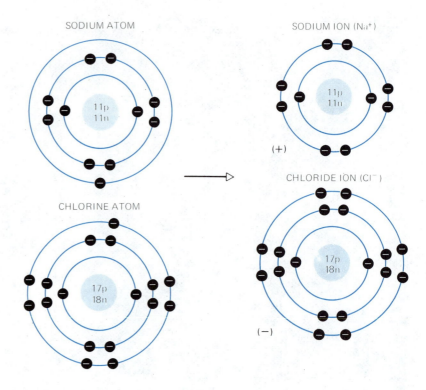

Fig. 9–3

Formation of an ionic bond. Sodium becomes stable if it loses the lone electron in its outer shell. Chlorine must add one electron to its outer shell of seven to achieve the stable number of eight. The sodium atom may transfer its outer electron to chlorine. After this transfer, sodium becomes an ion, with 11 protons but only 10 electrons, for a net charge of +1, and chlorine, with 17 protons and 18 electrons, has a net charge of −1; it is called a chloride ion.

9-2). The protons and neutrons in an atom are clumped together in a central structure, the **nucleus**. Because opposite charges attract one another, the negatively charged electrons are attracted to the positively charged protons in the nucleus, and this attraction holds the atom together. Each atom contains as many electrons as protons; therefore, the electrical charges cancel one another, and the whole atom is electrically neutral.

9–B Bonds Between Atoms

Although atoms are electrically neutral, many of them are not stable. To be stable, the outer layer, or **shell**, of the atom must be "filled" with a certain number of electrons—two for hydrogen and helium, and eight for atoms of the other elements we shall consider here. In a chemical reaction, atoms join, or **bond**, to

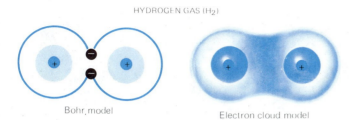

HYDROGEN GAS (H₂)

Bohr model

Electron cloud model

Fig. 9–4

Two different ways of representing covalent bonding between two hydrogen atoms. Each hydrogen atom has one electron and tends to gain another to complete an outer shell of two. By sharing their electrons, two hydrogen atoms both fill their electron shells. The "electron cloud" model uses density of color to indicate the proportion of time spent by the electrons in various places; the electrons in a covalent bond spend most of their time between the nuclei that are sharing them.

other atoms in such a way that each atom achieves greater stability by ending up with a filled outer shell. The behavior of the electrons, not of the nucleus, forms the bonds between atoms.

Three types of bonds are important in living things:

1.   An **ionic bond** forms when one atom gives up one or more electrons to another atom. As a result of this atomic give-and-take, each atom ends up with a stable set of electrons. For example, a sodium atom has one electron in its outermost shell, but it is more stable without this outer electron. An atom of chlorine, on the other hand, needs one electron to stabilize its outer shell; it can accept an electron from sodium (Fig. 9-3). After an electron has passed from a sodium to a chlorine atom, sodium has one more proton than it has electrons and so has a net positive charge of +1. It is now called a sodium **ion** (a charged particle) instead of an atom. The chlorine atom, with an extra electron, will have a negative charge of −1; it is called a chlor*ide* ion. The charged sodium and chloride ions are attracted to each other, and form crystals of sodium chloride, familiar to us as table salt.

eye-ON-ick

SO-dee-um

EYE-on

KLOR-een

2.   A **covalent bond** forms when two atoms share a pair of electrons—one electron from each atom (Fig. 9-4). Two pairs of electrons may be shared between two atoms to form a double covalent bond (Fig. 9-5).

coe-VAIL-ent

When two atoms of the same element bond covalently, they share the electron pair equally and the electric charge is distributed symmetrically between them (Figs. 9-4 and 9-5). If one of the atoms is attached to a third atom, however, or if the two partners of the bond are not of the same element, one of the atoms will usually attract the electron pair more strongly than the other. In such a case, the covalent bond will be electrically asymmetrical, and it is said to be **polar** (Fig. 9-6). The polarity of bonds between atoms can be arranged in a continuous series, with **nonpolar** (electrically symmetrical) covalent bonds at one end. At the other end is an ionic bond, with one atom monopolizing all the electrons involved in the bond so that the bond is highly asymmetrical. Between these two extremes are polar covalent bonds, in which the shared electrons are attracted to both atoms, but are attracted to one more strongly than to the other.

3.   The third type of bond between atoms that is important in living things is the **hydrogen bond**. One partner to this bond is a hydrogen atom attached by a polar covalent bond to another atom (usually oxygen or nitrogen) in such a way

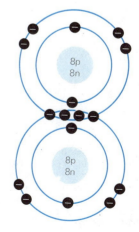

Fig. 9–5

A double covalent bond between two oxygen atoms. The atoms share two pairs of electrons, and each atom completes its outer electron shell.

Fig. 9–6

Two different representations of the polar molecule of hydrogen chloride (HCl). The electron pair that the atoms share in a covalent bond is attracted more strongly by the chlorine nucleus than by the single proton of hydrogen. As a result, the chlorine has a partial negative charge (δ^-), and the hydrogen atom has partly lost its electron, leaving it with a partial positive charge (δ^+).

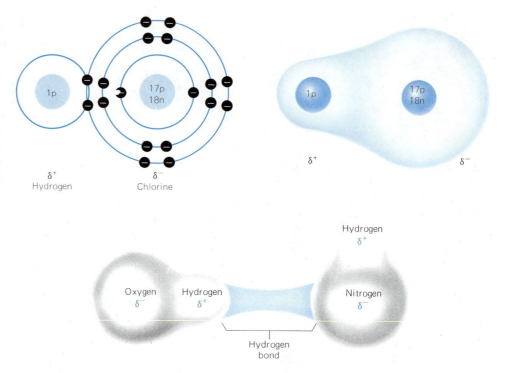

Fig. 9–7

A hydrogen bond (color) is a weak attraction between a polar-bonded hydrogen with a partial positive charge (δ^+) and a polar-bonded atom of nitrogen or oxygen with a partial negative charge (δ^-).

DIMETHYL ETHER

ETHYL ALCOHOL

Fig. 9–8

Structural formulae for two compounds with the same molecular formula (C_2H_6O). Ethyl alcohol, also known as ethanol, is the type of alcohol in alcoholic beverages. Each letter represents one atom (see Table 2–1 for symbols), and the lines between atoms represent covalent bonds.

that the hydrogen bears a slight positive charge. This slight positive charge is attracted to a third atom, bearing a slight negative charge (again, usually oxygen or nitrogen); this attraction is called a hydrogen bond (Fig. 9-7). A single hydrogen bond is very weak, but the billions of hydrogen bonds in an organism are strong enough to make them essential to life.

9–C Compounds and Molecules

Because each atom can gain, lose, or share only a particular number of electrons, the types of bonds that any particular atom can form are limited and predictable. In a **compound**, two or more atoms of different elements combine in specific proportions in a specific pattern of bonds. A compound has a definite composition and a definite set of properties, which are different from the properties of its component elements. Examples of compounds are sodium chloride (which we have already encountered), carbon dioxide, and water. A **molecule** is the smallest part of a compound that retains all the properties of the compound, just as an atom is the smallest part of an element that retains all the properties of that element. Ionically bonded compounds (such as sodium chloride) are usually spoken of as composed of ions rather than of molecules.

A molecular formula is a shorthand way of showing the kinds and proportions of different atoms in a substance. For instance, the formula H_2O, which stands for water, indicates that each water molecule contains two hydrogen (H) atoms and one oxygen (O) atom. (Each element is represented by one or two letters, as in Table 9-1.) Table salt, NaCl, contains one sodium ion (technically, Na^+ to show its positive charge) to each chloride (Cl^-) ion. Oxygen gas, O_2, consists of two oxygen atoms. When two different compounds contain the same numbers of each type of atom, only a **structural formula**, which shows the arrangement of atoms and bonds, will distinguish between them (Fig. 9-8). The structural formula for water is H—O—H, showing that each hydrogen atom is separately attached to the oxygen atom.

9–D Chemical Reactions

During a **chemical reaction**, molecules or ions interact to form new compounds. Reactions are often written in the form of equations. The equation for the burning of marsh gas (methane), for example, is:

$$CH_4 \quad + \quad 2\,O_2 \quad \rightarrow \quad CO_2 \quad + 2\,H_2O$$

methane   oxygen   carbon   water
                                   dioxide

The starting materials are written on the left, and the products on the right, after the arrow. The "2" before O_2 and H_2O represents the fact that two molecules of oxygen are needed to burn one molecule of methane, and two molecules of water result for each molecule of carbon dioxide produced.

Sometimes the arrows in a chemical equation point in both directions:

$$CO_2 \quad + \quad H_2O \quad \rightleftharpoons \quad H_2CO_3$$

carbon   water   carbonic
dioxide   acid

The double arrows mean that this reaction is reversible: it proceeds from right to left or from left to right, depending on the conditions.

9–E Water

It is impossible to imagine that life as we know it will be found on any planet that lacks abundant water, because water accounts for the bulk of material in most organisms.

Water (H_2O) is a most unusual compound. The molecule is polar, with a partial negative charge on the oxygen atom, and a partial positive charge on each hydrogen atom (Fig. 9-9). Because the partially negative oxygen on one molecule is attracted to the partially positive hydrogen on another, water molecules attach to one another by hydrogen bonds (Fig. 9-10). The structure of water molecules and the formation of hydrogen bonds between them endow water with a number of properties important to life:

1. *Water sticks to itself (cohesion) and to other substances (adhesion).* These properties are due to the polar nature of water molecules. The **cohesion** of water molecules to one another results in the **surface tension** which permits you to fill a glass of water slightly above its brim. Water also adheres strongly to any

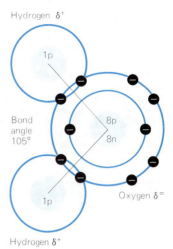

Hydrogen δ^+

1p

Bond angle 105°

8p
8n

1p

Oxygen $\delta^=$

Hydrogen δ^+

Fig. 9–9

A molecule of water. One atom of oxygen is covalently bonded to two atoms of hydrogen. The molecule is polar because it is bent at an angle, and the oxygen nucleus attracts the shared electrons to the point of the angle, giving the oxygen a partial negative charge, and leaving the hydrogen atoms at the open ends of the angle with partial positive charges.

Fig. 9–10

Hydrogen bonds (color) between water molecules. Oxygen atoms (color) bear partial negative charges, and so they hydrogen-bond to hydrogen atoms (black), bearing partial positive charges, on other water molecules.

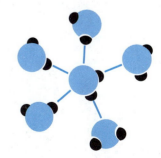

Fig. 9–11

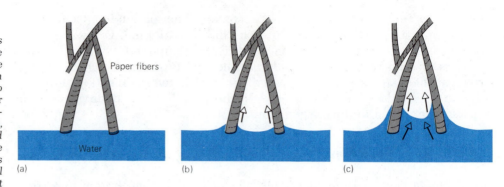

Capillarity of water. (a) Fibers in a piece of paper tissue touched to the top of a surface of water. (b) Adhesion between water and fibers allows water to creep along the fiber. (c) Water that has crept up the fiber coheres to other water molecules, pulling them along behind, and filling the space between the fibers with water. If the pores are small enough, cohesion will keep the water together until it has filled the space; it may then go on and fill spaces above.

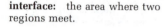

interface: the area where two regions meet.

electrically charged surface. This adhesion accounts for the **capillarity** of water—its ability to move upward into a slender glass tube or into the small pores of paper or soil. Cohesion between the water molecules means that the molecules adhering to the paper or soil or glass will pull other water molecules up behind them (Fig. 9-11).

2. *Water is a solvent.* More substances dissolve in water than in any other liquid. The polar nature of water permits it to dissolve other polar molecules (see Fig. 9-13).

Nonpolar (uncharged) molecules will not dissolve in water; instead they form interfaces with it like the interface formed when nonpolar oil and water stand in a jar together. Similar interfaces in living organisms are vitally important as the sites where many chemical reactions take place; thus the inability of water to dissolve nonpolar substances is also important.

3. *Water has high thermal conductivity.* Thermal conductivity is a measure of the ability of heat to spread throughout a substance. Water's high thermal conductivity is important in a living organism's body because chemical reactions go on all the time within the body, producing heat; however, this heat becomes evenly distributed throughout the watery body, eliminating the chances of local "hot spots."

4. *Water has a high boiling point.* It takes a great deal of heat to overcome the attraction between water molecules and change water from a liquid, in which all the molecules are held together, to a gas (water vapor) in which each molecule is separate. Fortunately for living organisms, temperatures on the surface of the earth seldom reach the boiling point of water.

5. *Water is a good evaporative coolant.* Many land animals and plants cool their bodies by expending the body's heat energy to change water from a liquid to a gas. Sweating in humans and panting in dogs are examples of evaporative cooling in animals.

6. *Water has a high freezing point.* The freezing point of water is, perhaps, higher than is ideal for life, since living organisms in many areas do encounter temperatures that freeze water. Water has the peculiar property that its maximum density occurs at 4°C, slightly above its freezing point of 0°C. As water cools from 4° to 0°C, it expands; therefore, an ice crystal is bigger than the volume of water it replaces. If ice crystals form within an organism they may destroy its delicate internal structure and cause death. Winter wheat and some insects are among organisms with natural antifreezes that prevent ice from forming. Other

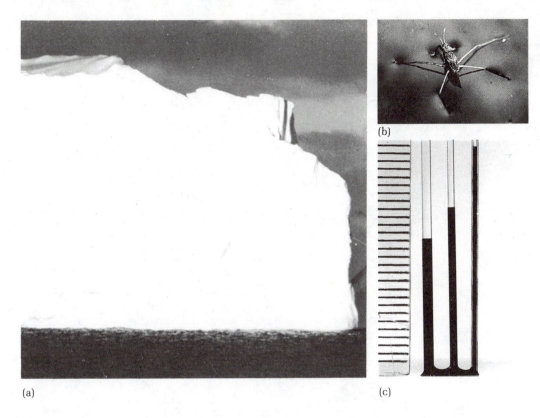

(a)

(b)

(c)

Fig. 9–12

Properties of water. (a) An iceberg in the Antarctic Ocean. Ice floats on top of water because ice is less dense than water; this property provides protection from freezing to water-dwelling organisms in cold climes. (U.S. Navy) (b) Surface tension, a result of the cohesion of water molecules, permits an insect to walk on water. (Biophoto Associates) (c) Capillarity: three glass tubes of different diameters have been placed in the ink puddle. The ink rises farthest up the narrowest tube. (Biophoto Associates)

organisms, like some plants, have tissues that are not damaged by ice crystals. Still other organisms, such as mammals and birds, keep their bodies at high enough temperatures so that they don't freeze.

The low density of ice has an advantage for aquatic organisms: the ice floats on top of the water, forming a blanket of insulation between the water and the cold air. Organisms living below the ice are thus protected from freezing.

9–F Dissociation

Many substances come apart, or **dissociate**, into ions when they dissolve in water. An **acid** is a substance that releases hydrogen ions (H^+) when it dissociates in water. Hydrogen chloride (HCl) dissociates into hydrogen ions (H^+) and chloride ions (Cl^-) in water to produce hydrochloric acid. A **base** (sometimes called an alkali) is a substance that releases hydroxyl ions (OH^-) in water or that can accept hydrogen ions in solution. The base sodium hydroxide (the active ingredient in some drain cleaners) dissociates into sodium and hydroxyl ions (Na^+ and OH^-) in solution (and gives off a lot of heat at the same time, as anyone who has used a drain cleaner containing it knows). A **salt** is a substance that gives off neither hydrogen nor hydroxyl ions when it dissociates. Our old friend sodium chloride (Na^+Cl^-) is an example.

AL-kuhl-eye

hi-DROX-ide

hi-DROX-ill

The acidity or alkalinity of a solution can be indicated by a number known as **pH**. The pH scale goes from 0 to 14. A pH of 7 is neutral (neither acidic nor basic);

STRAIGHT CHAIN

BRANCHED CHAIN

CHAIN WITH DOUBLE BOND

6-CARBON RING
WITH DOUBLE BONDS

6-CARBON RING

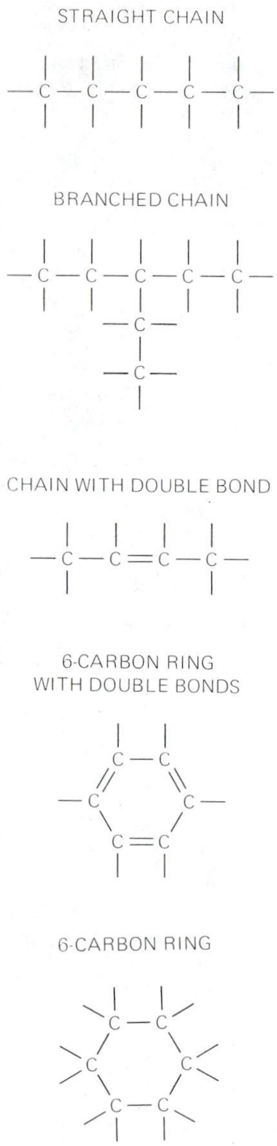

Fig. 9–15

Carbon atoms can bond to-gether in many different ways; a few examples are shown here. The unconnected lines sticking out from the carbon atoms (represented as C) can bond to atoms of other elements, commonly with hydrogen, oxygen, nitrogen, or sulfur.

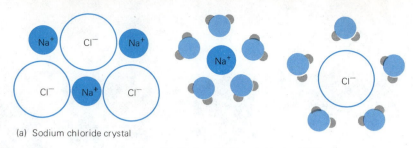

(a) Sodium chloride crystal

(b) Sodium (Na^+) and chloride (Cl^-) ions dissolved in water

Fig. 9–13

Sodium chloride (NaCl) dissolving in water. (a) Sodium (Na^+) and chloride (Cl^-) ions are attracted to one another because of their opposite electrical charges. When NaCl comes into contact with water, positively-charged sodium ions attract the partial negative charges on the oxygens of water. Similarly, negatively-charged chloride ions attract the partially-positive hydrogens of water molecules. All the tiny electrical tugs from water molecules pull Na^+ and Cl^- away from each other.

Fig. 9–14

The pH scale, with some familiar markers.

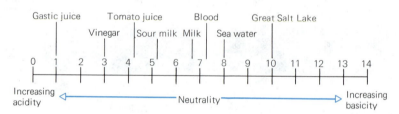

pH values below 7 are acid and above 7 are alkaline (Fig. 9-14). Water is considered neutral because it gives off H^+ and OH^- in equal amounts when it dissociates. The pH scale is logarithmic; pH = $-\log_{10}$ (concentration of H^+ ions in the solution)—thus a pH of 5 is ten times more acidic than a pH of 6.

9–G Carbon

Biological chemistry is based on large, carbon-containing molecules; these were once thought to be made only in living organisms and are, therefore, called **organic molecules**. Only the simplest carbon compounds, such as carbon dioxide (CO_2) are considered inorganic. All substances that do not contain carbon are also classed as inorganic, although many inorganic substances, such as water, table salt (NaCl), and carbon dioxide, are found in living things and may be vital to life.

The properties of carbon atoms make them uniquely suited to be the basis of the chemistry of life. A carbon atom can form four covalent bonds, and carbon atoms often join with each other to form long chains, sometimes with other carbon chains branching off from them. The ends of carbon chains may join, forming ring structures (Fig. 9-15). Carbon is the only element that can form enough different, complex, stable compounds to make up the variety of molecules found in living things.

9–H Building Biological Molecules

Organisms form various small organic molecules called **monomers**, the building blocks or subunits of larger molecules. Monomers may be linked together to form **polymers,** also known as **macromolecules** (macro = big). Organic polymers produced by living things include wool, silk, rubber, and cotton, which have been used by humans for thousands of years. Since about 1900, chemists have been synthesizing the artificial organic polymers we know as "plastics" by joining small organic monomers, such as dimethyl ether (Fig. 9-8) or ethyl acetate, in various ways—a case of art, or at least industry, imitating nature.

MON-oh-murz
POLL-ih-murz
MAC-roh-MOLL-ick-yules

DIE-meth-ill EE-thur
ETH-el ASS-it-ate

An organism's body builds the large molecules it needs by joining monomers together. The polymers so formed may be broken down into their original monomers. This is what happens in the digestive tract of an animal: macromolecules in its food are broken down into small molecules that can be absorbed into the body. Old molecules are also broken down, and their components used to build new macromolecules in all parts of the bodies of animals and plants.

All living organisms contain four main classes of organic molecules: carbohydrates, lipids, proteins, and nucleic acids. Each has characteristic monomers that combine to form polymers. In the rest of this chapter, we shall discuss each group briefly.

9–I Carbohydrates

Carbohydrate monomers are the simple sugars, called **monosaccharides**, such as **glucose** and fructose. Simple sugars contain carbon, hydrogen, and oxygen atoms in a 1 : 2 : 1 ratio. There are from three to nine carbon atoms in a monosaccharide; the most common ones contain five or six. When a monosaccharide containing five or more carbon atoms dissolves in water, it may take on a ring structure (Fig. 9-16). Monosaccharides may be broken down by the body to yield energy, or built up into larger polymers.

MON-oh-SACK-ahr-ides
GLUE-cose

Two monosaccharides may bond to form a **disaccharide** (Fig. 9-17). Sucrose (table sugar), maltose (malt sugar), and lactose (milk sugar) are familiar disac-

die-SACK-ahr-ide

SUE-krose

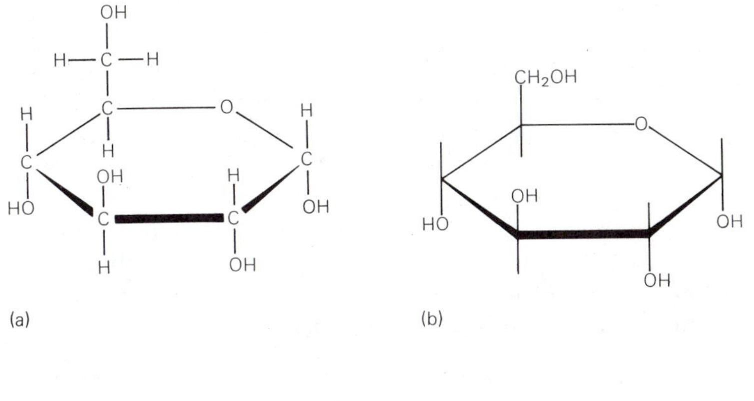

(a)   (b)

Fig. 9–16

Two ways of showing a molecule of glucose, a common monosaccharide. (a) When it is dissolved in water, as in the body of a living organism, the molecule is most frequently in the shape of a ring. (b) A "shorthand" way of showing the same molecule; carbon atoms are assumed to be at all corners unless another atom is shown (here, all corners have carbon except the upper right, which is occupied by an oxygen atom). Lines above and below the ring are assumed to be bonded to hydrogen unless another group (such as the several OHs and the CH_2OH at top left) is shown.

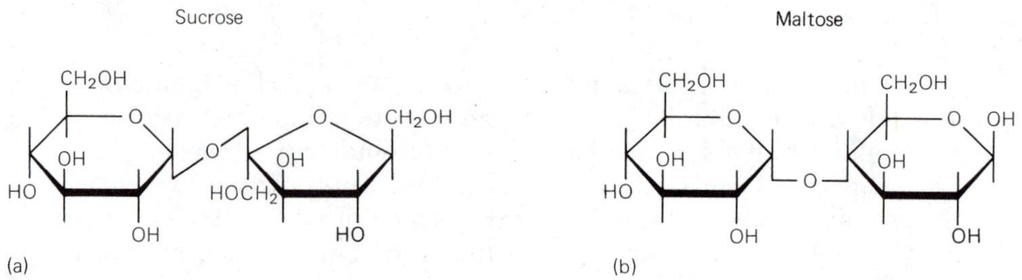

Sucrose   Maltose

(a)   (b)

Fig. 9–17

Two common disaccharides. (a) Sucrose, or table sugar, is formed by the joining of a glucose (left) and a fructose (right) monomer. (b) Maltose forms from two glucose monomers.

charides. (Despite the claims of health food enthusiasts, note that all these sugars have been extracted from living organisms and it is hard to see how one can be more "natural" than another.)

GLIKE-oh-jenn

A large number of monosaccharides may join together to form the polymers known as **polysaccharides**. A common polysaccharide in animals is **glycogen**, a polymer of glucose stored mainly in the liver and muscles. When glucose is needed for energy, it is released by breaking glucose units off glycogen molecules. Plants store glucose as the polymer **starch** (as in potatoes). Plants also form another important polysaccharide, **cellulose**, composed of glucose monomers but with a different arrangement of bonds (Fig. 9-18). Cellulose stiffens and supports the plant body (consider celery, which is rich in cellulose). The human digestive tract is not equipped to break the kinds of bonds that link glucose units in cellulose. Cellulose in the human diet, therefore, provides valuable bulk ("roughage") which stimulates the intestines to keep things moving along.

Sugary foods are quick-energy foods, since sugars are easily digested into a form that the body can use for energy. However, extra sugar, like any other excess food, is rapidly converted into fat.

9–J Lipids

LIP-idds

Unlike carbohydrates, with their relatively well-defined chemical composition and structure, lipids are varied in structure and in proportions of elements. The common property of lipids is that they are nonpolar, which means that they dissolve in nonpolar liquids, such as chloroform and ether, but are essentially insoluble in water.

Because they are insoluble in water, lipids are vital components of the membranes that divide watery compartments from one another in living organisms. They are also used as the main energy stores in animals since, unlike carbohydrates, they can be stored in concentrated form (without water).

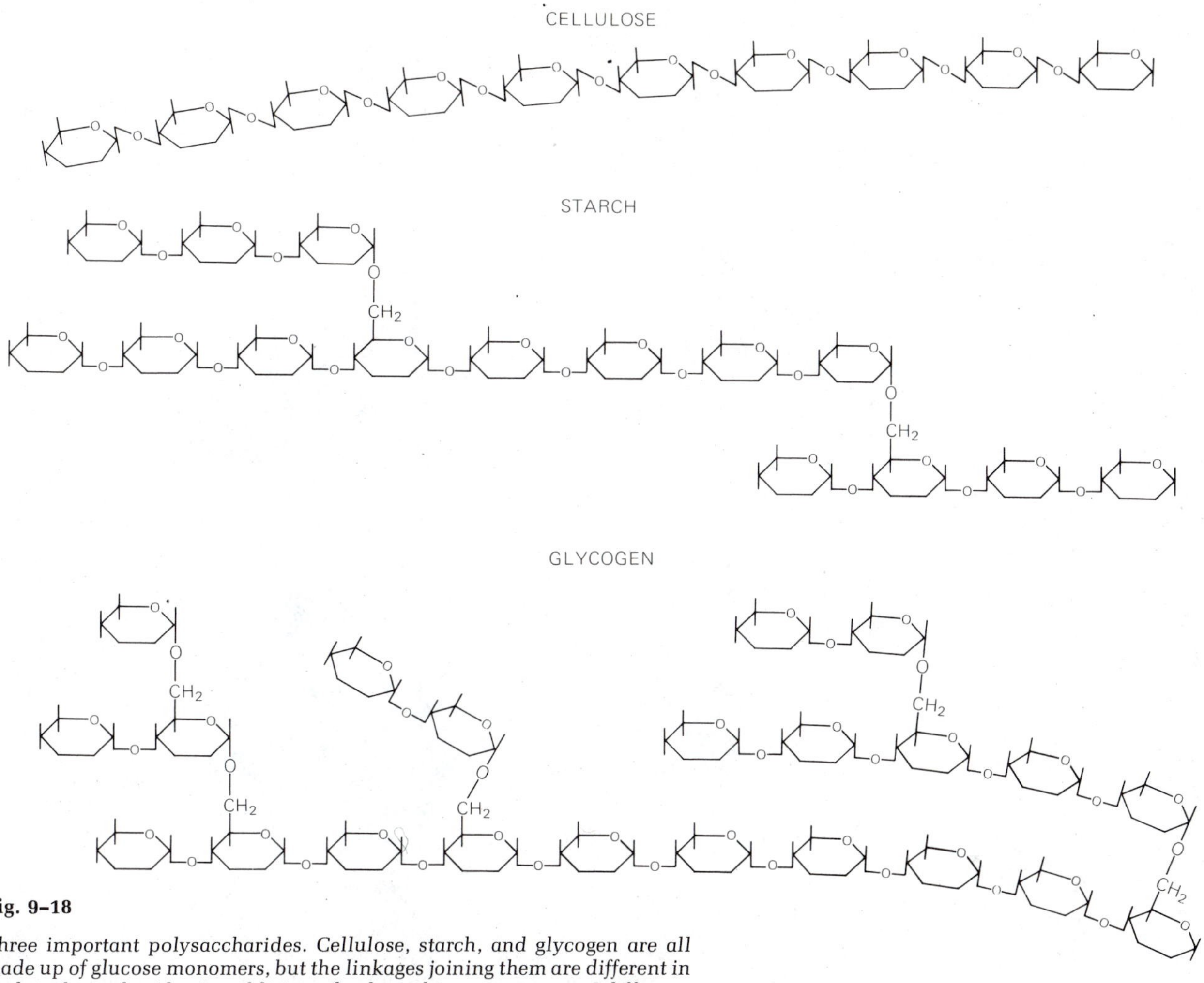

CELLULOSE

STARCH

GLYCOGEN

Fig. 9–18

Three important polysaccharides. Cellulose, starch, and glycogen are all made up of glucose monomers, but the linkages joining them are different in each polysaccharide. In addition, the branching patterns are different: cellulose molecules are unbranched, and glycogen branches more frequently than starch.

Small lipid molecules can combine to form somewhat larger ones, but polymers of the other three classes of organic molecules are all much larger than the largest lipid molecules.

Fatty acids are small lipid molecules, each with a **carboxyl group** (—COOH) at one end, and a long chain of carbon and hydrogen atoms (Fig. 9-19). A **saturated** fatty acid has no double bonds between any of its carbon atoms, whereas an **unsaturated** fatty acid has one or more carbon-carbon double bonds. It is called unsaturated because it could hold more hydrogen atoms if one bond in each pair of double bonds were broken and attached to a pair of hydrogen atoms instead.

car-BOX-ill

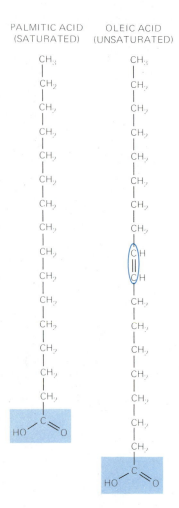

PALMITIC ACID (SATURATED) OLEIC ACID (UNSATURATED)

Fig. 9–19

Two fatty acids. The water-soluble carboxyl groups (−COOH) are in colored boxes. The long carbon-hydrogen chain at the other end of the molecule is not soluble in water. The double bond that makes oleic acid unsaturated is ringed.

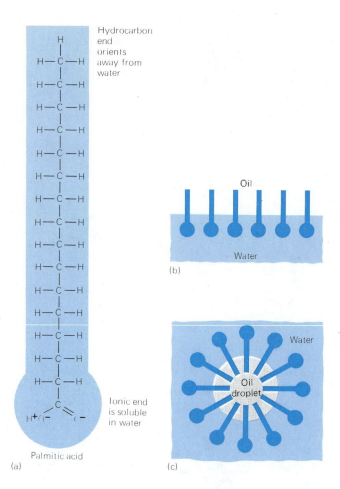

Fig. 9–20

Fatty acids have polar ends. The carboxyl group (ionic end) of a fatty acid dissociates in water (ionizes) and so is soluble in water, whereas the long hydrocarbon chain is insoluble in water. (b) As a result, the molecules orient themselves at interfaces between water and nonpolar organic substances such as oil, or between water and air, with their ionic ends in the water. (c) Soaps and detergents contain modified fatty acids; they clean by removing droplets of oil or other substances from dirty surfaces. Soap molecules surround an oil droplet, in which the soap's hydrocarbon chains dissolve. This leaves the water-soluble carboxyl groups sticking out into the water. The water dissolves the whole tiny droplet so that it floats and can be washed away.

try-GLISS-er-ides
GLISS-er-awl

The carboxyl end of a fatty acid molecule is polar, and will dissolve in water, whereas the carbon-hydrogen chain is nonpolar and will associate with nonpolar organic substances. Thus these molecules tend to lie along the interfaces between nonpolar and watery areas, making them important parts of the membranes that divide living systems into compartments (Fig. 9-20).

Triglycerides are molecules formed by joining three fatty acid molecules with a molecule of glycerol, an alcohol (Fig. 9-21). **Fats** are triglycerides that are solid at room temperature; oils are triglycerides that are liquid at room tempera-

GLYCEROL   THREE FATTY ACIDS

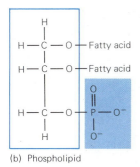

(a) Triglyceride

PHOSPHATIDIC ACID   CHOLESTEROL

(b) Phospholipid

Carbon-carbon
double bond

(c) Steroid

Fig. 9–21

Examples of different types of lipids. (a) A triglyceride. The glycerol part is boxed with a colored line; it is attached to three fatty acids. (b) In phospholipids, phosphate groups (solid-color box) are substituted for some fatty acids in the triglyceride-like structure. (c) A steroid, cholesterol, built on a multi-ring structure; steroids, including sex hormones, have the ring structure (colored), with different groups attached.

ture. Oils generally contain more unsaturated fatty acids than do fats. Animals living in cold areas, such as fish in arctic waters, usually contain more unsaturated triglycerides than do animals from warmer climes. This helps to keep their bodies flexible at low environmental temperatures.

Other lipids include phospholipids, molecules similar to triglycerides but containing the element phosphorus, and steroids, including cholesterol and many important hormones (Fig. 9-21).

FOSS-foh-LIP-idds

STEER-oids

9–K Proteins

Proteins perform a variety of jobs in the body of an organism (see Table 9-2). Familiar proteins include **enzymes**, which speed up chemical reactions, and the **structural proteins,** which make up things like hair, fingernails, and silk.

Proteins are composed of carbon, oxygen, hydrogen, nitrogen, and sulfur. The monomers of proteins are **amino acids**. Each amino acid has a carboxyl group (—COOH) and an amino group (—NH$_2$), both attached to the same carbon atom, and one of a variety of possible side chains, which is also attached to the

uh-MEAN-oh

TABLE 9-2   **SOME FUNCTIONS OF PROTEINS**

Protein	Function
Enzymes	
Amylase	Converts starch to glucose
DNA polymerase I	Repairs DNA molecules
Structural proteins	
Keratin	Hair, nails, horns, hoofs
Collagen	Tendons, ligaments, cartilage
Hormones	
Insulin	Regulates glucose usage
Vasopressin	Decreases water loss via urine
Contractile proteins	
Actin, myosin	Muscle contraction
Storage proteins	
Casein	Nutrient protein in milk
Ferritin	Stores iron in spleen and egg yolk
Transport proteins	
Hemoglobin	Carries oxygen in blood
Serum albumins	Carry fatty acids in blood
Toxins	
Cardiotoxin	Poison in cobra venom; blocks muscle contraction

Fig. 9–22

Some amino acids. The parts in the colored area are the side chains, which are different for each kind of amino acid. The parts shown on white are the same for all amino acids. Note that some side chains contain sulfur (S), nitrogen (N), or ring structures. In all, there are twenty amino acids that may be used to make up proteins.

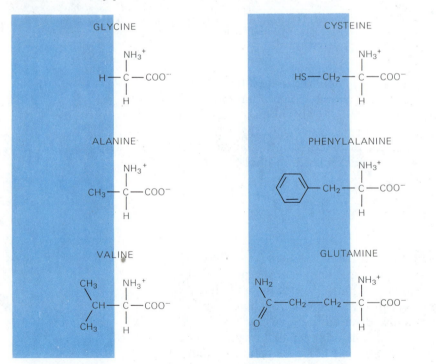

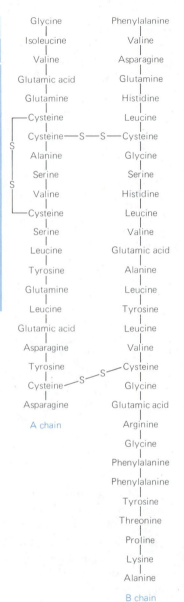

Fig. 9–23

The amino acid sequence of a small protein, insulin. The protein is made up of two polypeptide chains, A and B, which are joined by "sulfur bridges," attachments between sulfur-containing cysteine amino acids.

same carbon (Fig. 9-22). The different arrangements of the 20 common amino acids found in proteins account for the vast diversity of living things.

A long string of amino acids is called a **polypeptide**. Most polypeptides contain 100 to 300 amino acids. Some proteins consist of one polypeptide chain; others are composed of two or more polypeptides in a specific arrangement. For example, the hemoglobin molecules that make blood red contain four polypeptides each.

POLL-ee-PEP-tide

To function properly, a protein must be bent into a specific shape (see Fig. 9-23). The three-dimensional structure of a protein molecule is determined by the sequence of amino acids in its polypeptide chains. In other words, if you could take a polypeptide chain by either end and stretch it out and then let go, it would fold up into the same structure each time, and the structure would be different for each kind of polypeptide.

Most chemical reactions in living things are controlled by enzymes, protein molecules that act as catalysts. A **catalyst** is a substance that increases the rate of a chemical reaction that would normally occur at a slower rate. The catalyst is not permanently changed by the reaction.

CAT-uh-list

As an example of the action of an enzyme, let us consider a reaction that is familiar to most cat owners—the breakdown of urea from a cat's urine into carbon dioxide and ammonia, which gives its characteristic odor to a cat's litterbox in need of cleaning:

you-REE-uh

$$H_2N{-}CO{-}NH_2 \ + \ H_2O \ \rightarrow \ CO_2 \ + \ 2\,NH_3$$

urea   water   carbon   ammonia
                                dioxide

This reaction is catalyzed by the enzyme urease, which is produced by bacteria that settle out of the air and grow and reproduce in the litterbox. At room temperature, a molecule of urease can catalyze the breakdown of about 30,000 molecules of urea in 1 second. Without a catalyst, this reaction would take about 3 million years. Thus, in the presence of urease, the reaction happens at more than a trillion times its "normal" rate. Some enzymes work more quickly than urease, some more slowly.

YOUR-ee-ase

Enzymes are very specific in that each catalyzes reactions involving only one or a very few kinds of molecules. This is true because an enzyme actually binds to its **substrates**, the compounds upon which it acts. An enzyme has an **active site** whose shape and chemical composition are such that it can bind only to certain substrates. When an enzyme catalyzes a reaction, it binds its substrate molecules closely together, with the parts of the molecules that will react next to one another (Fig. 9-24). A substrate becomes slightly altered when it binds to an enzyme. For instance the enzyme may attract electrons in such a way as to put tension on the bonds holding the molecule together. This will make the molecule more reactive and is thought to be how an enzyme speeds up a reaction.

Various factors affect the rate of enzyme activity. For instance, because proteins contain electrically charged chemical groups that help to determine their folding patterns, the pH of a solution must be right for an enzyme to work. In an acid solution, negative groups on the enzyme may combine with H^+ ions

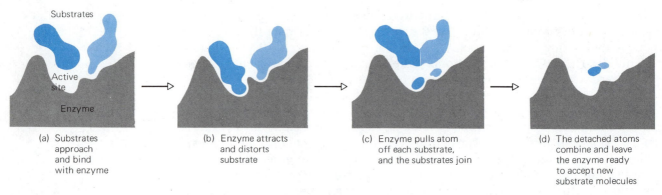

(a) Substrates approach and bind with enzyme

(b) Enzyme attracts and distorts substrate

(c) Enzyme pulls atom off each substrate, and the substrates join

(d) The detached atoms combine and leave the enzyme ready to accept new substrate molecules

Fig. 9–24

A diagram to show how an enzyme-catalyzed reaction between two substrates might look. Note that both the enzyme and substrate shapes are somewhat changed when the substrates are bound to the enzyme's active site. The enzyme probably loosens bonds in the substrate molecules, allowing them to release some of their atoms more easily and re-bond to each other.

in the solution and become neutralized. Only if an enzyme bears the appropriate charged groups on its surface will it fold up and work properly. Most enzymes work best when they are in an environment near pH 7 (neutral). There are exceptions, however, such as the digestive enzymes that work at the very acid pH (as low as pH 2) caused by the hydrochloric acid in the human stomach. These enzymes bear the electrical charges needed to bind their substrates only when H^+ surrounds them and combines with certain groups on the enzymes.

Temperature also affects the rate of enzyme reactions. When the temperature is higher, molecules move around faster, collide harder and more often, and so are more likely to react than at a lower temperature. The same is true of any reaction, whether it is catalyzed by an enzyme or not. On the other hand, high temperatures (usually above about 60°C) **denature** proteins by permanently destroying their three-dimensional structure so that they can no longer function. Heating preserves food by destroying the enzyme activity of organisms that cause decay. At the opposite extreme, chemical reactions proceed slowly at low temperatures; refrigeration also preserves food by slowing enzyme action.

Some enzymes need specific minerals or small organic molecules, which are not themselves catalysts, bound onto them before they will function. These are known as enzyme **cofactors** or **coenzymes**. Many vitamins are coenzymes. We need very little of them because cofactors are not destroyed in a chemical reaction, but, like enzymes themselves, can be used over and over again.

9–L Nucleic Acids

new-CLAY-ick

day-OX-ee-rye-bow-new-CLAY-ick

The **nucleic acids** include the largest molecules synthesized by organisms. There are two kinds of nucleic acids: **DNA**, or **deoxyribonucleic acid**, which contains hereditary instructions for the sequence of amino acids in polypeptides; and **RNA**, or **ribonucleic acid**, which participates in protein synthesis (see Chapter 15).

NEW-clee-oh-tide

Nucleic acid monomers are called **nucleotides**; each is composed of a phosphate group, a five-carbon sugar, and a single- or double-ring base containing nitrogen (Fig. 9-25). Besides their importance as the monomers that make up

TABLE 9-3   CHEMICAL COMPOSITION (EXCLUDING WATER) OF SOME COMMON BACTERIA

Type of molecule	% of total dry weight	Comments
Small molecules	10	Inorganic ions, monomers, coenzymes
Polysaccharides and lipids	16	Protective outer wall and membrane; some glycogen stored inside bacterium
DNA	4	One or two molecules per bacterium; each molecule is about 1 mm* long and highly folded; the bacterium itself is only about 0.002 mm long
RNA	20	About 3000 different kinds
Proteins	50	About 2500 different kinds: about 1/3 structural protein, 2/3 enzymes

*mm = millimetres; there are about 25 mm in 1 inch (see "metric system" end paper of this book)

Fig. 9–25

A representative nucleotide, a monomer of the nucleic acids. The five-carbon sugar is attached to a phosphate group, containing the element phosphorus (P), and to a base in the form of a single or (as here) double ring with nitrogen at several of its corners.

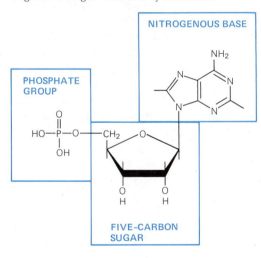

NITROGENOUS BASE

PHOSPHATE GROUP

FIVE–CARBON SUGAR

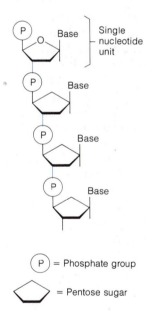

P = Phosphate group

= Pentose sugar

Fig. 9–26

Nucleic acids are made up of nucleotides linked with the phosphate group of one attached to the sugar of the next.

nucleic acids, several nucleotides have other functions. For instance, the nucleotide **ATP (adenosine triphosphate)** supplies the energy for many chemical reactions in all organisms.

add-EN-oh-seen
try-FOSS-fate

When nucleotides are polymerized into nucleic acids, the phosphate group of one nucleotide is linked to the sugar of the next, forming a long backbone of alternating sugar and phosphate groups, with the bases sticking out at one side (Fig. 9-26). The structure of nucleic acids is such an exciting illustration of the miracle of life that we have given this story its own chapter (Chapter 14).

SUMMARY

More than 100 different chemical elements are known, each with a unique set of chemical properties. Living things need about 20 of these elements. The special properties of the elements found in living systems are essential to life, growth, and reproduction.

Water is the most abundant substance in living things, and is absolutely necessary for life as we know it. Water dissolves polar and ionic substances; creeps into small spaces by capillarity; disperses heat throughout the body; and vaporizes from the body surface, carrying heat away from the body.

Organic polymers are composed of simple carbon-based monomers. It is possible to synthesize an endless variety of organic polymers by combining monomers in various ways. The four main groups of organic substances made by living organisms are carbohydrates, lipids, proteins, and nucleic acids.

The monomers of carbohydrates are monosaccharides (simple sugars). Some carbohydrates are important energy stores which can be broken down to release energy at need. The polysaccharide cellulose helps to support the bodies of plants.

Unlike members of the other groups, lipids are heterogeneous in structure, their common property being their nonpolar nature. Lipids are important energy storage molecules, and they make up the biological membranes that separate watery compartments in the bodies of organisms. Some lipids are hormones.

Proteins are by far the most common of the organic substances found in living organisms. Amino acids are the monomers of proteins. The specific three-dimensional structure of a protein is vital to its correct functioning. Enzymes are proteins adapted to fit specific substrates; they enable organisms to carry out reactions rapidly at relatively low temperatures. Hundreds of kinds of enzymes speed up the chemical reactions necessary to life. Proteins also form structures such as hair and fingernails, act as hormones and as poisons, and transport or store various substances.

The nucleic acids, DNA and RNA, are made up of nucleotides; they are involved in the inheritance of information. Some nucleotides, such as ATP, are important in their own rights; ATP supplies the energy for most chemical reactions in living organisms.

TABLE 9-4 **SUMMARY OF THE MAJOR GROUPS OF BIOLOGICAL MACROMOLECULES AND THEIR MONOMER SUBUNITS**

Monomers	joining together ⇌ breaking apart	Polymers
Monosaccharides		Polysaccharides
Glycerol, fatty acids, and other less common compounds		Triglycerides or related compounds
Amino acids		Polypeptides
Nucleotides		Nucleic acids

SELF-QUIZ

For multiple choice questions, choose the *one best* answer.

1. What is the difference between a sodium atom and a sodium ion?

2. A positively charged ion has:
   a. more protons than electrons
   b. more electrons than protons
   c. equal numbers of protons and neutrons
   d. equal numbers of protons and electrons
   e. more neutrons than electrons

3. What kind of bond involves the sharing of electrons between atoms? (covalent, ionic)

4. A water molecule is held together by (covalent, ionic, hydrogen) bonds.

5. A carboxyl group dissociates when it is surrounded by water; thus,

$$X-C\begin{matrix}O \\ \diagdown OH\end{matrix} \rightarrow X-C\begin{matrix}O \\ \diagdown O^-\end{matrix} + H^+$$

   This means that a molecule with a carboxyl group must be a(n):
   a. acid   c. salt
   b. base   d. nucleotide

6. Write the chemical formulas for:
   a. water _____
   b. table salt _____
   c. carbon dioxide _____
   d. oxygen gas _____

7. Water is able to dissolve ionic substances because:
   a. it contains ions   c. it is ionically bonded
   b. it is polar   d. it contains oxygen
                            e. it is covalently bonded

8. Which of the following is *not* made up of monosaccharides?
   a. sucrose
   b. starch
   c. glycogen
   d. cellulose
   e. hemoglobin

9. Complete the following table:

SUMMARY OF THE MAJOR GROUPS OF ORGANIC COMPOUNDS.		
Group	Monomers (subunits)	Importance to living organisms
Carbohydrates	a.	b. 1. 2.
c.	d.	1. Biological membranes 2. Energy storage 3. Hormones
e.	Amino acids	f. 1. 2.
g.	h.	i. 1. 2.

10. You have a solution of an enzyme. You add half the solution to each of two test tubes containing identical amounts of the enzyme's substrate. After waiting for a few moments, you test both tubes and find that the substrate in tube A has changed, but that in tube B has not altered. Suddenly you notice that tube B has been sitting on a hotplate switched to "high." The enzyme in tube B probably did not work because it had been:
   a. broken down into amino acids
   b. catalyzed
   c. denatured
   d. dehydrated

QUESTION FOR DISCUSSION

1. Refer to self-quiz question 10. How would the outcome in tube A be affected if you had added more enzyme to tube A at the beginning of the experiment? If you had added more substrate?

10 THE LIFE OF A CELL

OBJECTIVES When you have studied this chapter, you should be able to:

1. Describe the different uses of a light microscope, a transmission electron microscope, and a scanning electron microscope.

2. Give the function of each of the following structures, and state whether each would be found in plant, animal, or prokaryotic cells: cell membrane, nucleus, nuclear membrane, nuclear area, ribosome, mitochondrion, cell wall, chloroplast, vacuole, lysosome, endoplasmic reticulum, Golgi complex, cilium, flagellum, mesosome.

3. List at least three features that would enable you to tell whether a cell came from a plant or an animal.

4. Define the following processes, and state or identify the characteristics that distinguish them from one another: diffusion, facilitated diffusion, active transport, endocytosis, exocytosis, osmosis.

5. Describe the structure of the cell membrane, and relate its structure to its ability to exchange substances with the cell's environment.

6. Describe the main differences between a eukaryotic and a prokaryotic cell.

7. State the function of mitosis in cell division and briefly describe how mitosis occurs.

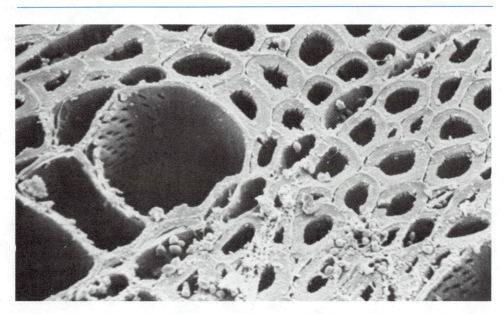

Biophoto Associates.

I took a good clear piece of Cork and with a Pen-knife sharpen'd as keen as a razor . . . cut off . . . an exceeding thin piece of it, and placing it on a black object Plate . . . and casting the light on it with a deep planoconvex Glass, I could exceedingly plainly perceive it to be all perforated and porous . . . these pores, or cells, were not very deep, but consisted of a great many little Boxes, separated out of one continued long pore by certain Diaphragms . . . Nor is this kind of texture peculiar to Cork only; for upon examination with my Microscope, I have found that the pith of an Elder, or almost any other Tree, the inner pulp or pith of the Cany hollow stalks of several other Vegetables: as of Fennel, Carrets, Daucus, Bur-docks, Teasels, Fearn . . . & c. have much such a kind of Schematisme, as I have lately shewn that of Cork.

With these words, written in 1665, the Englishman Robert Hooke first reported the existence of cells. The cells that Hooke saw in cork were really empty, dead **cell walls**, devoid of **cytoplasm**, the jellylike substance that living cells contain.

SITE-oh-plasm

After Hooke's discovery, those who used microscopes observed cells in all kinds of plants. Similar structures were found in animals, but they were harder to see because animal cells lack the thick cell walls of plant cells. Observers also reported the existence of many tiny organisms, each consisting of only one cell.

Then, 150 years later, in 1838 and 1839 the botanist Matthias Schleiden and the zoologist Theodor Schwann originated what we now call the **cell theory**. This theory states (1) that cells are the fundamental units of life; and (2) that all organisms are made up of one or more cells. In 1855, Rudolf Virchow added a third statement, (3) that cells arise only by division of other cells.

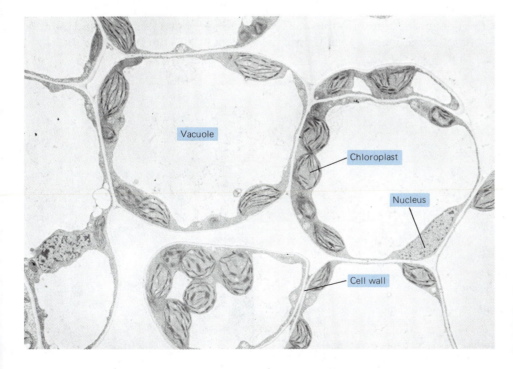

Fig. 10–1

Cells from a spinach leaf as they appear at low power with the electron microscope. (Biophoto Associates)

TABLE 10-1 **UNITS OF SIZE**

1 metre (m) = 100 centimetres (cm) = 39.4 inches
1 cm = 10 millimetres (mm)
1 mm = 10^{-3}m* = 10^3 micrometres (μm)†
1 μm = 10^{-6}m = 10^{-3}mm = 10^3 nanometres (nm)
1 nm = 10^{-9}m = 10^{-6}mm
The prefix "centi" means one-hundredth
              "milli" means one-thousandth
              "micro" means one-millionth
              "nano" means one-billionth

*$1/10 = 0.1 = 10^{-1}$; $10^{-3} = 1/1,000$; $10^6 = 1,000,000$
†Micrometres were formerly known as microns, denoted by the Greek letter μ, pronounced "mew."

In this chapter we shall examine the structure of cells, how the parts of a cell work, and how cells reproduce. We begin by considering microscopes, which enable us to see structures too small to be seen with the naked eye.

10–A Microscopes

Light microscopes with two main lenses were first invented in the sixteenth century. Surprisingly, the magnifying power of a microscope is not the main factor that determines how small an object can be viewed. The most essential feature is **resolving power**, the microscope's ability to distinguish the separateness of two objects that are close together. A particle smaller than the resolving power of the microscope is invisible to the viewer. The resolving power of a light microscope is limited by the wavelength of light: the shorter the wavelength, the better the microscope's resolution. The shortest wavelengths of visible light, in the violet range, have a resolving power of about 200 nm (see Table 10-1).

Fig. 10–2

A nineteenth century microscope, very similar to a modern microscope except that it uses natural instead of electric light to illuminate the specimen. (Biophoto Associates)

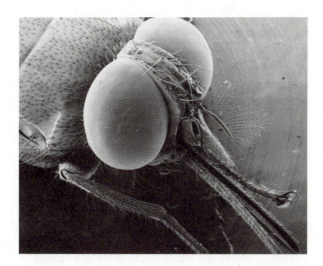

Fig. 10–3

A scanning electron micrograph of the head of a tsetse fly, a carrier of African sleeping sickness. (Biophoto Associates)

Transmission electron microscopes (EMs) were invented in the 1930s by people looking for something with a wavelength shorter than that of light. Although electrons act as particles, a beam of electrons also acts as if it travels in waves which, for all intents and purposes, act like light waves. An electron microscope is made in essentially the same way as a light microscope except that electromagnets instead of glass lenses are used to focus an electron beam instead of light rays. One major disadvantage of an electron microscope is that the electron chamber must be under high vacuum since electrons are easily deflected and absorbed by molecules of gas in the air. No living thing can exist in a high vacuum because all its water evaporates; therefore, only dead material can be examined under a transmission EM. Furthermore, the material viewed must be sliced very thin so that electrons can pass through it. The beam of electrons then makes an image of the specimen on a photographic plate.

In the **scanning electron microscope**, invented in the 1950s, electrons are bounced off the surface of the specimen and photographed as they return. The scanning EM has been used to take some exquisite photographs of the surface features of living things (for example, Figure 10-3). The resolution of a scanning EM is less than that of a transmission EM and the vacuum needed is less intense; hardy specimens, such as certain insects, can be viewed alive under the scanning EM.

TABLE 10-2 **SOME BIOLOGICAL SIZES**	
Nerve cell	Up to 2 m long (but very thin)
Average body cell of an animal	10–20 μm in diameter
Average body cell of a plant	30–50 μm in diameter
Chloroplast of a flowering plant	5–10 μm long
Mitochondrion	Up to 7 μm long
Escherichia coli (bacterium)	2 μm long
Ribosome	25 nm in diameter
DNA molecule	2 nm thick
Hydrogen atom	0.1 nm in diameter

In any microscope, a visible image is produced because some parts of the specimen absorb or deflect more light or electrons than other parts. To increase the contrast in the final image, most specimens are stained. Colored stains are used for the light microscope, and stains made of substances which absorb electrons (such as lead) are used for the transmission EM. Gold is frequently used to coat the surfaces of specimens for the scanning EM.

10–B Understanding Cell Organization

Let us choose any organ of the human body—heart, brain, liver, or jawbone, for instance—and, like Hooke, slice off a thin, transparent section. When we place our appropriately stained section under the microscope, we see that it does indeed consist of many tiny compartments, or cells. Furthermore, not all the cells are alike; they differ in size and shape, and in the amount of stain they take up (Fig. 10-4). We shall probably see cells of the same type grouped together, forming a **tissue**. An example is epithelial tissue, a thin sheet of closely-packed cells covering an organ; another is muscle tissue, which makes up most of the heart. We discover that the organ we are studying is made up of many tissues, each made up, in turn, of particular types of specialized cells. Higher magnification reveals various structures, collectively called **organelles**, within each cell.

Every cell has two different roles in the body. It carries on its own basic life processes, and it also contributes something to the economy of the body as a whole. A muscle cell in the heart is specialized to help pump blood. Since it is

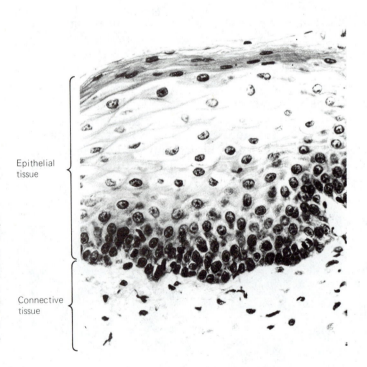

Epithelial tissue

Connective tissue

Fig. 10–4

Section through the human palate, showing two kinds of tissues. The flattened cells of epithelial tissue (top) protect underlying tissues. Surface cells slough off continuously and are replaced by new cells produced in the dark layer beneath. Below this dark layer of epithelium lies connective tissue containing many fibers. (Biophoto Associates)

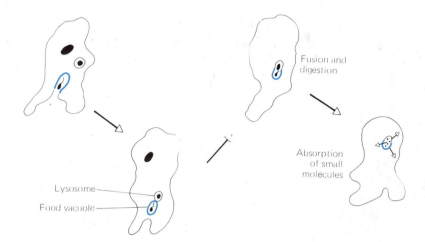

Fusion and digestion

Absorption of small molecules

Lysosome

Food vacuole

Fig. 10–5

An amoeba eating. The amoeba engulfs its prey to form a food vacuole. The vacuole fuses with a lysosome full of enzymes, which digest the prey. The small molecules formed by digestion are absorbed into the cytoplasm.

deep within the body, it cannot capture its own food or obtain oxygen from the air, but must rely on other specialized cells, such as those of the alimentary canal, lungs, and blood, to bring it the gases and nourishment it needs.

Some organelles are found in all cells. These are the organelles which every cell needs to survive. Others, like the contractile strands in a muscle cell, are there because of the cell's specialized role in the body and are not found in every cell.

We now examine the organelles found in multicellular organisms; these organelles are also found in many **unicellular** (one-celled) organisms, such as amoebas and paramecia. Bacterial cells are quite different, and we shall discuss them at the end of the chapter.

10–C The Life of a Cell

One of the foremost tasks facing a living thing, be it a cell or an organism, is obtaining the materials it needs: food, water, and oxygen. In addition, it must rid itself of waste products such as carbon dioxide. The exchange of substances between a cell and its environment is controlled by the **cell membrane**, a thin covering over the entire cell. Once inside the cell, small molecules, such as fatty acids, amino acids, oxygen, and sugars, dissolve in the cytoplasm and are available to the various organelles in the cell.

Sometimes the cell membrane forms a pouch around a large piece of food outside the cell and then pinches off, so that the pouch is inside the cytoplasm (Fig. 10-5). (This does not leave a hole in the cell membrane because the membrane is very flexible and seals itself instantly.) The pouch containing the food, called a **food vacuole**, can now fuse with another membrane-enclosed sac called a **lysosome**, an organelle which contains digestive enzymes. The enzymes digest the food and the small molecules released cross the membrane of the lysosome and enter the cytoplasm. Sometimes lysosomes digest the cell itself. This is what happens when a tadpole changes into a frog. Lysosomes in the

LIE-so-soam

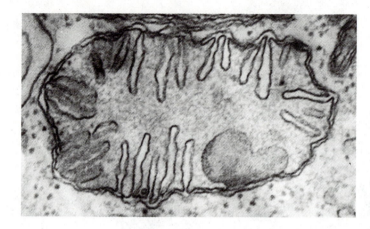

Fig. 10–6

A mitochondrion. Note the outer membrane surrounding the whole organelle and the deeply folded inner membrane. (Biophoto Associates)

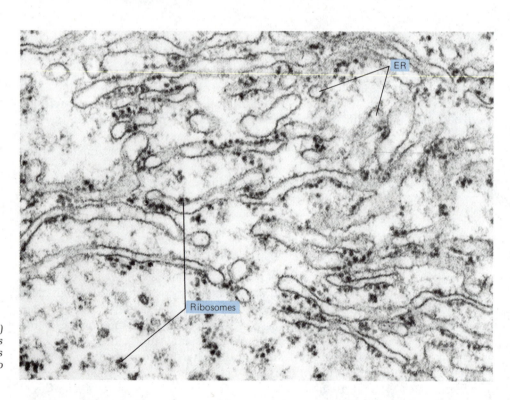

Fig. 10–7

Endoplasmic reticulum (ER) with and without ribosomes attached. Some free ribosomes are also visible. (Biophoto Associates)

tadpole's tail digest the tail, and the molecules released are absorbed back into the body.

my-toe-KON-dree-uh (-on)

Some of the cell's food molecules are broken down during **cellular respiration** to yield energy needed for the cell's various activities. **Mitochondria** (sing.: **mitochondrion**) are organelles that take in organic molecules and oxygen, use them in respiration, and release the end products (carbon dioxide and water)

uh-DEN-oh-seen
try-FOSS-fate

back into the cytoplasm (Fig. 10-6). Respiration also forms ATP (adenosine triphosphate), an energy-yielding molecule that leaves the mitochondria and powers chemical reactions in the cytoplasm and cell membrane (see Chapter 12).

Sugar molecules not needed for respiration may be stored in a cell. Liver and muscle cells store glucose as the polysaccharide glycogen; cells in plants store glucose as another polysaccharide, starch. Unneeded lipid molecules may be broken down and sent to the mitochondria to be respired, or may be stored as **lipid droplets** in the cytoplasm.

lipid (LIP-idd):   nonpolar organic substance, unable to dissolve in water.

Amino acids taken into a cell may be built up into proteins. Organelles called **ribosomes** are the sites of protein synthesis (Fig. 10-7). Each ribosome is made up of one large and one small subunit. Ribosomes are often found attached to a system of membranous tunnels and sacs known as the **endoplasmic reticulum (ER)**, and cells that manufacture much protein often contain a great deal of ER with many ribosomes attached (Fig. 10-7).

RYE-bo-soams

en-doe-PLAZ-mic
reh-TICK-you-lum

Endoplasmic reticulum divides the cytoplasm into compartments. This separates different chemical processes going on in the cytoplasm at the same time and reduces the interference of these processes with one another. Some enzymes work efficiently only if they are attached to a membrane; thus, the ER is also important as a "worktable."

Products made by a cell are often used outside the cell membrane. For instance, many digestive enzymes are made in the pancreas but sent to the small intestine. In these cases, proteins synthesized on ribosomes pass through the membranes of the endoplasmic reticulum and become enclosed in sacs pinched off from the ER membranes. These sacs may flatten out and become stacked up like so many pancakes, in a characteristic formation known as a **Golgi complex** or **Golgi apparatus** (Fig. 10-8). When it is time for the cell to export the proteins, the sacs fuse with the cell membrane and then open up, releasing their contents to the outside of the cell. How cells produce, package, and export some proteins, and how they know which proteins to keep, are exciting areas of current cell research.

GOLL-jee

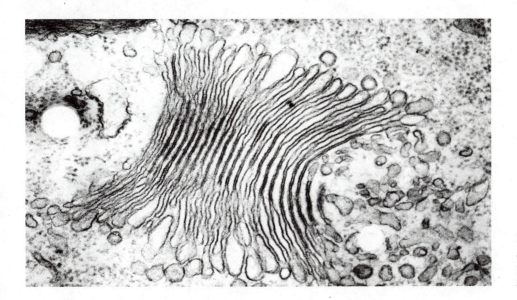

Fig. 10–8

The Golgi apparatus of a Euglena, a one-celled organism. (Biophoto Associates)

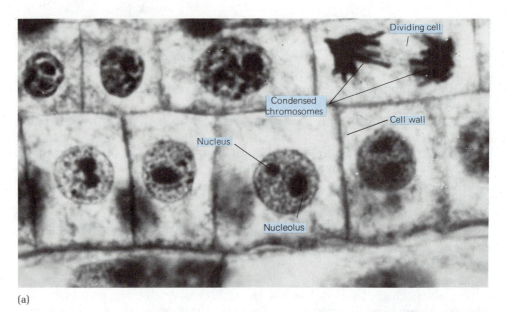

(a)

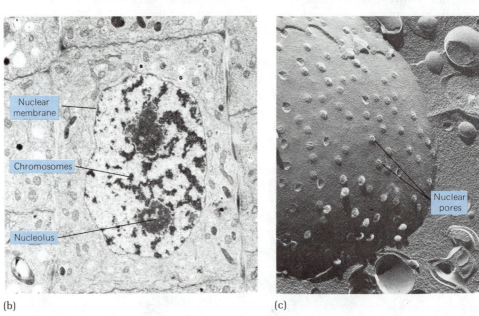

(b)

(c)

Fig. 10–9

Nuclei and chromosomes in onion root cells. (a) The nucleus is dividing in one cell, and here the chromosomes are condensed and look like rods. In the other cells, all that can be seen of the chromosomes are spots in the nuclei. (b) Electron micrograph of chromosomes and nucleoli in a cell that is not dividing. (c) Part of the outside of a nucleus to show the pores in the nuclear membrane. (Biophoto Associates)

10–D The Cell Nucleus

KROAM-oh-soams

When we look at an animal cell under the light microscope, the most obvious structure is usually the **nucleus**; a plant cell nucleus is often less obvious but is visible with proper staining. The nucleus contains DNA, the genetic material of the cell, combined with a great deal of protein in structures called **chromosomes**. The chromosomes appear as short, thread-like structures just before and during division of the nucleus, which precedes cell division (Fig.

10-9a). When the nucleus is not dividing, the chromosomes stretch out in a tangled mass. DNA directs the formation of RNA molecules, which pass out of the nucleus into the cytoplasm and participate in the synthesis of the cell's proteins. Some RNA becomes part of the ribosomes; this RNA is made in one or more dark-staining areas of the nucleus called the **nucleolus** (pl: **nucleoli**) (Fig. 10-9b).

new-KLEE-oh-lus

new-KLEE-oh-lie

Some of the RNA made in the nucleus bears instructions for the order of the amino acids in proteins. A cell's DNA contains instructions for making any protein that could be made in any cell in the entire organism, yet each cell makes only a fraction of these proteins. In addition, some cells change the assortment of proteins they make from time to time. The nucleus is thus seen as the cell's control center because the RNA it produces determines what proteins are made on the ribosomes in the cytoplasm. However, control is a two-way street; substances in the cytoplasm move into the nucleus and influence the DNA in such a way that it may start or stop the manufacture or export of particular types of RNA, depending on the chemical composition of the cytoplasm at the time.

The nucleus is surrounded by a double membrane, the **nuclear membrane** or **nuclear envelope**. The nuclear membrane is perforated by relatively large pores that allow materials, even fairly large particles such as ribosomal subunits, to pass between the nucleus and the cytoplasm (Fig. 10-9c).

10–E How Is a Plant Cell Different from an Animal Cell?

The organelles mentioned so far can be found in at least some cells of both plants and animals because all cells need them if they are to survive for any length of time. However, if we examine a section of the body of an unknown organism, we will be able to tell that it came from a plant rather than an animal by the following clues:

1. *Presence of a cell wall.* A plant cell is surrounded by a thick **cell wall**, which lies just outside the cell membrane. The cell wall is composed of cellulose and other fibers, making it a tough but flexible structure that contributes to the structure and support of the plant body (Fig. 10-11). Each cell wall holds its cell in shape, and since the cell walls of adjacent plant cells are cemented together, the cell walls in the body of a plant hold all the cells in position. Many cell walls are reinforced with materials that give additional strength to the cellulose, especially in woody plants (for instance, lignin is a major component of wood). The flexibility of the cell wall is also necessary, since it permits a plant to bend in the wind instead of breaking.

2. *Presence of plastids.* **Plastids** are organelles found only in plant cells. The most familiar plastids are green **chloroplasts**, the structures that make many plant cells green. Chloroplasts contain the green **pigment** (colored substance) chlorophyll, needed for photosynthesis.

Many plant cells have other types of plastids, such as those which contain the red, yellow, and orange pigments that color many flowers, fruits, and autumn leaves. Colorless plastids store starch or fats and are especially numerous in such structures as potatoes, roots, and seeds.

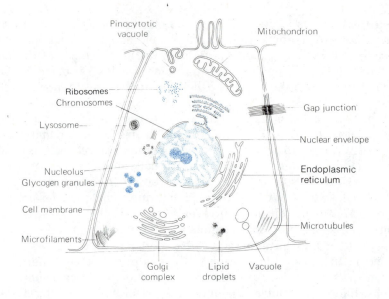

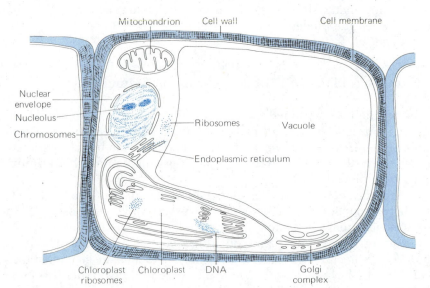

Fig. 10–10

Diagrams showing the structures characteristic of (a) an animal cell, and (b) a plant cell. Compare the plant cell with Figure 10–1.

VACK-you-oal

3. *Presence of large vacuoles.* A **vacuole** is a membrane-bound sac containing a more or less clear fluid. Cells of both plants and animals may contain vacuoles, but plant cells have especially large vacuoles. These vacuoles generally take up most of the cell, and, as a result, the nucleus, chloroplasts, mitochondria, and other organelles in the cytoplasm are crowded around the edges of a plant cell.

A plant cell may store red, blue, or purple pigments, and food molecules, salts, and other substances in its central vacuole. The vacuole is also a conve-

Fig. 10–11

Cellulose fibers in a plant cell wall.
(Biophoto Associates)

nient place to store toxic (poisonous) substances so that they will not disturb the cytoplasm and the other organelles. For instance, some acacia trees store cyanides in their vacuoles. These cyanides do not hurt the plant when they are stored in an intact vacuole, but any animal nibbling on the tree destroys the cell; this breaks the vacuole and releases the cyanide, which poisons the animal! (The plant cells have been destroyed by the animal already, so the plant is no worse off for the cyanide release.)

uh-KAY-shuh

10–F The Cell Membrane

Problems in chemical composition.   The ions and molecules of any substance are in constant, random motion and they tend, on average, to move from an area where they are more concentrated to an area of lesser concentration; that is, they move "down their **concentration gradient**." This process is known as **diffusion**.

Many substances can diffuse into or out of cells. In order to remain alive, a cell must have a chemical composition that varies only within narrow limits, and these limits do not correspond to the chemical composition of an organism's physical environment. The problem would not be so bad if a cell could perfect its own chemical composition and then seal itself off from the environment. However, this is impossible, for the cell must continually take in new food and oxygen molecules and expel wastes. Thus the cell must maintain a lively, but well-regulated, commerce with its environment. Control of this commerce is the task of the cell membrane.

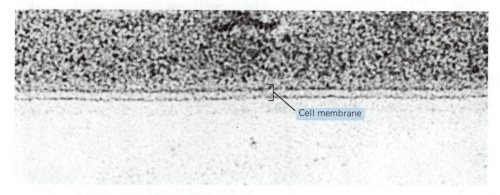

Cell membrane

Fig. 10–12

This electron micrograph shows the two-layered structure of the cell membrane. (Biophoto Associates)

Structure. The cell membrane (the covering of the cell) is too thin to be seen with the light microscope, but its existence was deduced long before the transmission electron microscope showed that it really is there. At the turn of the century, H. Overton discovered that the rate of penetration of many substances into red blood cells was directly proportional to their **solubility** (ability to dissolve) in lipid. Hence, the cell membrane is said to be highly **permeable** to lipid-soluble substances. Overton proposed that the cell membrane contained a great deal of lipid; molecules would dissolve into the cell membrane and emerge on the other side, in the interior of the cell.

PER-mee-uh-ble

PER-mee-uh-BILL-ih-tee

But the rule of lipid solubility did not explain all the observed permeability properties of the cell membrane. Water and a number of small, water-soluble (polar) molecules penetrate cells much more rapidly than one would expect from their very low lipid solubility.

The cell membrane is also more permeable to molecules with no electrical charge than it is to ions. For example, potassium and chloride ions, which are as small as water molecules, penetrate only one ten-thousandth as fast. Cell membranes also distinguish between positively and negatively charged ions, and in general cells are more permeable to the positively charged ions.

In 1925, E. Gorter and F. Grendel obtained lipid from cell membranes by bursting and removing the cell contents of red blood cells. They spread the lipid in a layer one molecule thick on the surface of a tray of water. The area of water covered by this lipid layer was twice as large as the surface area they had calculated for the original blood cells. Consequently, they suggested that the cell membrane was two lipid molecules thick. Investigations of the surface tension and flexibility of the cell boundary indicated that protein was also a component of the cell membrane.

Finally the transmission electron microscope revealed the cell membrane as a double-layered structure about 10 mm (0.00001 mm) thick (Fig. 10-12).

Various models of cell-membrane structure have come and gone as more and more data have accumulated. The most recent, the **fluid-mosaic model**, is shown in Figure 10-13. Molecules in the two layers of lipids are arranged so that their nonpolar, water-repelling, ends are in the middle of the membrane, and the polar, water-soluble, ends are facing the watery interior and exterior of the cell. Various protein molecules are embedded in the membrane: some of them rest on

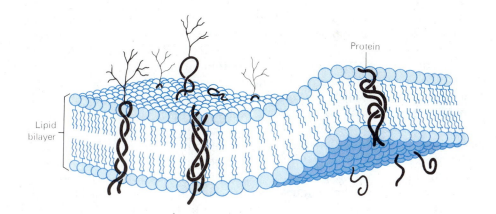

Fig. 10–13

The structure of the cell membrane according to the fluid-mosaic model.

the outer or inner surface of the lipid portion of the membrane; others extend completely through the membrane. The model is called "fluid-mosaic" because the protein molecules are of many different types (mosaic) and many of them move about (fluid) in the membrane rather than remaining in one position.

Function.   The cell membrane is **selectively permeable**—that is, it allows some kinds of substances to pass through more readily than others. We have seen that lipid-soluble substances may simply dissolve across the cell membrane. The movement of ions and small organic monomers, such as glucose and amino acids, however, is much more rapid than we would expect of polar molecules dissolving through a thin layer of lipid. There is evidence that these substances are ushered into (or out of) cells by **carrier molecules** in the cell membrane. Many of the protein molecules in the cell membrane act as carriers for various substances.

glucose: a simple sugar very commonly used as an energy source in cells.

In **facilitated diffusion** a carrier in the cell membrane combines with the molecule or ion on one side of the membrane and releases it on the other side, after diffusing the short distance across the cell membrane. The cell does not supply the energy needed for this to occur, except to make the carrier molecule. A carrier molecule makes the membrane more permeable to the substance transported.

A common example of facilitated diffusion allows most cells to take up glucose faster than glucose can diffuse through the lipid of the cell membrane. In the human body, for instance, liver and muscle cells, and red blood cells, can take up glucose by means of carrier molecules. Facilitated diffusion may also move glucose (or other substances) out of cells; when the blood glucose level drops, the cells of the liver give up glucose to the blood, using the same carrier molecules that take up glucose when the blood is glucose-rich.

In contrast to facilitated diffusion, **active transport** moves substances *against* their concentration gradients; that is, it moves substances from areas where they are less concentrated to areas where they are more concentrated. Since this is opposite to the direction of normal diffusion, the cell must expend energy.

SO-dee-um po-TASS-ee-um

Perhaps the most widely studied active-transport process is the so-called **sodium-potassium pump**. We do not yet know just how the pump works, but it uses energy to move sodium ions out of the cell and potassium ions in.

The observation that there is more sodium outside than inside a cell might mean only that the cell membrane is not permeable to sodium. How do we know that there is such a thing as the sodium pump? First, cells can be poisoned so that they cannot make ATP—the molecule that supplies energy for the pump. Such poisoned cells lose their ability to keep sodium out and potassium in; they leak potassium to the outside, and their internal sodium content rises. Second, if cells containing radioactive sodium are transferred to a medium containing non-radioactive sodium, they gradually lose the radioactive sodium to the outside and at the same time take in nonradioactive sodium. This indicates that sodium leaks into and out of the cell. So it is clear that the cell membrane is not impermeable to sodium.

How do carrier molecules work? It seems that carriers are proteins and related substances in the cell membrane. When a molecule to be transported combines with the carrier, the carrier changes shape. This shape change may open a temporary pore in the membrane (the pore can be visualized, not as a real hole, but as a patch of membrane with temporarily altered shape such that the substrate molecule can pass through easily). In the case of active transport, the energy from ATP may help to detach the transported molecule from its carrier on one side of the membrane, or may cause a change in the shape of the carrier so that it can pick up a different molecule for the return trip.

The cell membrane can take up or expel not only individual molecules or ions but also large molecules or particles composed of many molecules. This ability is due to the fluid nature of the cell membrane. In the process of **en-docytosis** (endo = inside; cyt = cell), part of the membrane flows out and engulfs a particle outside the cell. The edges of the membrane fuse together, creating a membranous sac around the particle. The sac detaches from the cell membrane, leaving the sac in the interior of the cell, where it may fuse with a lysosome that will digest its contents. **Exocytosis** (exo = outside) expels the contents of a membranous sac, such as part of the Golgi apparatus, to the exterior of the cell; it works by a reversal of endocytosis.

EN-doe-sigh-TOE-sis

Vital as water is to a living cell, as far as we know a cell has no special way of transporting water in or out. Water seems to travel through the cell membrane quite freely by the process known as osmosis.

Fig. 10–14

Osmosis in an artificial system. A tube containing glucose solution is covered with a membrane permeable to water but not to glucose and placed in a jar of water. Although water can move both ways across the membrane, glucose molecules in the tube interfere with movement of nearby water molecules, and so more water enters the tube than leaves. The height of the solution in the tube rises until the solution exerts enough pressure to force water out of the tube as fast as it enters.

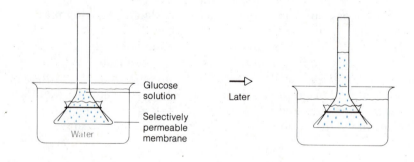

Glucose solution

Selectively permeable membrane

Water

Later

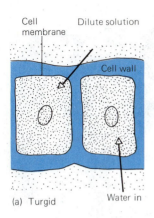

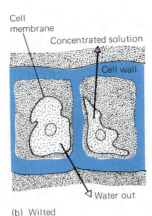

Fig. 10–15

Turgor and wilting in plant cells. (a) Plant cells surrounded by a dilute solution tend to gain water by osmosis across the cell membrane and remain turgid. (b) Plant cells surrounded by a concentrated solution lose water by osmosis and the plant wilts.

Osmosis is a special case of diffusion—the diffusion of water through a selectively permeable membrane, such as a cell membrane. Since the concentration of a solution depends on the amounts of substances dissolved in water, water will move from a more dilute solution (which has a *greater* concentration of *water*) to a more concentrated solution (with proportionately *less water*).

Osmotic movement of water depends on two main factors: (1) the total concentration of all particles dissolved in water on each side of the membrane; (2) the pressure exerted on each solution. All other things being equal, the solution with the higher total concentration of all dissolved particles will tend to gain water from a less concentrated solution across a selectively permeable membrane (Fig. 10-14). However, there may come a point when the water that has entered the more concentrated solution has built up so much pressure that water is forced out by pressure as quickly as it comes in.

Because cells have no direct control over their water content, they control their gain or loss of water by controlling the concentration of dissolved substances in the cell. To take up more water, for instance, a cell takes in more salt ions, glucose molecules, or other dissolved molecules; this raises the concentration of dissolved particles inside the cell. Water then follows by osmosis in its tendency to equalize its own concentration on the two sides of the membrane.

If the environment outside the cell has a higher concentration of dissolved substances than the cell itself, or if the environment is dry air, a cell will lose water and shrink, as happens when plants wilt on a dry, warm day. As water leaves the large plant cell vacuoles that store most of the cells' water, the cells sag inside their cell walls (Fig. 10-15). When a wilted plant is placed in water, however, water re-enters the cells and fills the vacuoles. The cells become **turgid**, or swollen with water, and press against their cell walls with an outward pressure known as **turgor pressure**. The cell walls can expand only to a certain extent, and then they press back; the pressure exerted by the cell walls prevents the cells from bursting by forcing water out as fast as it enters.

Placed in water or a very dilute medium, many animal cells will swell up and burst, because they have no cell walls (Fig. 10-16). However, animal cells that routinely come into contact with water, such as the cells lining your

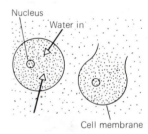

(a) An animal cell in a dilute solution may take in water until it swells and bursts

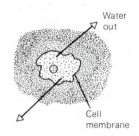

(b) An animal cell in a concentrated solution loses water and shrivels up

Fig. 10–16

Osmotic behavior of an animal cell.

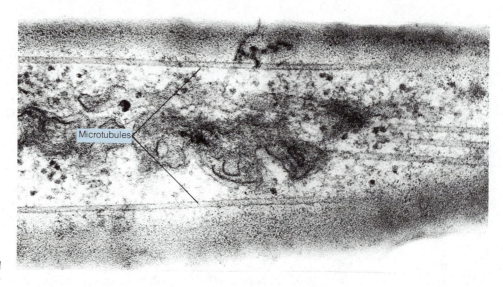

Microtubules

Fig. 10–17

Microtubules in a one-celled organism. (Biophoto Associates)

digestive tract, have special means of dealing with water. When you drink water, it is absorbed and distributed slowly, so that the cells in the interior of your body do not pop.

10–G Cell Movement

The world of living cells is filled with constant motion. In a pondweed cell, the green chloroplasts circle round and round the central vacuole; an amoeba changes shape as cytoplasm pours first this way and then that, forming and retracting its armlike extrusions. A heart cell, isolated on a microscope slide, puts out long filaments and moves around the slide.

Cell movement is still poorly understood, but recent advances in electron microscope technique have allowed us to see delicate structures that were destroyed by older methods of preparation (Fig. 10-17). A variety of cell movements requires two types of cell structures: microtubules and microfilaments.

my-cro-TOO-byules

Microtubules are long, slender tubes. They seem to be assembled at need from protein subunits that are always present in the cytoplasm, and they are broken down again into the reusable subunits. Microtubules form internal skeletons in cells of both plants and animals, and they sometimes act as tracks for the movement of various organelles.

SILL-ee-uh

flah-JELL-uh

Microtubules are also found in **cilia** and **flagella** (sing.: **cilium** and **flagellum**), threadlike projections jutting out from the cell surface (Fig. 10-18). Cilia are generally shorter and more numerous than flagella, but examination with the electron microscope has revealed that both have the same basic internal structure: nine pairs of microtubules arranged in a circle, with two single tubules at the center and a basal body at its base (Fig. 10-18). Cilia can move a small organism through a liquid medium, or move particles past a surface; for example, cilia on the cells lining the air passages to your lungs sweep it clear of mucus

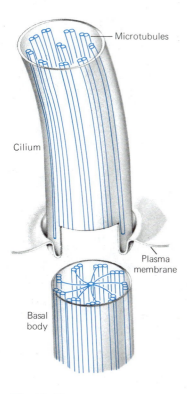

Fig. 10–18

Part of a cilium with its basal body.

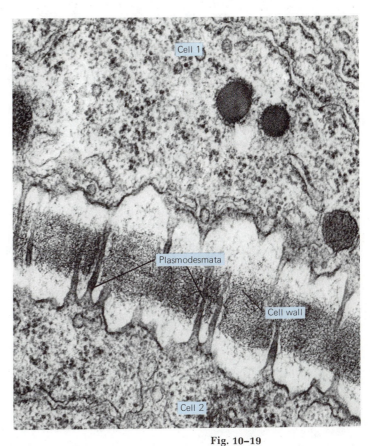

Fig. 10–19

Plasmodesmata between two plant cells. (Biophoto Associates)

and particles of dust. Flagella occur on many unicellular organisms and on sperm cells.

Microfilaments are protein fibers even more slender than microtubules (Fig. 10-17). They may help to cause cytoplasmic streaming, as in the pondweed cells mentioned above, or the movement of an amoeba. Like microtubules, microfilaments may exist as subunits or as assembled filaments, depending on conditions and needs in the cell.

10–H Cell Communication

For an organism to function, the activities of its cells must be coordinated. Communication between cells can be carried out by messenger substances (for example, hormones) that pass through the spaces outside cells, or by direct transfer of substances from one cell to another.

Plasmodesmata (sing.: **plasmodesma**) are channels through which the cytoplasm of one plant cell touches that of its neighbor (Fig. 10-19). **Gap junctions** are similar connections between animal cells. Small ions and molecules can pass freely across gap junctions. These junctions are particularly common

PLAZ-mo-dehz-MAH-tuh
(PLAZ-mo-DEHZ-muh)

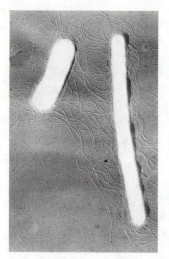

Fig. 10–20

Two rod-shaped bacteria. The flagella-like threads of this particular species are used for locomotion. (Biophoto Associates)

TABLE 10-3   **COMPARISON OF EUKARYOTIC AND PROKARYOTIC CELLS**

Structure	Eukaryotic cell	Prokaryotic cell
Cell wall	Absent in animals; present in plants	Present (different chemical composition from those of plants)
Cell membrane	Present	Present
Nucleus	Surrounded by a membrane	Nuclear area, not surrounded by a membrane
Chromosomes	Linear; with proteins	Circular; no protein
Endoplasmic reticulum	Usually present	Absent
Ribosomes	Present	Present (different type from those of a eukaryote)
Golgi complex	Present	Absent
Lysosomes	Present in many cells	Absent
Mitochondria	Present	Absent
Vacuoles	Present in most plant and some animal cells	Absent
Cilia and flagella	Present in all except higher plants	Flagella of different structure present in some bacteria

between cells in early embryos (for reasons unknown) and in the heart, where they speed the transmission of electrical impulses from one cell to the next (see Chapter 23).

10–I Prokaryotic Cells

PRO-karry-OT-ic

Before the invention of the electron microscope, it was assumed that all cells are variations on the same basic pattern. Cells of plants and animals and of some unicellular organisms were large enough so that it could be seen that they contained many of the same kinds of organelles. Cells of bacteria, however, are so small that their internal structure could not be discerned.

The electron microscope has revealed that the basic structure of a bacterial cell is different from that of the cells we have examined so far. The cells discussed above are called **eukaryotic** cells, and are believed to have arisen later in evolution than the **prokaryotic** cells of bacteria.

YOU-karry-OT-ic

Prokaryotic cells lack many of the organelles surrounded by membranes that are typical of eukaryotic cells: bacteria have no mitochondria, endoplasmic reticulum, chloroplasts, lysosomes, or Golgi complex. More importantly, they do not have a nucleus bounded by a nuclear membrane (Fig. 10-21). This difference is the basis of the terms prokaryotic and eukaryotic: the Greek word *karyon* means nucleus; hence prokaryotic (pro = before) means "before nucleus," and eukaryotic (eu = good) means "good nucleus." Prokaryotic DNA occurs as one circular molecule folded up in a **nuclear area**. The DNA of prokaryotes does not have proteins attached to it. Prokaryotic cells do have ribosomes, but these are smaller than the ribosomes in the cytoplasm of eukaryotic cells.

MEE-so-soam

Prokaryotic cells have cell walls and cell membranes, but the only internal membrane usually found is the **mesosome**, which is attached to the cell membrane. Its function is little understood. Various theories hold that it plays a role in

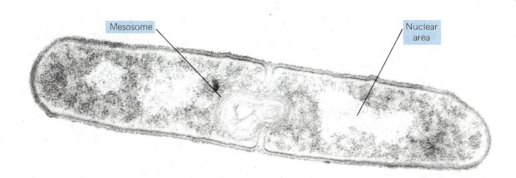

Fig. 10–21

A bacterium dividing in two. Note the nuclear area and the mesosome. (Biophoto Associates)

Fig. 10–22

Division in a bacterium. The DNA replicates and then separates into the two daughter cells as they are formed.

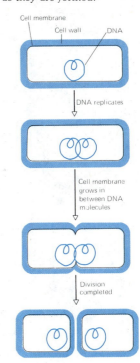

producing new cell-wall material after cell division; that it is involved in duplication of the DNA prior to cell division; or that it is involved in cellular respiration. Photosynthetic bacteria contain other membranes equipped with the various molecules needed to carry out photosynthesis, but these are not organized into chloroplasts, as are the photosynthetic membranes in the eukaryotic cells of plants.

10–J Cell Division

A cell reproduces by dividing to form two daughter cells. A cell's organelles must be duplicated before a cell divides; otherwise each cell division would leave the daughter cells with fewer and fewer organelles. Some organelles, such as chloroplasts and mitochondria, reproduce themselves by splitting in two; so long as a cell starts life with one of each it can produce as many as it needs. In addition, every cell must start life with some ribosomes, which it can use to make the proteins needed for more ribosomes, endoplasmic reticulum, and many other organelles. The DNA of a cell must be duplicated with great precision before a cell divides because the DNA contains the information that a cell uses to make its proteins. If each daughter cell does not inherit a complete copy of its parent's DNA, it will be unable to make all the kinds of proteins that it may need. A cell's DNA must be duplicated and the copies distributed to the daughter cells at cell division if this fate is to be avoided.

Cell division in prokaryotes. A bacterium contains only one DNA molecule, and it is attached to the cell membrane. Before division, the bacterial DNA is replicated, forming two identical DNA molecules, each attached to the cell membrane. When the cell divides, the cell membrane grows between the two molecules of DNA so that one ends up in each daughter cell (Fig. 10-22).

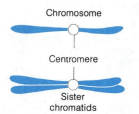

Chromosome

Centromere

Sister
chromatids

Fig. 10–23

Before cell division in a eukaryote, each chromosome replicates to produce two sister chromatids joined at the centromere.

PROPHASE

Sister chromatids visible. Nucleolus and nuclear membrane disappear.

Spindle

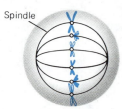

METAPHASE

ANAPHASE

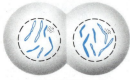

TELOPHASE

Fig. 10–24

The stages of mitosis in a eukaryote. The chromosomes are colored.

Cell division in eukaryotes. The cells of eukaryotes face a much more complex problem because they contain many chromosomes which are not identical. A more complicated mechanism is needed to ensure that each daughter cell receives a complete set of chromosomes: this mechanism is called mitosis.

Mitosis is a nuclear division that produces daughter nuclei with the same number of chromosomes as were present in the parent nucleus (Fig. 10-24). Since nuclear division is usually followed by cell division, the term mitosis is often used more loosely to refer to both mitosis and the cell division that follows. The mysterious dances of the chromosomes as they separate into two identical sets during mitosis have been observed for over a century, and yet there is still much we do not understand about the miraculously precise choreography of chromosome movement.

Before mitosis can occur, the chromosomes of the parental nucleus must **replicate** (duplicate). Replication of each chromosome produces two **sister chromatids**, which remain attached to each other for a time by a structure called a **centromere**. The chromatids eventually become chromosomes when the centromere divides and frees them as two separate structures (Fig. 10-23).

Mitosis is a continuous process but, to make it easier to describe, it is divided into four stages, which are defined by the appearance of the chromosomes under the light microscope (follow the stages in Figure 10-24):

1. **Prophase** gives the first indication that the nucleus is about to undergo mitosis. Distinct, thread-like sets of sister chromatids, each a chromosome that has replicated, become visible in place of the tangled mass of DNA and protein present between mitoses. This condensation of the chromosomes is an impressive process, comparable to coiling a thin strand some 200 m long into a cylinder 1 mm wide by 8 mm long. In most cases, the nuclear membrane disappears during prophase.

2. **Metaphase** is the stage of preparation for division. It can be recognized by the appearance of the **mitotic spindle**, made up of microtubules. Each set of sister chromatids moves to the middle of the spindle and attaches to microtubules.

3. **Anaphase** is the stage of separation of the chromosomes. First the centromeres divide, separating each set of sister chromatids to form two identical chromosomes. One of these chromosomes moves to each end, or pole, of the mitotic spindle.

4. In **telophase**, the last stage, the chromosomes, which are now at the poles of the spindle, begin to uncoil and form a mass of DNA once again. The nuclear envelope reappears around each set of chromosomes. Telophase is usually accompanied by division of the cytoplasm to form two new cells with one nucleus each. In animal cells, the cell membrane moves inward and finally pinches apart, leaving two separate cells. In plant cells, a partition forms across the center of the cytoplasm and each daughter cell builds a cell wall on its own side.

Agents that interfere with mitosis are used to produce **tetraploid** cells, which have double the number of chromosomes in the original cell. Such an agent is colchicine, a chemical extracted from the autumn crocus, which binds with a microtubule protein and prevents the mitotic spindle from functioning. This prevents the chromosomes from separating into two groups, and the nucleus ends up with twice as many chromosomes. Tetraploid seeds can be obtained by allowing a part of a plant treated with colchicine to flower and produce seeds. Tetraploid plants are usually larger and more vigorous than the original parent plant; many domesticated fruits, vegetables, and flowers are tetraploids produced either naturally or by deliberate human manipulation.

SUMMARY

Each cell must carry out all the processes of life: obtain food, release energy, produce and eliminate wastes, and reproduce itself. In addition to carrying out these functions, each cell of a multicellular organism also carries out some specialized function as its contribution to the body's overall economy.

Important structures found in eukaryotic cells are:

1. The cell membrane, which regulates the movement of substances into and out of the cell and keeps the cell's chemical content within the narrow limits required for life. The membrane's lipid layer admits lipid-soluble molecules; its carrier molecules provide for the entry and exit of many polar molecules and ions. When large particles must move into the cell, the membrane surrounds them and pinches off, forming a sac inside the cell. Substances can be discharged from many cells by the opposite process of exocytosis.

2. The nucleus, containing the genetic material in the form of the DNA of the chromosomes. A nucleolus is an area in the nucleus where parts of ribosomes are made. A nuclear envelope with relatively large pores surrounds the nucleus.

3. Ribosomes, the sites of protein synthesis.

4. Mitochondria, which produce most of the cell's energy supply of ATP molecules.

5. Lysosomes, sacs of enzymes involved in digesting food and worn-out cellular structures.

6. Endoplasmic reticulum, a system of membranes dividing the cell into compartments and forming the surface upon which many chemical reactions take place.

7. Golgi complex, the area where proteins are packaged for export from the cell.

8. The cell wall, a porous, protective, and supportive structure outside the cell membranes of plant cells.

9. Plastids, found in plant cells. The most important plastids are the chloroplasts, which carry out photosynthesis.

10. Vacuoles, especially prominent in plant cells.

11. Microtubules and microfilaments, structures concerned with movement of the cell and its organelles.

12. Cilia and flagella, organelles projecting from the cell and concerned with cell locomotion or with movement of external substances past the cell surface.

The prokaryotic cells of bacteria also have cell walls, cell membranes, cytoplasm, ribosomes, and DNA; their only internal membrane system is usually a mesosome, although photosynthetic membranes occur in some.

In eukaryotic cells, cell division is preceded by mitosis, a nuclear division which distributes identical sets of chromosomes to each daughter nucleus.

SELF-QUIZ

Match the structures listed on the right with the property given on the left.

1. Site of protein synthesis.
2. Used to propel a cell through a fluid, or to move a fluid past the surface of a cell
3. Rigid protective covering of some cells
4. Apparatus for exporting cell products
5. Carries on photosynthesis
6. Large fluid compartment in plant cells
7. Internal membranous structure in prokaryotic cells
8. Regulates passage of chemicals into and out of cells
9. Involved in cellular respiration
10. Contains genetic material of eukaryotic cell

a. cell membrane
b. cell wall
c. chloroplast
d. cilia
e. endoplasmic reticulum
f. flagella
g. Golgi complex
h. mesosome
i. mitochondrion
j. nucleus
k. ribosome
l. vacuole

For each item listed below, check whether it is found in cells of animals, cells of plants, prokaryotic cells, or cells of all types of organisms.

	Animal	Plant	Prokaryote	All
11. Ribosome	___	___	___	___
12. Nuclear area	___	___	___	___
13. Cell wall	___	___	___	___
14. DNA	___	___	___	___
15. Mitochondrion	___	___	___	___

16. To view the fine surface structure of a cell, a biologist would use:
    a. a light microscope
    b. a transmission electron microscope
    c. a scanning electron microscope
    d. an x-ray machine

17. Two students are operating on a frog. They constantly apply a salt solution to the organs exposed to the air, but they notice that the organs are beginning to shrink. The students consult the laboratory manual, and find that they misread the instructions and used a solution containing 9% salt when they should have used a 0.9% salt solution, of a concentration equal to that inside the frog's cells.
    a. Explain why the frog succumbed (a biologist's polite way of saying "died") during this surgery.
    b. What process was involved?
    c. Did this process involve carrier molecules?

18. The Golgi complex discharges substances from the cell by fusing its membranous sacs with the cell membrane and the subsequent opening of what was the interior of the sac to the outside environment. This is an example of:
    a. exocytosis
    b. endocytosis
    c. active transport
    d. facilitated diffusion

19. Arrange the following events of mitosis in chronological order:
    a. chromosomes condense; nuclear envelope disappears
    b. spindle forms
    c. chromosomes replicate
    d. chromosomes move to opposite poles of the spindle
    e. chromosomes line up at the middle of the cell
    f. cytoplasm divides

QUESTIONS FOR DISCUSSION

1. What useful applications of the scanning electron microscope can you think of?

2. Would you expect the cells of your hair follicles to contain more ribosomes than a cell that stores fat? Why?

3. Hydrogen cyanide (HCN) and carbon monoxide (CO) are poisons that penetrate cell membranes readily. Can you think of an explanation for the fact that no cells have evolved adaptations to keep these molecules out?

4. It has been said that animals, as we know them, could not exist if they had cell walls. Why not?

5. What is the advantage to cells of keeping spindle protein subunits on hand in the cytoplasm, rather than making them anew from amino acids each time they are needed?

6. Tobacco smoke reduces the activity of cilia in the air passages. How does this contribute to "smoker's cough" and lung disease?

Essay:
ORIGIN OF MITOCHONDRIA AND CHLOROPLASTS

Mitochondria and chloroplasts have several features in common. Each is bounded by an outer membrane, separating it from the cytoplasm, and each contains an inner membrane with a complex arrangement of folds. Each also contains its own DNA, RNA, and ribosomes, and therefore makes at least some of its own proteins; and both mitochondria and chloroplasts reproduce by dividing in two.

These and other observations led to the interesting theory that chloroplasts and mitochondria evolved from free-living prokaryotic cells. According to the theory, these prokaryotes took up permanent residence in some other prokaryotic cell to the mutual benefit of both parties. During the course of evolution, the engulfing cell underwent other changes and eventually became a eukaryotic cell. The evidence in favor of this theory is as follows:

1. Some prokaryotes can grow inside eukaryotic cells today.

2. Plastids and mitochondria are about the size of prokaryotic cells.

3. Like prokaryotes, plastids and mitochondria contain DNA which is circular and not complexed with proteins.

4. Plastids and mitochondria synthesize at least some of their own proteins. Their ribosomes are about the same size as prokaryote ribosomes, which they also resemble chemically.

5. Some antibiotics block protein synthesis on the ribosomes of prokaryotes, mitochondria, and plastids, but not on cytoplasmic ribosomes of eukaryotic cells.

6. Eukaryotic cells deprived of their mitochondria or plastids cannot replace them. Apparently these organelles arise only by division of existing plastids or mitochondria.

Those who disagree with the theory that plastids and mitochondria originated as prokaryotes engulfed by another prokaryote argue as follows:

1. Although the protein-synthesizing machinery of mitochondria is like that of prokaryotes, it is not identical.

2. Plastids require proteins coded by genes on the nuclear chromosomes. Thus, if plastids were originally prokaryotes, some of their genes must have moved into the cell nucleus.

3. Other membranous structures, such as nuclear membranes and the Golgi apparatus, do not appear to be descended from engulfed prokaryotes; presumably they arose as extensions of existing membranes.

4. It is not surprising that the DNA and ribosomes of mitochondria and plastids resemble those of prokaryotes. After all, prokaryotes were probably the ancestors of eukaryotes. Cytoplasmic ribosomes and nuclear DNA have merely evolved differences from their prokaryote ancestors faster than those of mitochondria and plastids.

This controversy can never be settled conclusively. Even advocates of the theory that mitochondria and plastids originated as free-living prokaryotes disagree about how they got into the first eukaryotic cell. Some feel that they arose as parasites; others argue that the first eukaryotic cell formed by the fusion of several prokaryotic cells. It is also possible that organelles originated as prey which were resistant to digestion by cells that ate them. In any event, this controversy has stimulated a lot of interesting work on mitochondria and plastids and the division of labor in eukaryotic cells.

11 PHOTOSYNTHESIS

When you have finished this chapter, you should be able to:

1. Explain why we say that life on earth is powered by the sun.

2. Discuss the role of ATP in the energy economy of a cell.

3. Describe what was known about photosynthesis before 1900 and explain the experiments on which that knowledge was based.

4. Describe or sketch the structure of a chloroplast and point out or describe the location of the photosynthetic pigments, the electron transport chain, the "acid battery," and the enzymes of the dark reactions.

5. Name and recognize the raw materials and end products of the light reactions and of the dark reactions of photosynthesis.

6. Name the most important photosynthetic pigment.

7. Summarize the major steps in the light and dark reactions.

8. State why water and carbon dioxide are used up during photosynthesis and where the photosynthetic waste product, oxygen, comes from.

U.S.D.A.

Fig. 11–1

Energy enters living systems when plants make their own food by photosynthesis. Then herbivores like this elk obtain energy by eating plants, and carnivores eat the herbivores.

Pogo, hero of the comic strip which bears his name, once declared that he did not like bombs because they "put everything too everywhere." What he did not realize is that bombs are only a faster way to arrive at an inevitable end result. Left to its own devices, everything will eventually go everywhere anyway. This is because all molecules have a certain amount of energy of motion, which makes them move randomly until they are more or less evenly distributed through space—a process known as diffusion (see Section 10-F). **Entropy** is a measure of randomness, and the entropy of the universe constantly and inexorably increases.

Like everything else, living organisms are unstable; their molecules tend to diffuse away, and if this process goes too far the organism dies. But, as we saw in Chapter 10, organisms control the exchange of substances between living cells and their environment, and they can even accumulate materials from outside themselves and grow. Organisms accomplish these feats by using energy to push molecules and ions against their natural tendencies to become randomly ordered. In so doing, the organism increases the entropy of its surroundings to such an extent that the entropy of the organism plus its surroundings increases even though the organism's own entropy decreases.

In this chapter and the next, we address two important questions:
1. How do organisms obtain energy?
2. How do organisms expend energy?
The answer to these questions is that some organisms capture the sun's energy and store it as chemical energy in food molecules, and all organisms break down food molecules to release the energy they need for everyday purposes.

In this chapter we consider photosynthesis, the process whereby plants store the energy from the sun in food molecules. In Chapter 12, we shall consider cellular respiration, the process whereby all organisms break down food molecules and release the energy they need.

11–A Energy and Life

Energy occurs in many familiar forms, such as heat, light, electrical energy, and the chemical energy stored in food and fuel molecules. The behavior of energy is described in the **laws of thermodynamics**. The first law of thermodynamics states that energy can neither be created nor destroyed; it can only be transformed from one form into another. If the first law tells us that we cannot "win" by creating energy, the second law of thermodynamics says that we cannot break even: each time energy is converted from one form to another, some of the energy's ability to do useful work is lost as useless heat. This is important because it is not ultimately the quantity of energy available to us that matters but how much useful work it does. Everyday experience tells us that energy is converted into forms that are not useful: there is plenty of energy in the heat leaking from a poorly insulated house or in the exhaust from a car, but it is of little use.

THUR-moe-die-NAM-ics

Each time an organism uses a quantity of energy in one of its life processes, it loses some of this energy in a useless form. Living organisms face the problem of obtaining a constant supply of energy to make up for these losses and to power processes such as movement, growth, and reproduction.

During the course of evolution, some organisms became capable of trapping the energy of sunlight and converting it to the energy of chemical bonds in food molecules by the process called **photosynthesis**. Since sunlight is the most abundant energy source on earth, photosynthetic organisms quickly came to outnumber the various bacteria that use other sources of energy to make food molecules. With minor exceptions, the living world is now utterly dependent on sunlight to supply all its energy needs. Photosynthetic organisms use sunlight to make their food molecules; organisms that cannot carry out photosynthesis use food that ultimately comes from photosynthetic organisms.

11–B ATP

We have already met (Section 9–L) the small but vital energy-rich nucleotide **ATP**, or **adenosine triphosphate**. ATP has much the same role in a cell's energy flow that money has in a human economy. A cell needs ATP, as you or I need ready cash, to pay for the necessities of life. Because many enzymes use the energy from ATP to perform chemical reactions, a cell needs ATP to move, produce heat, excrete waste products, carry out active transport, and synthesize new protein molecules, and for many other transactions. If it has ATP, a cell can even build food molecules, which are a form of stored energy—a savings account which can be broken down again as needed to make the ready cash of ATP. The central position of ATP in the economy of life can be illustrated like this:

$$\text{Solar Energy} \rightarrow \text{ATP} \rightleftharpoons \text{Food Molecules}$$
$$\downarrow$$
$$\text{Growth, Reproduction, Movement, etc.}$$

Photosynthesis is the capture of the sun's energy; respiration is the break-down of food molecules. Both processes are really just means to an end: the synthesis of ATP.

An ATP molecule is said to have two **high-energy phosphate bonds**, shown as wavy lines in Figure 11-2. Breaking these bonds releases a great deal more energy than is released by breaking most other covalent bonds. A cell usually releases energy from ATP by breaking off the last of the three phosphate groups; this yields a molecule of **ADP (adenosine diphosphate)** and a free inorganic phosphate group (abbreviated as P_i):

covalent (coe-VALE-ent) **bond:** a linkage between two atoms that share a pair of electrons.

$$\text{ATP} \rightarrow \text{ADP} + P_i + \text{energy}$$

It has been known for years that one way a cell can make ATP is by transferring a phosphate group from another molecule to ADP (Fig. 11-3). Most of a cell's ATP, however, is made by a process discovered in the 1960s, called **chemiosmosis**, which uses energy stored in a sort of "battery" within mitochon-

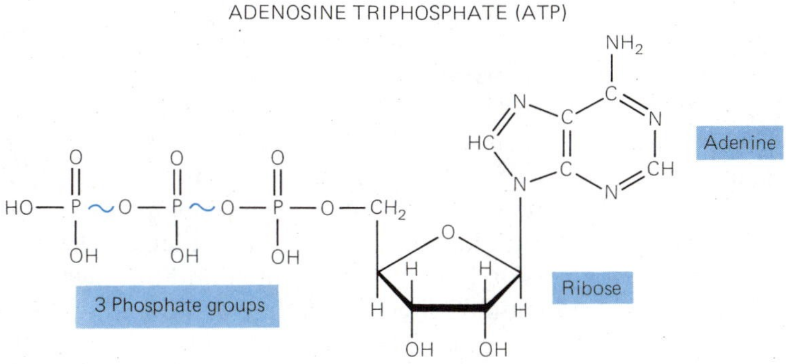

ADENOSINE TRIPHOSPHATE (ATP)

Fig. 11–2

Adenosine triphosphate (ATP). ATP is a nucleotide because it is made up of phosphate groups, a sugar (ribose), and a nitrogenous base (adenine). Each of the colored squiggle bonds releases a lot of energy when it is broken. The bond joining the end phosphate group is the one most frequently broken as ATP is used to power an enzyme-catalyzed reaction.

Fig. 11–3

One method of synthesizing ATP. An enzyme moves a phosphate group from another molecule (in this case diphosphoglycerate) and attaches it to adenosine diphosphate (ADP) to produce ATP.

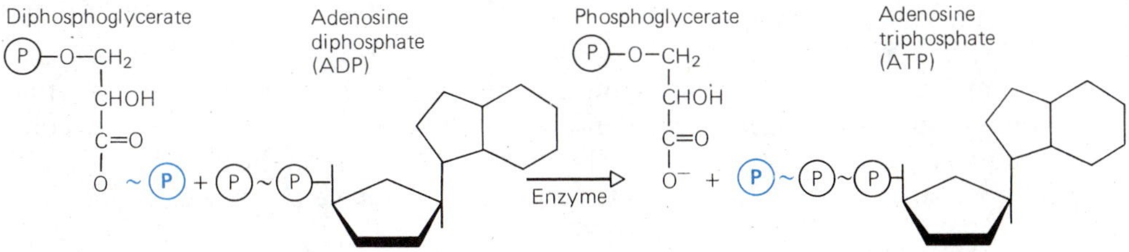

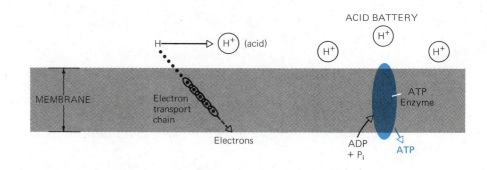

Fig. 11–4

Creating an acid battery full of hydrogen ions (H^+) that can be used to power ATP synthesis. The acid battery must be enclosed by an acid-impermeable membrane (gray). The electron transport chain in the membrane carries electrons away from hydrogen atoms, leaving H^+ (acid) behind. This allows an ATP-synthesizing enzyme (color), also in the membrane, to make ATP using ADP and P_i. In making ATP, the enzyme uses up some of the acid, and so acid must be made continually if ATP synthesis is to go on.

catalyze: to cause a chemical reaction without being permanently changed by the reaction.

Fig. 11–5

Joseph Priestley, who discovered that plants produced something that animals needed. We now know that the "something" is oxygen, but Priestley interpreted his results incorrectly, hampered by the views of the old phlogiston theory. Phlogiston was a hypothetical substance supposed to exist in all flammable objects. (Biophoto Associates, courtesy of the Linnaean Society, London)

dria and chloroplasts. This battery consists of a reservoir of an acid solution of hydrogen ions (H^+) trapped by a membrane that is impermeable to H^+. In this membrane are enzymes that catalyze the synthesis of ATP. These enzymes need energy to make ATP and this energy is supplied, in a way that is not yet completely understood, by the leakage of hydrogen ions out of the reservoir. The main task of mitochondria and chloroplasts is to keep the reservoir full of H^+. They do this by taking hydrogen atoms and removing their electrons, leaving the H^+ behind. A series of molecules called the **electron transport chain**, also located in the membrane around the reservoir, carries electrons away, preventing them from recombining with the H^+ (Fig. 11-4).

11–C Early Experiments on Photosynthesis

Until near the end of the eighteenth century, it was generally assumed that plants use water, and possibly minerals from the soil, to synthesize food. This belief was partly due to an experiment carried out by a Dutchman, Jean-Baptiste Van Helmont, in 1648. Van Helmont planted a willow tree weighing 2.3 kg in a tub containing 90.8 kg of dry soil. For 5 years he gave the plant only water, and then he weighed the soil and tree again. He found that the tree now weighed 76.8 kg, whereas the soil had lost only 0.06 kg in weight. Van Helmont concluded "that all vegetable matter immediately and materially arises from the element of water alone." (According to the ancients, there were four "elements:" water, fire, earth and air.)

Van Helmont had identified one chemical (water) which plants use to make more plant, but we now know that he was wrong in thinking that the tree's weight gain was due to water alone. What was another possible source of nourishment for his willow tree?

The Unitarian minister Joseph Priestley (Fig. 11-5) provided the answer to this question in 1771, although he did not realize it. Priestley showed that animals and plants altered the composition of the air around them in complementary ways. He burned a candle in a covered jar until the candle went out. If a living plant was then grown in the covered jar for several days, a candle could

(a)   (b)   (c)

Fig. 11–6

The results of Priestley's experiments. (a) A candle burning in an airtight jar went out. (b) A mouse died if kept in an airtight jar. (c) If a plant was placed in the airtight jar with the mouse, the mouse lived.

once more burn in the jar. Priestley concluded that "there was something attending vegetation, which restored the air . . . that had been injured by the burning of candles." Priestley also found that a live mouse changed the air in the same way that a candle did (Fig. 11-6). Priestley thought that the single substance air had been altered from one form to another and back again during his experiments. Today we know that animals and candles use up oxygen out of the air and that plants give off oxygen during photosynthesis. In 1782 Jean Senebier, a Swiss clergyman, showed that plants use up carbon dioxide when they produce oxygen. He suggested that the carbon in carbon dioxide was converted into plant material during photosynthesis; a gas in the air was a source of plant nutrition that Van Helmont was not aware of.

Jan Ingenhousz, a physician at the Austrian court, found that plants need sunlight if they are to produce oxygen. He put willow twigs under water in sunlight and watched bubbles of oxygen form on the leaves. By 1796, Ingenhousz could write a general equation for photosynthesis:

in a green plant,
in light

carbon dioxide + water ⟶ plant tissue + oxygen

Nineteenth-century workers, watching illuminated leaves through a microscope, saw that starch grains grew in the cells as photosynthesis went on. They suggested that carbohydrates were the immediate products of photosynthesis. This completed the basic picture of photosynthesis (Fig. 11-7).

Since carbon dioxide is the only raw material of photosynthesis that contains carbon, the carbon in the carbohydrate must come from carbon dioxide. Both water and carbon dioxide contain oxygen, however, and it was not until 1941 that workers showed that the oxygen given off in photosynthesis comes from water. To do this, they prepared water with "heavy oxygen"—oxygen with extra neutrons (see Section 9–A) in its nucleus. When a plant was given heavy-oxygen water and normal carbon dioxide, it gave off heavy-oxygen gas. When the experiment was done the other way around, with normal water but with carbon dioxide containing heavy oxygen, oxygen gas of the normal weight was released (Fig. 11-8).

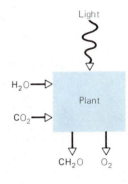

Fig. 11–7

Diagram showing what was known about photosynthesis by the end of the nineteenth century. It was known that plants exposed to light could make carbohydrate (CH_2O) using carbon dioxide (CO_2) and water (H_2O) as raw materials. Oxygen (O_2) was released as carbon dioxide was used up.

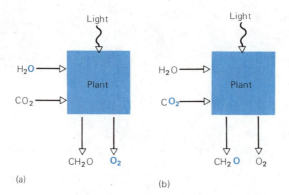

Fig. 11-8

Experiments to show whether water or carbon dioxide carries the oxygen that is released as oxygen gas during photosynthesis. Heavy oxygen (color) is used to identify the sources of oxygen. (a) When water containing heavy oxygen is given to a plant, the oxygen gas released contains heavy oxygen atoms. (b) When the carbon dioxide given to the plant contains heavy oxygen atoms, some of the heavy oxygen shows up in the carbohydrate, but none is found in the oxygen gas given off.

11-D The Two Sets of Reactions in Photosynthesis

The name photosynthesis covers the whole process from absorption of light energy by a plant to the production of carbohydrate. This includes a long and complex series of reactions which are being identified and fitted into the jigsaw puzzle by researchers.

As early as 1905, F. F. Blackman proposed that photosynthesis could be divided into two main parts. If plants were exposed to increasing intensities of light, the rate of photosynthesis (measured by the amount of oxygen given off) increased steadily, up to a point, and then stopped increasing. At this point the plant was absorbing as much light as it could use. The rate of photosynthesis could still be speeded up, however, if the temperature were increased.

Blackman proposed that this meant there were two series of reactions, the "light" reactions and "dark" reactions. The light reactions require light. The dark reactions do not actually require dark but are independent of light, and can be speeded up by increasing the temperature. Blackman proposed that the light reactions produced substances necessary for the dark reactions (Fig. 11-9). Discoveries half a century later showed that he was right. The dark reactions require ATP, and also hydrogen atoms extracted from water; these are produced in the light reactions. The question is, how do the light reactions produce these intermediates? To answer this, we must consider the structure of a chloroplast.

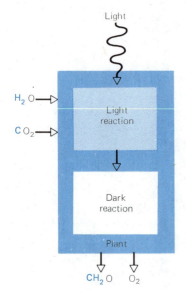

Fig. 11-9

Diagram incorporating Blackman's deductions about the light and dark reactions in photosynthesis. (The experiment in Fig. 11-8 had not yet been done—it came 35 years later.)

11-E Structure of Chloroplasts

One of the basic principles of biology is that the structure of a system is intimately related to its function. We shall see that the structure of a chloroplast organizes the chemicals that carry out photosynthesis in a way that is necessary to their function.

If we look at a photosynthetic plant cell under a microscope, we see that the green color of the plant is due to the green chloroplasts; the rest of the cell is almost colorless. Chloroplasts take their hue from the green **pigment** (colored substance) **chlorophyll**; chlorophyll appears green because it absorbs red and blue wavelengths of light, which are used in photosynthesis, and reflects the green wavelengths, which are then detected by our eyes (Fig. 11-10).

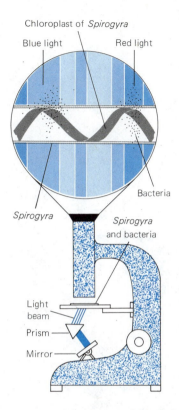

Chloroplast of *Spirogyra*
Blue light   Red light

Spirogyra

Bacteria

Spirogyra
and bacteria

Light
beam
Prism
Mirror

Fig. 11–10

An experiment performed by T. W. Engelmann, in 1883, which showed the wavelengths of light used in photosynthesis. The prism above the mirror breaks a beam of white light up into a spectrum: violet, blue, green, yellow, orange and red. On the stage of the microscope Engelmann placed a drop of water containing a strand of a small alga, Spirogyra, and some bacteria that need oxygen. Spirogyra has a single, long chloroplast, which Engelmann placed so that it lay with some parts in each area of the spectrum. Oxygen released during photosynthesis attracted bacteria; they gathered where red and blue light fell on the chloroplast. Engelmann concluded that red and blue light support higher rates of photosynthesis than do other colors of light. This corresponds to the colors absorbed by chlorophyll, showing that it is the most important photosynthetic pigment.

In addition to chlorophyll, chloroplasts contain yellow, orange, or brown pigments called **carotenoids**, which aid in photosynthesis by absorbing other wavelengths of light and passing their energy on to chlorophyll. These pigments are usually masked by the green chlorophyll, but when chlorophyll is destroyed in the autumn we can see the brilliant pigments that were previously masked.

carROT-en-oids

The photosynthetic pigment molecules are embedded in the innermost membrane of the chloroplast, which forms a system of tunnels and sacs that contain the acid battery (Fig. 11-11). Outside this membrane is the **stroma**, the "ground substance" of the chloroplast, containing the chloroplast's DNA and ribosomes and the enzymes of the dark reactions, as well as ADP and ATP. Surrounding the stroma is a double membrane that separates the chloroplast from the cytoplasm of the cell.

11–F The Light Reactions

The unique part of photosynthesis is the use of light energy to make electrons move along the electron transport chain. As a result of this electron flow, light energy is converted into chemical energy in two forms: ATP and special hydrogen-laden carrier molecules. The process is not completely understood, but we have reached the point where we can outline its main features.

The special quality of the pigment chlorophyll lies in the fact that when it absorbs a unit of light energy, one of its electrons can reach "escape velocity"

1 μm

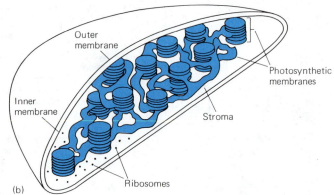

Fig. 11–11

(a) Electron micrograph of a chloroplast of tobacco. (Herbert W. Israel, Cornell University) (b) Diagram of the three-dimensional structure of a chloroplast. The acid battery accumulates inside the photosynthetic membranes, shaded in color. Embedded in these membranes are the pigments (chlorophyll and carotenoids), the electron transport chain molecules, and the enzyme that makes ATP (see Fig. 11–4).

and leave the chlorophyll molecule. This electron passes to various carrier molecules of the electron transport chain (discussed in Section 11–B). Where does this electron end up? It joins with another electron from the chain, and with a hydrogen ion (H^+) from water in the stroma (in any group of water molecules, some are dissociated to form H^+ and OH^-). The two electrons and the H^+ all attach to a **hydrogen carrier molecule**, in this case **$NADP^+$ (nicotinamide adenine dinucleotide phosphate**, a nucleotide), which is converted to NADPH:

nick-oh-TIN-uh-mide
AD-den-een
di-NEW-klee-oh-tide
FOSS-fate

$$2\ e^- \ + H^+ + \ NADP^+ \ \rightarrow \ NADPH$$

electrons   hydrogen   carrier with
carrier   hydrogen attached

Electrons activated by light energy have therefore been used to add hydrogen to a carrier to form NADPH during the light reactions. This NADPH is released into the stroma.

But the light reactions do more than this. The formation of NADPH leaves chlorophyll molecules lacking electrons. These electrons are replaced by breaking down the hydrogen atoms in water according to this reaction:

$$2\ H_2O \ \rightarrow \ 4e^- \ + 4H^+ + \ O_2$$

water   electrons   oxygen (waste product
of photosynthesis)

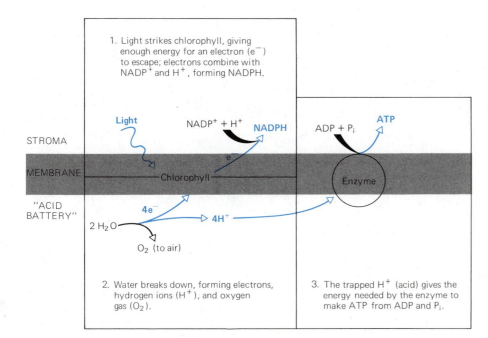

Fig. 11–12

Summary of the light reactions of photosynthesis. The pathway of energy is shown in color; it starts with light, and ends as chemical energy stored in NADPH and ATP. The membrane is the innermost membrane of the chloroplast, the stroma is above it, and the acid battery accumulates below it.

Two molecules of water break down to form one molecule of oxygen gas and the equivalent of four atoms of hydrogen, that is, four electrons, which can go to chlorophyll molecules that have lost electrons, and four hydrogen ions, which become part of the acid battery. The energy stored in the battery then powers the synthesis of ATP by enzymes that lie in the surrounding membrane. The ATP is released into the stroma. The oxygen gas produced during the splitting of water is the waste product of photosynthesis, and it diffuses out of the plant. It is a remarkable fact that all the oxygen in the air, so essential to animal life as we know it, originates in this way.

11–G The Dark Reactions

The light reactions release ATP and NADPH into the stroma of the chloroplast. This is where the dark reactions take place. They consist of "fixing" the gas carbon dioxide from the air into the form of carbohydrate (Table 11-1).

Each carbon dioxide molecule attaches to a five-carbon sugar molecule called ribulose diphosphate. The resulting six-carbon structure immediately splits into two three-carbon fragments, which subsequently undergo a series of reactions using the energy from ATP and the hydrogen from NADPH.

RIB-you-lows die-FOSS-fate

$$C_5 \quad + CO_2 + 2\,NADPH + 2\,ATP \rightarrow 2\,C_3 + 2\,NADP^+ + 2\,ADP + P_i$$

ribulose   three-carbon
diphosphate   fragment

TABLE 11-1 SUMMARY OF THE LIGHT AND DARK REACTIONS OF PHOTOSYNTHESIS, SHOWING ONLY RAW MATERIALS AND END PRODUCTS (NOTE THAT MANY OF THE END PRODUCTS OF THE DARK REACTIONS ARE RAW MATERIALS FOR THE LIGHT REACTIONS AND VICE VERSA.)

	Raw materials	End products
Light reactions	Light energy	
	Water	Oxygen
	ADP + P_i (from dark reactions)	ATP
	$NADP^+$	NADPH
Dark reactions	Ribulose diphosphate	
	Carbon dioxide	Sugars
	ATP (from light reactions)	ADP + P_i
	NADPH (from light reactions)	$NADP^+$

Fig. 11–13

Carbon fixation, in which carbon dioxide is incorporated into carbohydrate molecules. The six-carbon compound formed by joining carbon dioxide to ribulose diphosphate breaks into two identical molecules (shown oriented in opposite directions). The energy of ATP (formed by the light reactions) is used to attach a phosphate group to each molecule, and then the hydrogens carried by NADPH (also formed by the light reactions) are attached.

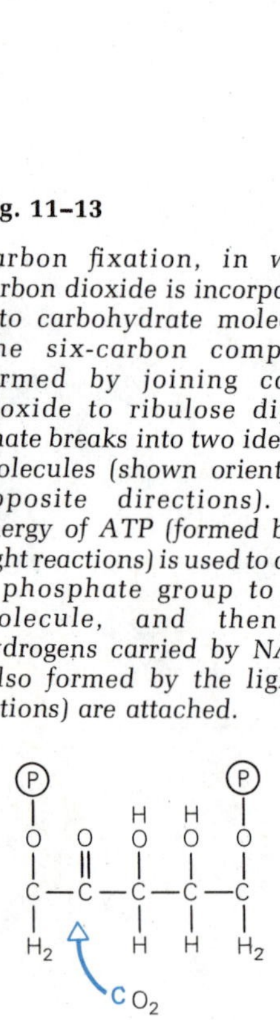

Ribulose diphosphate

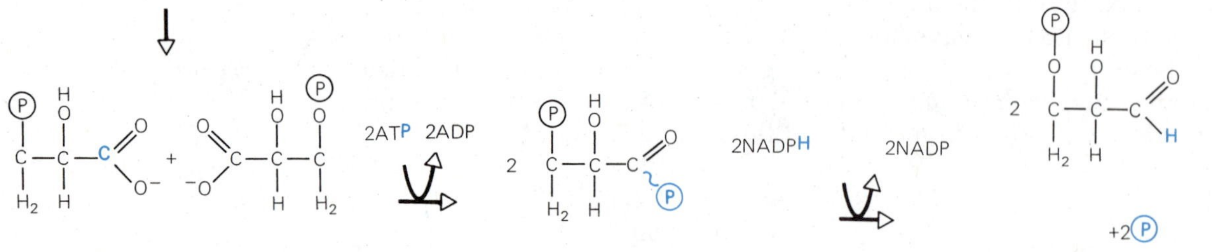

2 Phosphoglycerates

Diphosphoglycerate

Phosphoglyceraldehyde

Keep in mind that many of the same types of molecules are going through this series of reactions at any one time. When the three-carbon fragments reach a certain stage, several possibilities are open to them. Some go through a long series of reactions in which they attach to other molecules and eventually form more five-carbon ribulose diphosphate molecules. These can go back to the beginning of the reaction and join with more carbon dioxide, thus increasing the total amount of fixed carbon in the plant. Other three-carbon fragments go to the synthesis of various amino acids, by the addition of nitrogen-containing groups. Still others join together and form six-carbon sugars, such as glucose. Glucose, in turn, may go to form sucrose, starch, cellulose, and other substances.

Because some of the three-carbon fragments end up as parts of more ribulose diphosphate molecules, the whole process is a cycle (known as the **Calvin cycle** after its discoverer). Note that to make the equivalent of one new (six-carbon) glucose molecule, the cycle must go through six turns, each turn adding one carbon from a carbon dioxide molecule to the plant's store of fixed carbon.

Note also that the ADP, P_i, and $NADP^+$ released from the dark reactions return to the inner membrane, where the light reactions re-form them into ATP and NADPH. These molecules cycle back and forth between the light- and dark-reaction areas; there are few of them in the chloroplast, so that in the dark, the ATP and NADPH made in the light reactions are used up within minutes by the dark reactions. Photosynthesis then stops until light shines on the plant once more.

11–H Beyond the Dark Reactions

Plants are **autotrophs**, which means that they can make all their own organic molecules. Animals are **heterotrophs**, meaning that they must take in organic molecules as food. Although plants do not need organic molecules from the outside world, they do need water and many inorganic minerals (which we often supply in fertilizer) such as nitrate, phosphate, and sulfate, absorbed mainly through the roots. Some of these minerals are cofactors necessary to enzyme activity; others are parts of the proteins, hormones and other molecules that plants can build up using the carbohydrates produced in photosynthesis.

These carbohydrate monomers can be used to synthesize every sort of organic molecule that the plant needs. Large quantities of the polysaccharide cellulose are produced to form cell walls, and chemical energy for later use is stored in the form of the polysaccharide starch. Fats in plants are synthesized from sugars such as sucrose.

Plants use sulfates and nitrates absorbed by the roots to synthesize amino acids, most of which go to form the plant's enzymes. When an animal eats the plant, it will digest the plant's proteins into their component amino acids and use the amino acids to build up the animal's own particular proteins.

AUT-oh-troafs
HETT-er-oh-troafs

nitrate: NO_3^-
phosphate: PO_4^{3-}
sulfate: SO_4^{2-}

polysaccharide: large organic molecule formed by linking many simple sugar molecules.

SUMMARY

Living organisms require energy in order to maintain their chemical composition, grow, repair damage, and reproduce. Almost all organisms living today are directly or indirectly dependent on the sun to supply their energy needs.

The direct source of energy for most chemical reactions in living cells is

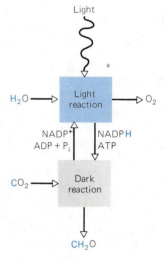

Fig. 11–14

Present concept of the overall scheme of photosynthesis. Figs. 11–12 and 11–13 show more detail of what is inside the boxes.

Fig. 11–15

The Calvin cycle (which includes the dark reactions of photosynthesis) and its connections with the light reactions and the formation of other kinds of molecules made within the plant body. The black arrows trace the path of energy from sunlight through temporary chemical intermediates (such as NADPH and ATP) to the more permanent storage of sugars, starch, fats, and so forth.

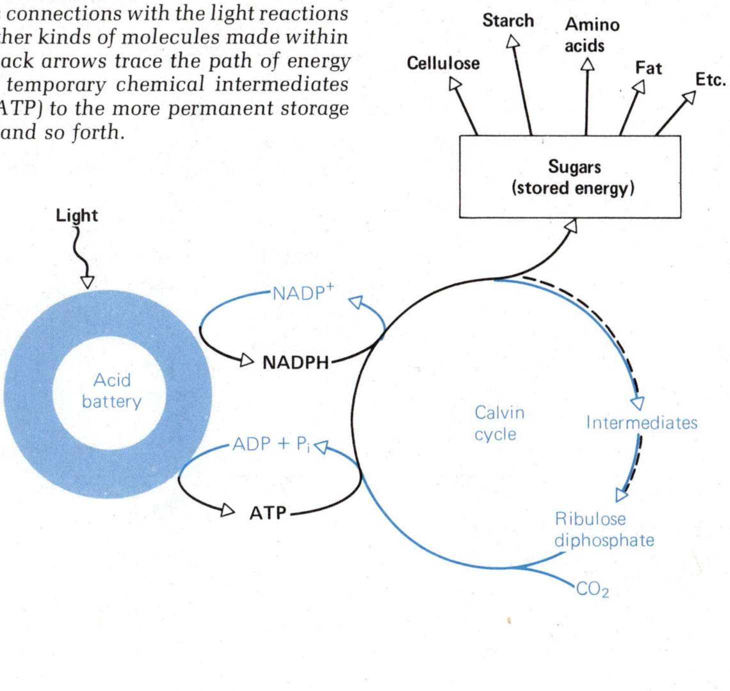

Cellulose   Starch   Amino acids   Fat   Etc.

Sugars (stored energy)

Light

NADP⁺

NADPH

Acid battery

ADP + P_i

ATP

Calvin cycle

Intermediates

Ribulose diphosphate

CO_2

ATP. Most ATP is formed by enzymes using the energy of a hydrogen ion-filled "acid battery" to power the joining of ADP and inorganic phosphate (P_i).

Early experimenters deduced the general equation for photosynthesis:

$$CO_2 + H_2O \xrightarrow[\text{chlorophyll}]{\text{light}} \text{carbohydrate} + O_2$$

In this century photosynthesis has been broken down into a series of light reactions, touched off by the absorption of light energy, and a series of dark reactions, which use chemicals produced by the light reactions to drive a set of light-independent reactions.

The light reactions set up and maintain an acid battery that powers the synthesis of ATP, and they also provide electrons used to join hydrogen to a hydrogen carrier, $NADP^+$, forming NADPH. Oxygen is released from water molecules whose hydrogens are trapped in the reservoir.

The dark reactions use energy from ATP and hydrogen from NADPH to incorporate CO_2 into carbohydrate molecules.

SELF-QUIZ

1. Match:
   ___a. Location of chlorophyll
   ___b. Location of dark reaction enzymes
   ___c. Location of ATP synthesis
   ___d. Location of acid reservoir

   i. Stroma
   ii. Outer two membranes of chloroplast
   iii. Innermost membrane of chloroplast
   iv. Inside inner membrane of chloroplast

2. Which of the following are raw materials of photosynthesis?
   a. CO_2 and H_2O
   b. NADPH and ATP
   c. CO_2 and O_2
   d. NaCl and ATP
   e. NADPH and H_2O

3. Priestley noticed that a mouse will live in a sealed jar if a live plant is also inside. What additional factor is necessary for the mouse's survival in this situation?
   a. chlorophyll
   b. sunlight
   c. hydrogen ions
   d. a chemical that releases CO_2
   e. a source of NADPH

4. The synthesis of ATP during photosynthesis is believed to be directly powered by energy from:
   a. NADPH
   b. ADP
   c. water
   d. H^+ trapped in a membrane-surrounded compartment
   e. energized chlorophyll electrons

5. The substances produced as a result of the light reactions which are vital to the dark reactions of photosynthesis are:
   a. CO_2 and H_2O
   b. ATP and NADPH
   c. ATP and H_2O
   d. O_2 and NADPH

QUESTION FOR DISCUSSION

1. If chloroplasts are treated with a detergent known to make membranes more permeable to ions, they stop making ATP. Explain why this is so.

Essay:
ECOLOGICAL ASPECTS OF PHOTOSYNTHESIS

When biologists measure the rates of photosynthesis of plants growing in their native habitats (Fig. 11-16), they find an enormous variation between different types of plants and in different locations. Part of this variation is because some plants have different dark reactions from the ones described in this chapter. Plants such as rice, corn, sugar cane, wheat, and oats produce a four-carbon instead of a three-carbon molecule in the dark reactions and so are called C_4 plants. At high temperatures, C_4 plants can produce sugar faster than C_3 plants, provided there is plenty of light and water. C_4 plants seem to have evolved this ability as an adaptation to tropical conditions where temperatures are high and there is plenty of light. A C_4 plant from Death Valley, California, was found to undergo photosynthesis most rapidly when the temperature was an incredibly high 47°C (117°F).

Some plants use yet a third pathway to fix carbon dioxide—a pathway known as crassulacean acid metabolism (CAM for short). The advantage of this pathway is that it preserves water. It is found in succulents (see Fig. 4-19) that live in dry areas and are perennially short of water. Plants always lose some water during photosynthesis since they have to open pores (called stomata) in their leaves in order to take in carbon dioxide, and water evaporates from these pores. CAM plants reduce water loss by closing their stomata during the hot day and opening them at night when it is cooler and water evaporates more slowly. They are then in the awkward position of having light only during the daytime, while their stomata open to take in carbon dioxide for photosynthesis only at night. CAM solves this problem by permitting the plants to fix CO_2 as organic acids at night and to use CO_2 released from these acids to make carbohydrate during the day when light is available.

Different plants are adapted to different light intensities. Plants such as corn and tomatoes, which need a lot of light, cannot be grown in a house without artificial light. Other plants, such as many plants of the forest floor (some of which have become popular houseplants), are "shade plants"; they do not photosynthesize very much even when there is plenty of light available.

Fig. 11–16

Measuring the rate of photosynthesis of a shrub in Death Valley. The shrub is enclosed in a tent, and the air surrounding it is circulated back to the mobile laboratory where its carbon dioxide content is measured continuously. The rate at which the plant uses up carbon dioxide measures its rate of photosynthesis. (Harold Mooney)

Fig. 11–17

C_3 and C_4 plants. (a) Potatoes, C_3 plants that are important crops. (b) Sugar cane, the plant in which the C_4 system was discovered. (Biophoto Associates)

12 CELLULAR RESPIRATION

When you have studied this chapter you should be able to:

1. Name the starting materials and the important end products of glycolysis, the tricarboxylic acid (TCA) cycle, and the electron transport chain.

2. Write a brief paragraph explaining how the three processes mentioned in objective 1 fit into the overall scheme of cellular respiration.

3. Explain why most organisms need oxygen, and state how carbon dioxide and water are produced as waste products of cellular respiration.

4. State where, in the cell, glycolysis, the TCA cycle, and electron transport occur in eukaryotes and in prokaryotes.

5. Compare and contrast the process of alcoholic fermentation by a wine yeast with the process of anaerobic respiration by a muscle.

6. Compare aerobic and anaerobic respiration with respect to the amount of ATP regenerated by each process.

7. Explain why we become fat when we eat more food than we need.

As we saw in Chapter 11, some useful energy is converted into a useless form each time energy is transferred from one molecule to another. All organisms, both photosynthetic and nonphotosynthetic, must constantly use energy, most often by releasing it from the energy-rich molecule ATP (adenosine triphosphate), the energy donor for many reactions that go on in a living organism. A cell contains only a small amount of ATP, and it continually renews its ATP supply by releasing energy from food molecules to drive the reaction that joins ADP and phosphate (P_i) together to form ATP.

In this chapter we shall examine the release of energy from food molecules, a process known as **cellular respiration**, in some detail. The process involves a long string of chemical reactions, with a cast of chemical characters whose polysyllabic titles would be the envy of any bureaucrat. We present them here to give you an appreciation for just one of the many complex chemical pathways that have arisen during the long eons of evolution. This pathway occupies a central spot in the chemical activities of most cells.

As we proceed, you will notice a number of similarities and differences between cellular respiration and photosynthesis. Whereas photosynthesis uses carbon dioxide and water to produce carbohydrate and molecular oxygen, cellu-

Fig. 12–1

Energy flow in living things. All organisms break down organic molecules to produce energy during cellular respiration.

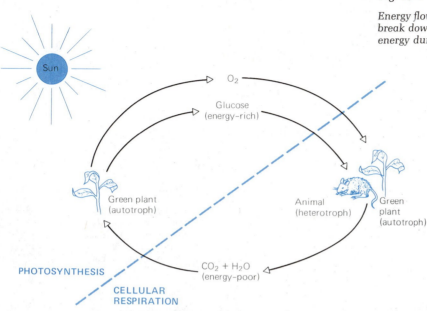

lar respiration breaks down carbohydrates, in the presence of oxygen, to carbon dioxide and water. This does not mean that respiration is photosynthesis in reverse. It is not; but in our discussion of respiration we will meet many molecules and reactions that are familiar from photosynthesis.

12–A Overview of Respiration

In this chapter we will follow the respiration of the food molecule glucose, a six-carbon sugar. Given a supply of glucose, most cells will produce energy using this sugar rather than other food molecules that may be present. Glucose is broken down through a long series of steps, and at some of these steps energy from the chemical bonds in the glucose molecule is used to synthesize ATP. Glucose may be respired either **aerobically** (using oxygen) or **anaerobically** (when no oxygen is available). The entire process of cellular respiration can be divided into three main sequences of reactions:

air-OH-bic; AN-air-OH-bic

1. **Glycolysis** consists of reactions that break down the six-carbon glucose molecule into two three-carbon molecules of a compound called pyruvate. Two important products are made along the way: ATP and hydrogen atoms, which attach to a hydrogen carrier molecule, **NAD⁺** (**nicotinamide adenine dinucleotide**), forming NADH (Fig. 12-2).

nick-oh-TIN-uh-mide
AD-den-een
die-NEW-klee-oh-tide

2. The **tricarboxylic acid cycle** or **TCA cycle**. During the TCA cycle, two-carbon groups broken off from pyruvate are dismantled, forming carbon dioxide. The important products of the cycle are more ATP and several hydrogen-laden carrier molecules, including NADH and also some **FADH₂**, the hydrogen-rich form of **flavin adenine dinucleotide (FAD)**.

try-car-box-ILL-ic

FLAY-vin
AD-den-een
die-NEW-klee-oh-tide

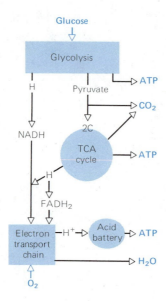

Glucose

Glycolysis

→ ATP

H   Pyruvate

→ CO₂

NADH   2C

TCA cycle

→ ATP

H

FADH₂

Electron transport chain   —H⁺→   Acid battery   → ATP

→ H₂O

O₂

Fig. 12–2

A summary of what happens during respiration. Important raw materials (in color) top and bottom, important end products in color on the right. Intermediate molecules linking glycolysis, the TCA cycle and the electron transport chain are in black.

SITE-oh-kroam

3. The **electron transport chain**, or **cytochrome chain**. In this series of steps, the NADH and FADH₂ molecules formed previously pass electrons from the hydrogen atoms they carry to a series of electron-transport molecules. As in photosynthesis, passing electrons along the chain sets up a reservoir of hydrogen ions (H⁺) by a process that is not yet completely understood. Energy trapped in the hydrogen reservoir "battery" is used to make ATP. Electrons leaving the electron transport chain join with oxygen and with still other hydrogen ions to make water.

glie-COLL-iss-iss

12–B Glycolysis

Glycolysis is a series of reactions found as part of both aerobic and anaerobic respiration. A small amount of ATP is produced during glycolysis, and many

invertebrates: animals without backbones.

bacteria, fungi, and small invertebrates can survive on the ATP made during glycolysis alone. Most of these organisms, however, as well as all so-called "higher" organisms, also possess the TCA and electron transport pathways, which produce more ATP than glycolysis does. These two pathways need oxygen to function, and most organisms use them to make ATP when oxygen is available. In organisms with the TCA and electron transport systems, the main function of glycolysis is to produce pyruvate molecules that can be processed and sent to the TCA cycle. Glycolysis is the only source of ATP under anaerobic conditions, in which the TCA cycle and electron transport chain cannot function.

The enzymes of glycolysis are dissolved in the cytoplasm. Follow the steps of glycolysis in Figure 12-3 as we mention them:

1. Glucose is a stable compound, and it does not react easily. Before the energy stored in glucose can be released, the molecule must be made more

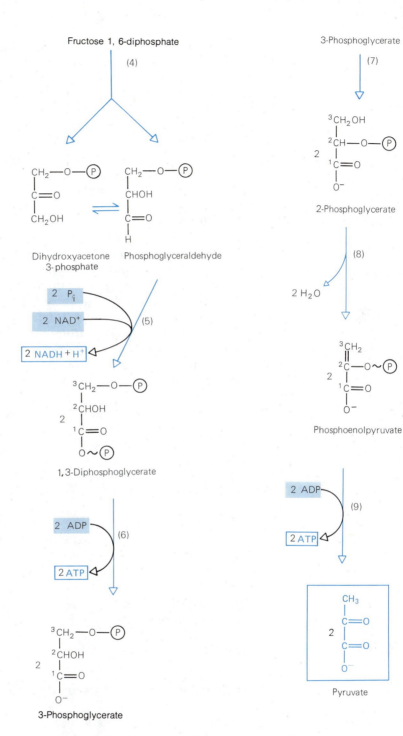

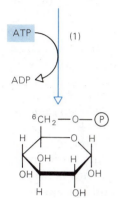

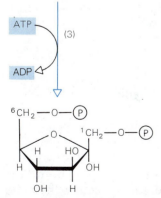

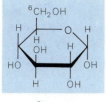

Glucose

Glucose-6-phosphate

Fructose-6-phosphate

Fructose 1,6-diphosphate

Fructose 1, 6-diphosphate

Dihydroxyacetone 3-phosphate   Phosphoglyceraldehyde

1,3-Diphosphoglycerate

3-Phosphoglycerate

3-Phosphoglycerate

2-Phosphoglycerate

Phosphoenolpyruvate

Pyruvate

Fig. 12–3

The stages of glycolysis. Important molecules entering the pathway appear on shaded backgrounds. Important products are boxed in with colored outlines.

reactive. This is done by transferring a phosphate group from ATP to glucose. The phosphate group attaches to the sixth carbon atom of glucose, forming glucose-6-phosphate. In this first step of glycolysis, an ATP is actually used up. This is like priming a pump; it allows a cell to reach a point where it can begin to make ATP.

2. Next, the molecule is rearranged, converting it to another six-carbon sugar, fructose-6-phosphate.

FRUCK-tose

3. In a second pump-priming reaction, another ATP donates a phosphate group, resulting in a molecule of fructose-1,6-diphosphate.

4. Fructose diphosphate splits into two three-carbon molecules, each with a phosphate group attached to one end. One of these three-carbon molecules, dihydroxyacetone phosphate, is eventually converted into the same form as the other, phosphoglyceraldehyde.

DIE-hi-DROX-ee-ASS-uh-tone
FOSS-foe-GLISS-er-AL-duh-hide

5. In the next step, each three-carbon molecule receives a phosphate group (P_i) that has been floating around in the cytoplasm. Each molecule also gives up two hydrogen atoms; these are picked up by the hydrogen carrier NAD^+, but the carrier quickly gives up one H^+. The three-carbon molecules are now molecules of diphosphoglycerate, and the squiggly bond ($\sim$) represents a high-energy bond with enough energy to form ATP from ADP and P_i when the bond is broken.

die-foss-foe-GLISS-er-ate

6. In the next step, each high-energy bond is broken, and the energy is used to make a molecule of ATP from ADP and P_i.

7. The remaining phosphate group of each phosphoglycerate molecule is transferred to the molecule's center carbon.

8. A molecule of water is removed from each phosphoglycerate molecule, forming phosphoenolpyruvate; the conversion changes the bond of the phosphate group into a high-energy bond.

foss-foe-EEN-all-PIE-rue-vate

9. The remaining phosphate group is removed from the phosphoenolpyruvate, and breaking the bond provides enough energy for the formation of ATP from ADP and P_i; the phosphoenolpyruvate has become a molecule of pyruvate.

In summary, glycolysis converts a six-carbon molecule, glucose, to two three-carbon molecules of pyruvate, with the production of four ATPs and two $NADH + H^+$. Since two ATPs have been used up, the net gain from glycolysis is two ATP and two $NADH + H^+$. As we shall see, the NADH eventually passes its hydrogen to the electron transport chain. Glycolysis has thus converted glucose into pyruvate, which can undergo further degradation, and has formed NADH and a small amount of usable energy in the form of ATP.

PIE-rue-vate

12–C Further Adventures of Pyruvate: The TCA Cycle

SIT-rick

The tricarboxylic acid cycle is also known as the **citric acid cycle** or the **Krebs cycle**, after Sir Hans Krebs, who first described the cycle in 1937 and received a Nobel Prize for this work in 1953.

Although glycolysis occurs in the cytoplasm, its products move into the mitochondria, where respiration is completed. Most of the molecules involved

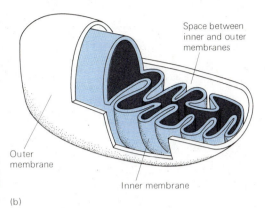

Fig. 12–4

A mitochondrion cut apart to show its structure. The acid battery forms in the space between the inner and outer membranes.

Space between inner and outer membranes

Outer membrane

Inner membrane

(b)

in the TCA cycle and the electron transport chain are located on or in the inner membranes of mitochondria (Fig. 12-4). In prokaryotic cells, which have no mitochondria, the reactions of respiration take place in the cytoplasm and at the cell membrane.

After pyruvate enters a mitochondrion in a eukaryotic cell, it reacts with a molecule known as coenzyme A and with NAD^+. The carbon and oxygens on one end of the pyruvate are removed as a molecule of carbon dioxide, leaving a two-carbon acetyl group, which attaches to coenzyme A. The resulting molecule is known as **acetyl coenzyme A**, or **acetyl CoA**. Meanwhile, the NAD^+ has picked up two hydrogens (Fig. 12-5). The acetyl CoA enters the TCA cycle proper.

ass-uh-TEEL coe-EN-zyme

The reactions of the TCA cycle are shown in Figure 12-6. Acetyl coenzyme A transfers its two-carbon acetyl group to the four-carbon compound oxaloacetic acid, forming a six-carbon compound, citric acid. In the transfer, coenzyme A is released and freed to pick up another acetyl group. A series of enzymes next removes two of the six carbons, in the form of carbon dioxide, from the citric acid molecule, and the remaining four-carbon compound is converted into a new molecule of oxaloacetic acid, ready to accept another two-carbon acetyl group from acetyl coenzyme A.

ox-uh-loe-uh-SEET-ic

PYRUVATE

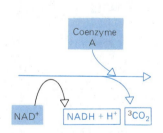

ACETYL COENZYME A
(ACETYL CoA)

Fig. 12–5

Pyruvate is converted into an acetyl group which combines with coenzyme A to form acetyl CoA. Acetyl CoA enters the tricarboxylic acid (TCA) cycle.

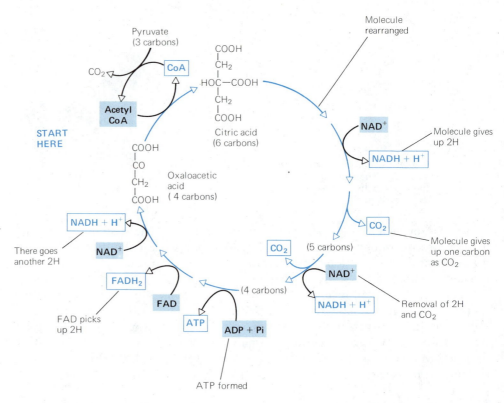

Fig. 12–6

The tricarboxylic acid cycle, simplified.

We will not go into the details of the TCA cycle, but you should note the following points:

1. The carbon dioxide which we breathe out is a waste product. For each molecule of pyruvate oxidized, all three carbon atoms are given off as molecules of carbon dioxide; one carbon dioxide is produced in the conversion of pyruvate to an acetyl group, and the other two in the TCA cycle.

2. Hydrogen atoms are removed at various stages of the TCA cycle and passed to the electron acceptors NAD^+ and FAD, forming $NADH + H^+$ and $FADH_2$.

3. One molecule of ATP is formed directly during each turn of the TCA cycle.

4. The TCA cycle may be summarized as

oxaloacetic acid + acetyl CoA + ADP + P_i + 3 NAD^+ + FAD →
oxaloacetic acid + CoA + 2 CO_2 + ATP + 3 NADH + 3 H^+ + $FADH_2$.

12–D Charging the Batteries: The Electron Transport Chain

One molecule of glucose has now been completely dismantled. Some of the energy so released has been used to make ATP, but most of it remains in the electrons carried by NADH (from glycolysis and the TCA cycle) and by $FADH_2$

Fig. 12-7

Electron transport and ATP synthesis in a mitochondrion. Hydrogen carried by NADH is accumulated between the mitochondrial membranes (shown in gray), forming an "acid battery." The enzyme that makes ATP uses the energy of the acid battery to form ATP from ADP and P_i. Meanwhile, the electrons that entered with the H^+ are transported by electron transport molecules in the inner mitochondrial membrane and eventually combine with oxygen, forming water.

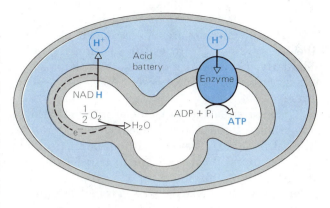

(from the TCA cycle). These hydrogen carrier molecules pass their hydrogens to the electron transport chain. Most of the molecules in the chain are **cytochromes**, which can accept only electrons. The H^+ parts of the hydrogen atoms are pushed into the hydrogen reservoir. Meanwhile, the electrons pass from one molecule to the next until they reach the end of the chain. The final electron acceptor is oxygen, which takes the electrons, plus some H^+ (which is *not* part of the reservoir; some H^+ is always present from the dissociation of water into H^+ and OH^-) and forms water:

$$\tfrac{1}{2}\,O_2 + 2\ \text{electrons} + 2\,H^+ \rightarrow H_2O$$

The electron transport chain is located in the inner membrane of the mitochondrion, and the acid battery reservoir builds up between the inner and outer membranes (Fig. 12-7). Enzymes in the inner membrane can use the energy stored in the battery to power the synthesis of ATP from ADP and P_i. Each time an ATP is made, the battery loses some of its charge. The battery is recharged by the continuous flow of electrons arriving via NADH and $FADH_2$.

The chemical reactions of glycolysis and the TCA cycle have been well established. What happens in the electron transport chain, however, has been the subject of intensive study but little definite proof—partly because complete electron transfer takes place only when the mitochondrial membranes are intact, and it is very difficult not to break the membranes when doing biochemical experiments. It took many years to develop the techniques now being used to study this pathway.

As yet, the evidence that the acid battery powers the synthesis of ATP (**chemiosmotic synthesis**) is not completely convincing, but it is more satisfying than earlier theories. Indeed, it has gained such acceptance that its originator, Peter Mitchell, received a Nobel Prize for this work in 1978. The only reasonable alternative to the chemiosmotic theory is that ATP is made in the electron transport chain by transfer of a phosphate group to ADP (thus forming ATP) from some other molecule, as happens in glycolysis (see Fig. 12-3). Despite 30 years of searching for such a phosphate-donating compound, none has been found. In addition, such a theory would not explain why ATP synthesis stops if the membranes of the mitochondrion are broken. In these circumstances, the chemical reactions of respiration proceed, but no ATP is formed. It is easy to see, though, that if the membranes are necessary to contain the reservoir of H^+, then breaking them allows the H^+ to escape, letting the power of the battery leak away. Similarly, the chemiosmotic theory explains why detergents, which make membranes more permeable, prevent ATP synthesis but not the reactions of the TCA cycle.

die-nie-tro-FEEN-ols

Several substances that make mitochondrial membranes leaky have medical importance. If the membrane leaks, the respiration of food still yields H^+, but the H^+ leaks out of the battery without being used to make ATP. A well-nourished cell can literally starve to death if it lacks the ATP to carry out the various reactions of its life processes. Dinitrophenols, yellow substances once used as food additives to make baked goods look as if they contained more eggs than they really did, make mitochondria leaky; so do large amounts of one of the thyroid hormones, thyroxin. Each of these substances was prescribed for a time as a cure for obesity, and each was abandoned after several patients ran out of ATP and died during treatment. Large quantities of unsaturated fatty acids have the same effect, and have been touted as cures for obesity in some dangerous fad diets of more recent years.

How much ATP is made from each glucose molecule during aerobic respiration? The best estimate is 36 or 38 molecules of ATP (depending on what enzymes are present in the particular cell). This figure is an estimate because biochemists are not yet satisfied that they know just how many ATPs are made per molecule of glucose by the chemiosmotic mechanism involving the acid battery.* As we have seen, two ATPs (net) are made during glycolysis, and two more are made directly in the TCA cycle (one per turn of the cycle; it takes two turns to dispose of the two pyruvate molecules produced by glycolysis). So about 89% of the total ATP derived from the breakdown of each glucose molecule is made with energy from the acid battery.

12–E Respiration Without Oxygen

Oxygen is necessary if the TCA cycle and electron transport chain are to function. When they have no oxygen, many cells can produce ATP by **anaerobic respiration**, also known as **fermentation**. The most common type of anaerobic respiration uses the pathway of glycolysis, which, as we saw before, yields a small amount of ATP—two ATPs per glucose, which is about 5% as much ATP as is produced by the aerobic respiration of one molecule of glucose.

We have seen that NADH is made in the glucose-to-pyruvate pathway (glycolysis). NAD^+, the carrier that picks up hydrogen atoms from molecules undergoing glycolysis, is in very short supply in a cell. When no oxygen is present to receive electrons from the electron transport chain, NADH cannot be converted back to NAD^+; as a result, the TCA cycle and electron transport chain

---

*When biochemists thought that ATP was made directly in the electron transport chain, they identified particular steps that were supposed to generate ATP, and older texts state that precisely 36 or 38 molecules of ATP are made for each molecule of glucose. Recent research has shown that the situation is much more complex than was once thought, but the figure 36 to 38 is still believed to be a reasonable guess of the average ATP yield per glucose molecule.

Fig. 12–9

Yeast cells. Small cells are budding off from some of the larger ones. (Biophoto Associates)

Fig. 12–8

Louis Pasteur (1822–1895). Pasteur was the first person to work out the process of alcoholic fermentation during winemaking. (Smithsonian Institution)

become backed up, and stop. Even glycolysis stops, because there is no NAD^+ left to pick up hydrogens that must be removed for glycolysis to proceed. In this situation, cells that can undergo anaerobic respiration use means other than the electron transport chain to rid NADH of its load of hydrogen. NAD^+ can then go back and accept more hydrogens from glycolysis, permitting glycolysis to proceed.

Anaerobic respiration proceeds by somewhat different pathways in different organisms. The first to be worked out was the process by which yeasts (one-celled fungi) ferment the sugars in grapes as they produce wine (Fig. 12-10). In this **alcoholic fermentation**, the pyruvate formed during glycolysis is dismantled into a molecule of carbon dioxide and a molecule of the two-carbon compound acetaldehyde. Acetaldehyde next accepts two hydrogen atoms from $NADH + H^+$, forming the two-carbon alcohol ethanol, also known as ethyl alcohol, the active ingredient in alcoholic drinks. After this hydrogen transfer, NAD^+ is freed to pick up more hydrogen atoms, allowing the process of glycolysis to continue.

asset-AL-duh-hide
ETH-uh-nol

Fermentation, which occurs when bread rises as well as during winemaking, continues until the yeast cells have used up all the sugar around them (when they eventually starve to death), or until they are killed by the heat of baking. Stoppering a wine bottle before fermentation has finished yields an effervescent liquid because carbon dioxide is still being given off. Such was the young wine that stretched and split the wine skins in the New Testament story. To make a fizzy wine like champagne you use a very strong bottle and cork it up before fermentation has finished. The carbon dioxide dissolves in the wine under pressure and is released as bubbles when the bottle is opened. Yeasts produce alcohol only when there is no oxygen present; when there is enough oxygen, they use aerobic respiration to break sugar down completely to carbon dioxide and water. When a vat of wine is fermenting rapidly, it produces carbon dioxide

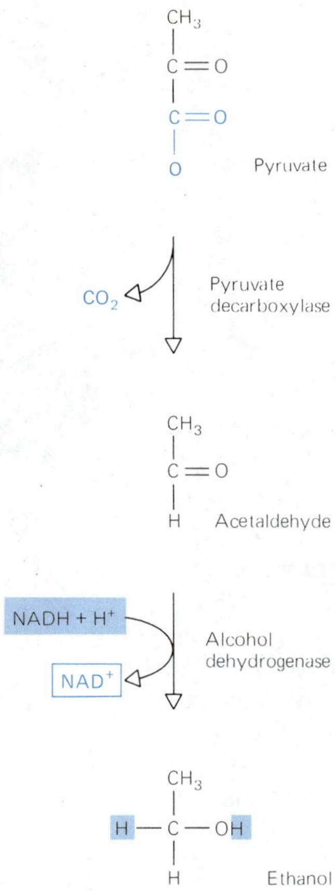

CH$_3$
|
C=O
|
C=O
|
O   Pyruvate

CO$_2$ Pyruvate
decarboxylase

CH$_3$
|
C=O
|
H   Acetaldehyde

NADH + H$^+$
NAD$^+$ Alcohol
dehydrogenase

CH$_3$
|
H—C—OH
|
H   Ethanol

Fig. 12–10

Alcoholic fermentation as it occurs in yeast. Pyruvate produced in glycolysis releases carbon dioxide to form acetaldehyde. This accepts electrons to produce NAD$^+$, which is needed to continue glycolysis and produce ATP when a cell is deprived of oxygen.

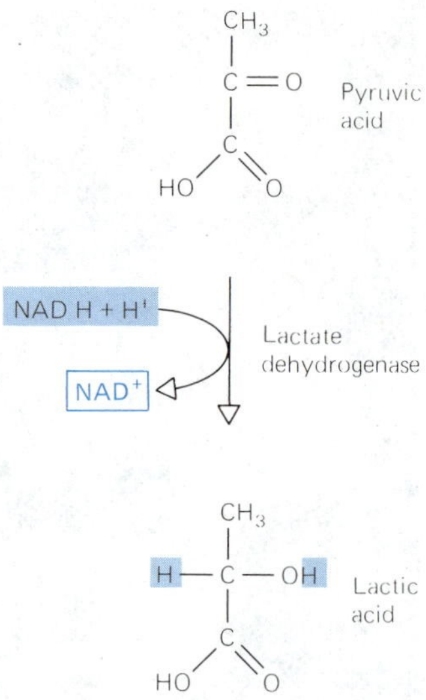

CH$_3$
|
C=O   Pyruvic
| acid
C
HO   O

NAD H + H$^+$ Lactate
NAD$^+$ dehydrogenase

CH$_3$
|
H—C—OH   Lactic
| acid
C
HO   O

Fig. 12–11

The conversion of pyruvic acid to lactic acid in muscle. When the cell lacks oxygen, NADH accumulates: pyruvic acid produced by glycolysis acts as a hydrogen acceptor and becomes lactic acid, releasing NAD$^+$ to accept more hydrogens produced during glycolysis. (Pyruvic acid is pyruvate [an ion] combined with H$^+$, and lactic acid is lactate ion combined with H$^+$.)

ass-EAT-ic

fast enough to drive off all the air over the wine. But when fermentation slows down, wine must be sealed up immediately. Otherwise aerobic bacteria, which are always present in the air, will convert the alcohol into acetic acid. Acetic acid is the acid in vinegar, and this is how wine vinegar is made.

Another familiar example of anaerobic respiration occurs in our own bodies during overexertion. Normally, our cells are supplied with oxygen and carry on aerobic respiration. During strenuous exercise, however, our muscles use up oxygen faster than the bloodstream can replace it, and in these conditions the muscles use anaerobic respiration to supply a small amount of ATP that enables them to keep going. In muscles the pyruvate is converted, not into carbon

LACK-tate

dioxide and alcohol (as in wine), but into a three-carbon compound, lactate, formed by adding two hydrogen atoms from NADH + H$^+$ to pyruvate. The NAD$^+$ so freed picks up more hydrogens, allowing glycolysis to proceed.

However, the body cannot tolerate a great buildup of lactate. When muscular activity stops after overexertion, we still keep breathing heavily for a time, to repay the "oxygen debt" we have incurred while respiring anaerobically. As oxygen reaches the muscles, the TCA cycle and electron transport chain can clear out the backlog of electrons waiting to be transferred, ultimately, to oxygen. Lactate is converted back to pyruvate, which may either proceed toward the TCA cycle, if the muscle still needs to replenish its supply of ATP, or may enter reverse glycolysis and re-form the muscle's supply of glucose, which in turn is stored as glycogen.

glycogen   (GLIE-coh-jenn): a polysaccharide, made by joining many molecules of glucose.

12–F Alternative Food Molecules

Presented with a smorgasbord of the commonly available food molecules, most cells will respire glucose to make ATP. However, since all organic molecules contain stored energy, any of them may be broken down to release the energy needed to synthesize ATP under the appropriate conditions.

Many carbohydrates other than glucose are processed by way of glycolysis. Polysaccharides may be broken down into their component sugar molecules, and various six-carbon sugars may be changed into the glucose or fructose found in glycolysis.

Fatty acids in the food or stored in the body do not go through glycolysis, which breaks down only carbohydrates, but go through a different pathway. The fat-breakdown pathway feeds into the respiratory pathway we have studied at the point where acetyl CoA forms. Fatty acids are broken down in two-carbon chunks that join to coenzyme A, forming acetyl CoA. After this, the acetyl CoA is indistinguishable from acetyl CoA made from the pyruvate at the end of glycolysis, and it proceeds through the TCA cycle, whose NADH and $FADH_2$ go on to the electron transport chain (Fig. 12-12).

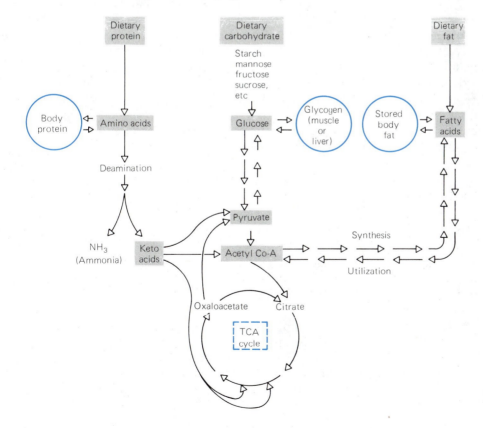

Fig. 12–12

Pathways for processing proteins, fats, and carbohydrates in the diet.

Excess amino acids taken in as food may also be used for energy, but protein that is already part of the body is not used until an advanced stage of starvation, after carbohydrate and fat reserves are depleted. Any amino acids not needed to form body proteins first have their amino (nitrogen) groups removed; the rest of the molecule contains carbon, hydrogen, and oxygen, the same elements found in carbohydrates and fats. Amino acids vary greatly in the structure of the molecule left after removal of the amino group. Some are like pyruvate and enter the respiratory pathway at that point; others enter as acetyl groups joined to CoA, or as one of the molecules in the TCA cycle.

From this we can see that the TCA cycle and electron transport chain are a sort of final common pathway for the breakdown of organic molecules to yield energy. Figure 12-12 shows how protein, carbohydrate, and fat pathways are linked and shows that many of these pathways also work in reverse. The TCA cycle supplies many intermediate molecules to various other pathways in the body's biochemical processes, and when the cell has enough energy, molecules flow through the TCA cycle into body-building processes. Note especially that it is possible to trace pathways from both carbohydrates and proteins to fat deposits; any excess food in the diet, no matter which group it comes from, can ultimately end up as stored fat, a "savings deposit," which for many of us seems to earn compound interest year by year. So we see that fad diets—such as those permitting us to eat all the food we want, provided that it's protein—cannot really help us lose weight; only eating less food, or using up more, will do that.

Fats are nonpolar, and thus they tend to repel water and arrange themselves as concentrated fat droplets. Fats also yield more than twice the energy of proteins or carbohydrates, weight for weight. These two properties have made it advantageous for animals to store most of their energy reserves as fat, although a small amount of carbohydrate, in the form of glycogen in the liver and muscles, is kept for a "quick energy boost" as needed. Plants store most of their energy reserves in the form of starch (carbohydrates). The —OH groups of carbohydrates make them polar, and they attract and hold much water. This makes them bulky to store, but since plants do not have to move around and drag this food supply along, they are not inconvenienced by the weight as an animal would be. Carbohydrates are also much easier to mobilize into the energy-release pathways than are fats. These properties seem to have selected for storage of carbohydrate energy reserves in plants.

SUMMARY

Organisms usually store and use energy in the form of ATP, and ATP must be replaced constantly for life to continue. The breakdown of food molecules, especially glucose, during cellular respiration releases energy that is used to re-form ATP.

Glycolysis, the first series of reactions in the breakdown of glucose, takes place in the cytoplasm. It produces a gain of two molecules of ATP and two molecules of NADH + H^+. The original six-carbon glucose becomes two molecules of a three-carbon compound, pyruvate.

After entering a mitochondrion, each pyruvate is next converted to one

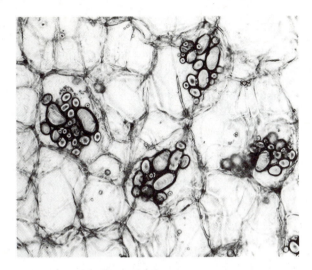

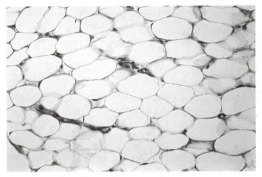

Fig. 12–13

Food storage. (a) Starch grains in the cells of a potato. (b) Adipose cells in a cat. The colorless material in the cells is fat. (Biophoto Associates)

molecule of carbon dioxide and one two-carbon acetyl group, which attaches to coenzyme A, forming acetyl CoA. Acetyl CoA enters the TCA cycle, becoming attached to a four-carbon oxaloacetate molecule to form a six-carbon citrate molecule. As the TCA cycle proceeds, two carbons are released as carbon dioxide, and several hydrogen carrier molecules pick up pairs of hydrogen atoms. One molecule of ATP is made during each turn of the TCA cycle. The molecules in the cycle are eventually converted back to oxaloacetate, which picks up more acetyl CoA.

The hydrogen-laden carrier molecules from glycolysis and the TCA cycle pass the hydrogen atoms' electrons to the electron transport chain in the inner mitochondrial membrane. The electrons eventually pass to oxygen, which accepts them, along with some H^+, to form water. Thus oxygen is the final electron acceptor in the chain. The H^+ from the hydrogen atoms accumulates in a hydrogen reservoir trapped between the membranes of the mitochondrion, forming an acid battery that supplies the energy needed by enzymes in the inner mitochondrial membrane to make ATP from ADP and P_i.

When there is not enough oxygen to accept electrons as fast as they come down the chain, some cells undergo anaerobic respiration, using only the reactions of glycolysis as a meager source of energy. Under anaerobic conditions, organisms such as yeasts convert the pyruvate resulting from glycolysis to carbon dioxide and alcohol, whereas muscles convert it to lactate. In both cases, NAD^+ is relieved of its hydrogen burden, and goes back to the pathway of glycolysis, where it picks up more hydrogen, thus allowing glycolysis to proceed.

Although glucose is the main food molecule used during respiration, any organic molecule can be broken down to release energy for use in ATP synthesis. The interconnecting breakdown pathways can also be reversed, and it is possible for excess carbohydrates and proteins in the diet to be converted to fat and stored along with any excess dietary fat.

SELF-QUIZ

1. The important end products of the TCA cycle include:
   a. carbon dioxide and oxygen
   b. carbon dioxide and $FADH_2$
   c. oxaloacetate and pyruvate
   d. oxaloacetate and ADP
   e. oxaloacetate, NADH, and ADP

2. When oxygen is available, NADH + H^+ from glycolysis proceeds to:
   a. the electron transport chain
   b. water
   c. the TCA cycle
   d. fat storage
   e. pyruvate

3. The role of oxygen in cellular respiration is:
   a. formation of carbon dioxide
   b. release of energy from the hydrogen reservoir
   c. conversion of pyruvate to acetyl CoA
   d. acceptance of hydrogen to form water
   e. combination with NADH + H^+

4. The electron transport chain is found in:
   a. the inner mitochondrial membrane in eukaryotes and the cell membrane in prokaryotes
   b. the inner mitochondrial membrane in eukaryotes and the cytoplasm in prokaryotes
   c. the outer mitochondrial membrane in eukaryotes and the cytoplasm in prokaryotes
   d. the cytoplasm in both prokaryotes and eukaryotes

5. A common feature of anaerobic respiration in both yeasts and muscle is formation of:
   a. a great deal of carbon dioxide
   b. alcohol
   c. NAD^+ from NADH + H^+
   d. acetyl CoA
   e. lactate

6. Complete aerobic respiration of glucose in muscle yields about 36 ATP; anaerobic respiration of the same molecule would yield _____ ATP. Therefore, the lack of sufficient oxygen temporarily causes the muscle to lose _____% of the energy it could derive from the glucose molecule.

7. True or False? Carbohydrates eaten in the food must be respired within a short time or the body will lose the chance to obtain the energy stored in the molecule. Explain.

QUESTIONS FOR DISCUSSION

1. Compare and contrast photosynthesis (Chapter 11) and cellular respiration. What similarities and differences do you notice in starting and end products, intermediate molecules, other molecules needed for the process to work, energy flow, etc.?

2. Why does a photosynthetic plant need to carry on cellular respiration? Do all its cells respire?

3. Amygdalin (Laetrile) has been widely touted as a cure for cancer. Digestive enzymes break amygdalin down to release cyanide, and some people who consumed overdoses of amygdalin have died of cyanide poisoning. Cyanide inactivates cytochromes. Why is it poisonous?

4. Why do foods that are rich in fat tend to be expensive (in dollars and cents) compared with carbohydrates?

5. Comment on the claim that other natural" sugars, such as fructose, are less fattening than glucose or sucrose.

ORIGIN OF LIFE 13

When you have studied this chapter, you should be able to:

1. Give two reasons why spontaneous generation is unlikely to occur on earth today.

2. State the main differences in composition between the atmosphere of the early earth and the present atmosphere.

3. Explain the importance of time in the theory of the origin of life presented in this chapter.

4. Trace the steps by which life may have originated on earth, from the formation of organic molecules through the origin of aerobic respiration; state what evidence we have that these steps were possible.

5. State the significance of the evolution of photosynthesis and aerobic respiration in the overall evolution of life.

6. List changes in the environment brought about by living organisms, and explain how the evolution of modern organisms depended on the change to modern environmental conditions.

If we could trace the ancestry of living organisms, we would find a long line of cells stretching back in time. Each cell came from division of a previously existing cell . . . but where did the first cell come from?

Most scientists today believe that chance chemical events, occurring over hundreds of millions of years, built up increasingly complex and lifelike clusters of chemicals; some of these eventually became cells. In this chapter we shall sketch the outlines of this theory and consider some of the evidence supporting it.

13–A Spontaneous Generation

For thousands of years people believed in **spontaneous generation**, the routine appearance of living organisms from nonliving matter. Aristotle wrote that frogs and insects formed in moist soil; others observed that stagnant watering troughs produced worms and algae, and spoiled meat produced maggots.

In 1668, the Italian Francesco Redi dealt the first important blow to the theory of spontaneous generation. Redi placed dead snakes and eels in glass jars, and covered some of the jars with the finest Neapolitan muslin, leaving the others

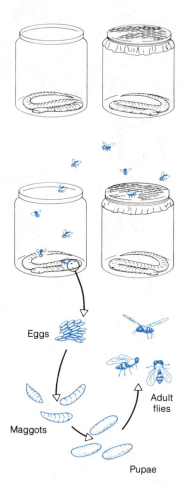

Eggs

Maggots

Pupae

Adult
flies

Fig. 13–1

Redi's experiment, 1668. Some jars containing dead animals were covered with muslin; others were left open. Maggots appeared in the open jars, but not in the covered ones. Redi observed that this was because flies entered the open jars and laid eggs that developed there (see the fly life history below the jars). Flies could not enter the covered jars so no maggots or flies appeared in these jars.

ANTONIUS A LEEUWENHOEK.

Regiæ Societatis Londinensis membrum.

Fig. 13–2

Antonius Leeuwenhoek, the Dutch microscope maker who found microorganisms and unicellular plants and animals everywhere he looked. (Biophoto Associates; Linnaean Society of London)

open to the air. Flies soon arrived and laid eggs on the dead animals in the open jars; the eggs hatched into maggots. No maggots appeared in the covered jars, but flies did settle on the muslin and lay eggs there. Similar experiments finally convinced most people that plants and animals come only from their parents' seeds or eggs.

Meanwhile, in the late 1600s, the Dutch lens grinder Antonius van Leeuwenhoek began building microscopes and using them to examine water, soil, and even his own body; everywhere he found different kinds of tiny "animalcules." He proposed that these animalcules, too, came from the reproduction of others of their kind.

Abbé Lazzarro Spallanzani agreed. In the eighteenth century, he set out to show that microorganisms do not arise spontaneously in the nutrient broths where they are sometimes found in abundance. He prepared flasks of a nutrient gravy, sealed and heated them, and found that most of the flasks remained free from living microorganisms. (We now know that some bacteria make resting spores that survive heat treatment and grow afterward—accounting for the appearance of microorganisms in some of his flasks.) On the basis of his experi-

microorganisms: small organisms such as bacteria and fungi.

ments, Spallanzani contended that microorganisms could not arise spontaneously. People with opposing viewpoints contended that Spallanzani had somehow made the air in the sealed jars unfit to support life, or that he had destroyed "vital molecules" needed by nonliving matter in order to become living.

The theory of spontaneous generation of microorganisms met its death in the mid-nineteenth century at the hands of Louis Pasteur in France and John Tyndall in England. They demonstrated that bacteria travel through the air, and that if air is purified before it passes into a flask of sterilized nutrient broth, no bacteria will appear in the broth. Pasteur drew the necks of his glass flasks out into S-shaped curves that could trap bacteria as the air passed freely into or out of the flask (Fig. 13-3). Tyndall sterilized the air entering his flasks by passing it through a flame or through absorbent cotton. By the late 1870s, all but a few diehards agreed that living organisms, of whatever size, come from reproduction of previously existing organisms . . . which brings us back to the question: Where did the first living organisms come from?

13–B Conditions for the Origin of Life

Louis Pasteur is often credited with disproving the theory of spontaneous generation. But Pasteur himself once remarked that his fruitless 20-year search for spontaneous generation did not lead him to believe it was impossible. What Pasteur showed was that life did not arise in his flasks under the conditions he used (sterilized nutrient medium, clean air) in the time he waited. He did not show that life could *never* arise from nonliving matter under *any* set of conditions.

Indeed, scientists today believe that life *did* arise from nonliving matter, but under conditions very different from those we see around us today, and over the course of hundreds of millions of years. Many scientists see the origin of life as an inevitable stage in the evolution of matter, and believe that it has probably happened many times, in many parts of the universe.

Under what conditions can life arise? There appear to be four basic requirements: the presence of certain chemicals and an energy source, absence of oxygen gas (O_2), and eons of time. Of the necessary chemicals, water is abundant on earth, and the required inorganic minerals abound in rocks and volcanic gases. The action of various energy sources in the production of organic molecules, molecules produced today mainly by living organisms, will be the topic of most of this chapter, so let us turn for a moment to the last two conditions.

Lack of oxygen gas.   Oxygen combines readily with organic molecules and destroys them; organic molecules exposed to oxygen on the early earth would not have lasted long enough to form more complex structures. This is one reason why spontaneous generation from organic matter does not occur today. (Another is that free organic molecules are usually absorbed and used as food by bacteria and fungi even before oxygen can damage them.) Geologists tell us that the oldest rocks on earth were formed in an atmosphere without oxygen. The earth's first atmosphere, composed largely of hydrogen gas (H_2), was soon replaced by

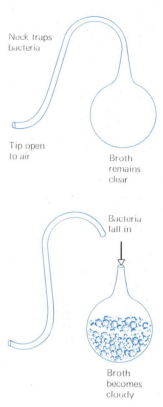

Neck traps bacteria

Tip open to air

Broth remains clear

Bacteria fall in

Broth becomes cloudy

Fig. 13–3

Pasteur's swan-necked flasks. Air could enter the flasks freely through the open tip, but it could not travel quickly enough through the curved neck to carry along bacteria, which are heavier than the air in which they travel. Any bacteria or other cells present were trapped at the bottom of the curve, while air molecules continued on into the main part of the flask. Only if the neck was broken off could bacteria enter the flask and putrify the nutrient broth.

TABLE 13-1	PROBABILITY THAT AN EVENT WILL NOT HAPPEN
Given this probability: in 1 year	0.999
then: in 2 years	0.998
in 3 years	0.997
in 4 years	0.996
in 1024 years	0.359
in 2048 years	0.129
in 4096 years	0.017
in 8128 years	0.000276

gases in which hydrogen was combined with other common elements: carbon (CH_4, methane), nitrogen (NH_3, ammonia), and oxygen (H_2O, water vapor). The oxygen in the atmosphere today comes almost exclusively from photosynthesis, carried on by living plants.

Time. The events involved in the origin of life were extremely unlikely, according to the laws of probability. Given enough time, however, even very improbable events are bound to occur. For example, if the probability that an event will occur in a year is one in a thousand, the probability that it will not occur is 999 in a thousand, or 0.999; the probability that it will not happen in 2 years is $(0.999)^2$; in 3 years, $(0.999)^3$. Table 13-1 shows that there is a very small probability that the event will not happen at least once in 8128 years, or conversely, there is a very high probability (0.9997) that it *will* happen. The events required for the origin of life were much less probable than 1/1000, but there was much more time available. We believe that the earth formed about 4.5 billion years ago; the rocks of 1 billion years later contain spherical particles that were probably forms of pre-life. The earliest known fossils of cells come from rocks formed about 1.6 billion years after the earth formed; these may have been photosynthetic. So, unlikely as living systems are, they had so much time to evolve that their origin was probably inevitable!

13–C Search for the Beginning of Life

We can never know exactly how life began. In 1924, a Russian, Alexander Oparin, published a theory of how life might have originated; in 1929 his ideas, not yet translated from the Russian, were echoed by the Englishman J. B. S. Haldane. Subsequent research has supported their predictions.

In 1953 Stanley Miller, then a graduate student, began to test this twentieth-century theory of the origin of life. Miller built a laboratory apparatus to represent a small-scale model of conditions on the early earth, including a primitive "atmosphere" of hydrogen-rich gases and an "ocean" (Fig. 13-4). Electrodes in the "atmosphere" chamber produced electrical discharges, representing lightning, a possible source of energy on the early earth (most chemical reactions require an energy input to proceed at appreciable rates). At the end of a week, the "ocean" contained many organic compounds. The presence of amino acids in the mixture caused special excitement, because proteins, made up of

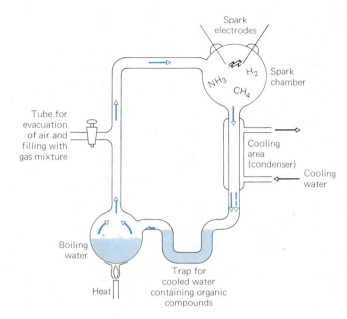

Spark
electrodes

Spark
chamber

NH_3 H_2
CH_4

Tube for
evacuation
of air and
filling with
gas mixture

Cooling
area
(condenser)

Cooling
water

Boiling
water

Heat

Trap for
cooled water
containing organic
compounds

Fig. 13–4

Miller's apparatus for imitating conditions on the early earth. The "atmosphere" is represented by ammonia (NH_3), methane (CH_4), and hydrogen (H_2) gases in a glass sphere. Sparks from electrodes represent lightning, a source of energy. The water represents the sea, heated to produce water vapor in the "atmosphere" and then cooled in a condenser to represent rainfall.

amino acids, are such important chemicals in living organisms (Section 9–K). This experiment was quite simple, and numerous high-school students have repeated it. Similar experiments in other laboratories, using many plausible variations on Miller's conception of the early environment, have added a large number of organic compounds to the ones found by Miller.

Such molecules are formed even today as volcanic gases and lava react with water. In addition, meteorites and the atmospheres of other planets also contain various organic compounds. All these lines of evidence support the idea that organic molecules could have, and did, form on the early earth without the aid of living things. Without oxygen to destroy them or bacteria or fungi to use them as food, these substances could have accumulated in the early oceans until, as Haldane said, the sea was like a "hot, dilute soup."

The next step was the formation of larger polymers from small organic monomers. Formation of polymers requires energy, and if water is present it is apt to break short polymers back into monomers as fast as they form. To be stable, a mixture of short-chain polymers must contain as little water as possible.

Sidney Fox produced chains of amino acids, called **proteinoids**, by applying heat—an energy source—to a dry mixture of amino acids. (This could have occurred on the early earth in tide pools as water evaporated on hot, sunny days.) Other workers have produced polymers from molecules in the atmosphere, using light or electrical discharges as energy sources.

13–D Formation of Aggregates

Short polymers are quite unstable and tend to break back up into monomers if water is added to them. However, longer chains tend to fold up and stabilize themselves by attraction between different parts of the molecules; rainfall or a high tide washing these larger polymers into the sea would not have destroyed them.

In fact, shaking different kinds of polymers in water may lead to the formation of larger structures: **coacervate droplets**. If fats are present, they coat the droplets with a "skin" like a lipid cell membrane. Oparin, who studied coacervate droplets extensively, proposed that the ancestors of the first living organisms were such droplets.

Depending on what polymers they are made of, coacervate droplets may absorb molecules from the surrounding fluid. In addition, molecules inside coacervate droplets form polymers more readily than do molecules in the surrounding fluid. As droplets take up substances, they expand until they become so large that they fragment.

Fox found that when he added water to his proteinoids, they formed **microspheres** (Fig. 13-5). Microspheres show many properties of living cells. They have a double layer of protein as an outside boundary; the boundary is selectively permeable, and admits polymers composed of nucleotides (the precursors of nucleic acids, which contain the genetic information of modern cells) quite readily; and the interior is divided into watery and lipid-like areas, separated by protein boundary areas. In addition, microspheres accumulate small molecules from their surroundings in high concentrations, catalyze chemical reactions, and produce buds that grow and eventually break off as new microspheres.

13-E Beginnings of Metabolism

As we have seen (Chapters 11 and 12), living cells carry out complex chemical changes consisting of many steps. The total of all the chemical reactions in a living cell is its **metabolism,** encompassing the respiration that releases energy; the use of energy to move, to build up polymers, and to accumulate substances from the environment; and the conversion of some substances to others for use, storage, or excretion.

Studies of microspheres and coacervate droplets show how the complex biochemical pathways of metabolism may have arisen. Before such aggregates became living organisms, they must have undergone long eons of chemical evolution, in which there was selection for those chemical combinations that lasted the longest time before disintegrating. If an aggregate contained molecules that catalyzed chemical reactions making it more stable, it would last longer. The raw materials needed for these reactions were probably abundant at first to the few aggregates that needed them. As these successful, stable aggregates grew and fragmented, however, and as new ones formed spontaneously, they would have used up the needed raw materials faster than these were produced by lightning, warm tide pools, and volcanoes. As a result, there would have been selection for the systems that were most efficient at competing for the now-scarce raw material, and for systems that could convert a second, abundant material to the first, now-scarce one. Eventually, competition among systems that could use the second substance would render it scarce also; then systems that could convert a third material to the second, and the second to the first, would be

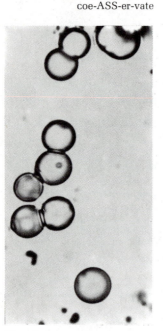

coe-ASS-er-vate

Fig. 13-5

Proteinoid microspheres, each 1-2 μm in diameter. (Sidney Fox)

muh-TAB-ull-ism

catalyze: to cause a chemical reaction without being permanently changed by the reaction.

at a competitive advantage. At each repetition of this sequence of events, metabolic pathways would have become longer. In addition, those systems whose catalysts could perform the various reactions fastest would have accumulated raw materials, grown, and reproduced faster than their fellows.

Let us consider a familiar metabolic pathway, cellular respiration, and see how it may have evolved. The important product of cellular respiration is ATP, which donates its energy to drive other chemical reactions in a cell. We can speculate that early aggregates absorbed ATP directly from the surrounding "soup." As the supply of ATP became low, systems that could use the energy released by breaking down other substances to make ATP would have been at a selective advantage. The process of anaerobic fermentation of glucose to pyruvate seems to be universal in modern organisms, and so biochemists believe that this pathway evolved early as a means of regenerating ATP. The tricarboxylic acid cycle and the electron transport chain are highly efficient means of extracting further energy from glucose, and they are found in all but a few primitive bacteria. These processes are thought to be relative latecomers in the evolution of means to make ATP.

anaerobic fermentation, tricarboxylic acid cycle, electron transport chain: see Chapter 12, Sections A-E.

13–F Beginnings of Reproduction

Although self-assembled aggregates have much in common with modern cells, they lack genetic information and so they cannot be considered as "living." **Genetic information** allows a living cell to make protein molecules to precise specifications and to pass on copies of this information to its daughter cells. In most modern organisms, the genetic information is contained in the order of nucleotide monomers in DNA molecules. DNA is a self-duplicating molecule (see Chapter 14). This property assures that each daughter cell receives identical genetic information when a cell divides. In addition, DNA directs the formation of RNA molecules, which in turn direct the linking of amino acids into protein molecules (discussed in Chapter 15). All of this is done by biochemical reactions which depend on shapes and electrical forces that fit molecules into precise positions for the reactions to take place. These masterpieces of molecular engineering have been selected from an astronomical number of trial-and-error chemical events over the course of billions of years.

Note: review the role of genetic information (genes) in natural selection in Section 3-A.

Molecules in nonliving aggregates show rudiments of this ability to "recognize," and fit together with, other molecules. Amino acids that have joined into proteinoids attract nucleotides, and vice versa. Certain sequences of amino acids tend to attract certain nucleotides; if these nucleotides then join to form a nucleic acid molecule, that molecule contains biological information in that its order of nucleotides was determined by the order of amino acids in the proteinoid, rather than at random. Evolution favored the aggregates in which the relationship between proteinoids and nucleic acids became more precise, because successful combinations would then produce successful offspring and become abundant. Finally, systems with predictable activities evolved: the modern cell's machinery for duplicating its DNA exactly and for making RNA and proteins with monomers arranged in precise sequences specified by the DNA.

13–G Further Developments

Once a reliable mechanism for reproducing genetic information evolved, the story of the origin of life was complete. The era of chemical selection was over, and the era of natural selection began.

For a long time, all organisms were **heterotrophs**, feeding on organic molecules in the primordial "soup" or on the less fortunate of their own number. They probably met their energy requirements by anaerobic fermentation of glucose to pyruvate (described in Section 12–B).

The next important event in the evolution of life was the development of **autotrophy**, the ability to make food from inorganic molecules. In a world where the supply of environment-made nutrients was severely depleted by competition, this opened up a whole new era of self-sufficiency.

The most familiar autotrophs today are green plants, but photosynthesis as we know it is one of the more advanced forms of autotrophy, which has outcompeted earlier evolutionary "experiments" in do-it-yourself food production. The more ancient forms of autotrophy, known as **chemosynthesis**, involved trapping energy not from sunlight, as in photosynthesis, but from some inorganic chemical reactions that release a great deal of energy. Many groups of bacteria living today practice different forms of chemosynthesis. There are also bacteria showing different degrees of ability to trap light energy; studying these bacteria allows us to reconstruct possible steps of the pathway to photosynthesis as practiced by higher plants today.

Why do we believe that autotrophy arose relatively late in evolution? First, we have seen that there was probably food available for heterotrophs, produced by nonbiological means, long before anything "alive" came into being. Second, photosynthesis requires a wide variety of molecules in a precise array on the surfaces of membranes inside cells (see Chapter 11); such an orderly process surely represents a relatively advanced state of evolution.

Most of the oxygen in the air today has come from photosynthesis. Oxygen destroys organic compounds, and when a large quantity of oxygen appeared in the atmosphere soon after photosynthesis evolved, it posed a threat to the continuation of life. The destructiveness of oxygen was probably the selective pressure that brought about the next important evolutionary advance, aerobic respiration.

Like other metabolic pathways, aerobic respiration undoubtedly evolved in steps. By turning again to studies of living bacteria, we can trace its evolution. Aerobic respiration probably began with cells that combined their wastes, such as the pyruvate resulting from glycolysis, with molecular oxygen, neatly disposing of two toxic substances at once. (Bioluminescent organisms—for example, fireflies and certain bacteria—use just such a process to produce their flashes of light.) As time went on, organisms evolved ways of interpolating more and more steps into the process of respiration. This provided more precise control over the release of energy from the waste products of glycolysis and better recovery of the energy released. The modern tricarboxylic acid cycle and electron transport chain permit a high recovery of ATP in proportion to the amount of energy invested. The ability to make vast amounts of ATP per molecule of glucose

heterotroph (HET-er-oh-troaf: hetero = other, troph = feeder) an organism that must obtain its food from a source outside its body.

KEY-moe-SIN-the-sis

aerobic respiration: breakdown of food molecules to release energy, using oxygen to combine with the breakdown products.

respired (anaerobic glycolysis yields two ATPs per glucose molecule, whereas aerobic respiration yields an estimated 36 to 38) made it possible for organisms to grow and reproduce much faster, and to evolve new enzymes and structures. Although these more complex structures required a great deal of energy to operate, they gave aerobic organisms superior competitive abilities and allowed them to outstrip their anaerobic neighbors. Today anaerobic organisms grow only where there is not enough oxygen to support aerobic life.

13–H Organism and Environment

When we study evolution we are used to thinking that the environment sets certain boundaries and that organisms must function within these boundaries or perish. But organisms also change their environments, both immediately and over the billions of years of evolutionary time. We have already seen that heterotrophic organisms consumed the nutrients from the original primordial soup, and that autotrophs added molecular oxygen to the air and so set the stage for the evolution of aerobic respiration.

Oxygen also formed the ozone layer of the atmosphere: ozone (O_3) forms when ultraviolet light from the sun hits oxygen gas (O_2). This ozone layer acts as a filter, preventing much of the ultraviolet light from reaching the earth's suface. Ultraviolet light is very destructive to proteins and nucleic acids; ages ago, when it penetrated in full force through the atmosphere, life was confined to the water, which shields its inhabitants by absorbing the ultraviolet rays.

With the ozone layer established, organisms could start moving onto land. The first colonists were plants, which found abundant light and minerals and, at first, little competition. Terrestrial plants evolved and flourished, and they were soon joined by terrestrial animals, which ate them. The trees and grasses now clothing much of the land continue to add oxygen to the atmosphere; in addition, they change the patterns of water flow from the land to the seas and speed the formation of soil from rock. Thus organism and environment have molded one another during the history of life on our planet.

Fig. 13–6

Bioluminescence. This is a Portuguese man-of-war, a big, transparent, stinging jellyfish. Parts of its sail, above water, and of its tentacles, below, contain cells that produce light. (Biophoto Associates)

SUMMARY

Until the nineteenth century, people believed in "spontaneous generation": living organisms arose spontaneously, constantly being formed from nonliving substances when conditions were favorable. The work of Pasteur and Tyndall laid to rest the belief that organisms arise spontaneously on earth nowadays. However, most scientists now believe that life originally arose in this way.

The conditions under which life began, however, were very different from those on earth today. The atmosphere of the primitive earth was probably composed largely of hydrogen compounds, with virtually no oxygen gas. A hydrogen-rich atmosphere would have been conducive to the formation and stabilization of organic compounds, which gradually formed polymers, and then aggregates, and evolved systems of metabolism, information transfer, and reproduction, eventually becoming living organisms.

Some important events during the early history of life were the evolution of photosynthesis and aerobic respiration. Organisms have changed their environment from an earth of barren water and rock in a hydrogen-rich atmosphere to one of teeming oceans and verdant landscapes in an atmosphere containing a great deal of oxygen. Each environmental change caused by organisms exerted selective pressures to adapt to the new environment, which in turn changed the environment even more. Thus, living organisms and their environment have shaped each other during the evolution of life on earth.

SELF-QUIZ

1. Spontaneous generation is unlikely to occur on earth today because:
   a. there are very few active volcanoes
   b. any chemicals on the road to life would be devoured by organisms
   c. there is not enough ultraviolet light to furnish the necessary energy
   d. no one has the patience to discover the right formula

2. The present atmosphere differs from the early atmosphere of the earth in containing more of the element _____ and less of the element _____:
   a. hydrogen (See Table 9-1 for a list of elements)
   b. ammonia
   c. methane
   d. oxygen

3. Number the following in the order in which they are believed to have evolved:
   ___ aerobic respiration
   ___ proteinoids
   ___ photosynthesis
   ___ organic monomers
   ___ fermentation

4. Oparin believes coacervate droplets could be the precursors of life because:
   a. they can pass on identical copies of their genetic information
   b. they carry on respiration
   c. they can accumulate substances, catalyze chemical reactions internally, divide, and grow
   d. all of the above
   e. none of the above

5. The evolution of photosynthesis presented life with both an opportunity and a threat. Explain what these were.

6. One change brought about by living organisms was depletion of the nutrients formed by nonbiological means in the primordial "soup." How did this influence further evolution of life?

QUESTIONS FOR DISCUSSION

1. Are the laboratory simulations that were presented in this chapter (Miller's reducing atmosphere simulation, Oparin's coacervate droplets, and Fox's microspheres) experiments (see Section 1–A)? Are there any controls?

2. Suppose that a scientist claimed to have produced life from nonliving materials under laboratory conditions. What criteria must such an "organism" meet before you would agree that it was truly living?

3. Two very important evolutionary events that happened after the evolution of aerobic respiration were the origin of eukaryotes and, probably much later, the evolution of multicellular organisms. How do eukaryotes differ from prokaryotes? (See Chapter 10, including Essay.) How do multicellular organisms differ from unicellular ones? What is the advantage of each of these events, and how might each have come about?

SUGGESTED READINGS

Cells and their chemistry

Berns, M. W. *Cells.* New York: Holt, Rinehart, and Winston, 1977. Well-illustrated introduction to cells and their activities.

Calvin, M. *Organic Chemistry of Life;* Readings from *Scientific American.* San Francisco: W. H. Freeman, 1973. A collection of articles on biologically important organic compounds; readable and very well illustrated.

DeRobertis, E. D. P., F. A. Saez, and E. M. F. DeRobertis. *Cell Biology,* 7th ed. Philadelphia: W. B. Saunders, 1980. An excellent text describing recent work on the features of cells; frequent summaries of information and conclusions.

Kennedy, D., Ed. *The Living Cell.* San Francisco: W. H. Freeman, 1965.

———. *From Cell to Organism.* San Francisco: W. H. Freeman, 1967.

———. *Cellular and Organismal Biology.* San Francisco: W. H. Freeman, 1974. These three volumes edited by Kennedy are collections of articles on cell structure and function that originally appeared in *Scientific American;* with introductory essays.

Hill, J. W. *Chemistry for Changing Times,* 2d ed. Minneapolis: Burgess, 1975. An excellent and entertaining "chemistry for poets."

Respiration and photosynthesis

Hinkle, P. C., and R. E. McCarty. "How cells make ATP." *Scientific American.* March 1978. Presents the hypothesis of the chemiosmotic mechanism for ATP synthesis and compares how it is thought to work in bacteria, mitochondria, and chloroplasts.

Krebs, H. A. "The history of the tricarboxylic acid cycle." *Perspectives in Biology and Medicine,* **14:**154, 1970. An engaging account of how the TCA cycle was orked out by Sir Hans Krebs, who won a Nobel Prize for this work.

Stryer, L. *Biochemistry.* San Francisco: W. H. Freeman, 1975. A standard biochemistry textbook with excellent descriptions of cellular respiration and photosynthesis.

Origin of life

Calvin, M. *Chemical Evolution.* New York: Oxford University Press, 1969. A discussion of the chemical evolution that preceded life on earth and the possibility that such evolution is occurring on other planets.

Farley, J. *The Spontaneous Generation Controversy from Descartes to Oparin.* Baltimore: Johns Hopkins University Press, 1977. A historical account of philosophical and scientific controversy over the origin of life; full of fascinating tidbits of information and interesting ideas about the development of scientific thought.

Keosian. J. *The Origin of Life,* 2nd ed. New York: Reinhold, 1968. Presents many arguments that have been advanced against the chemical origin of life as viewed by the main body of scientists. Summarizes experimental data and discusses chemical reaction mechanisms in non-biological synthesis of organic compounds.

Ponnamperuma, C. *The Origins of Life.* New York: E. P. Dutton, 1972. Excellent illustrations and photographs.

Wald, G. "The origin of life." *Scientific American,* August 1954. Interesting discussion of the role of time in probability.

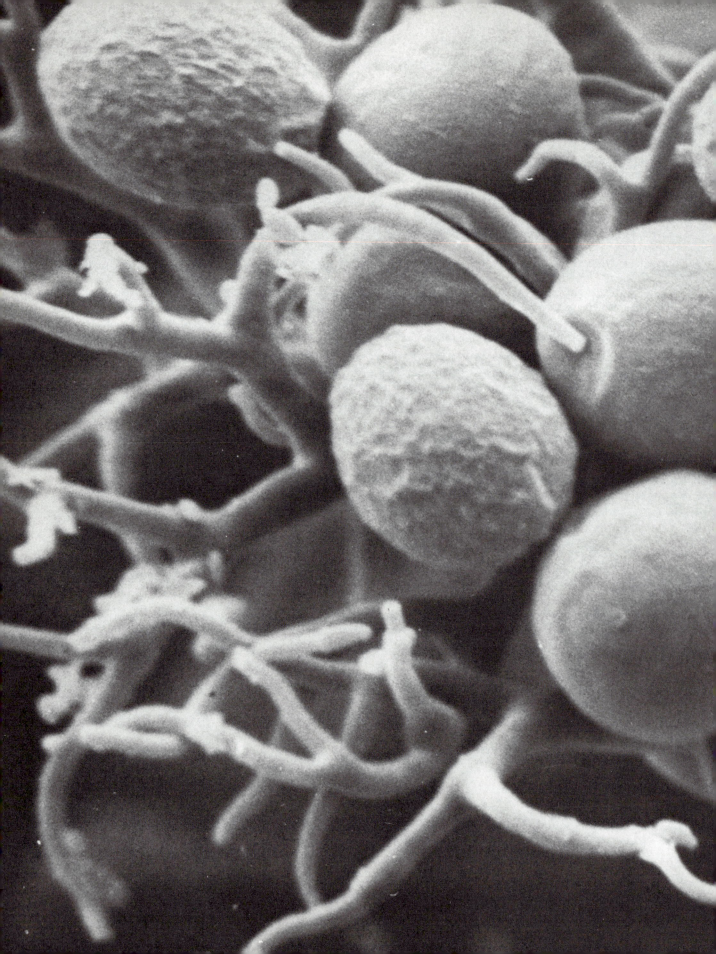

Part Three

BIOLOGICAL INFORMATION AND ITS EXPRESSION

14 DNA

OBJECTIVES   When you have studied this chapter, you should be able to:

1. Describe and explain the evidence that DNA is the genetic material, using these studies as evidence: (a) bacterial transformation; (b) infection of bacteria by bacteriophages; (c) the quantity of DNA found in tissues of different species and in body cells and reproductive cells of one species; (d) comparison of the base composition of DNA in cells from members of the same and different species.

2. Describe the structure of a nucleotide.

3. Describe the structure of a molecule of DNA, and explain why the number of adenine bases in the molecule equals the number of thymine bases and the number of guanine bases equals the number of cytosine bases.

4. Describe the experiments of Meselson and Stahl and explain how they provided evidence that DNA replicates in a semiconservative fashion.

5. Describe the structure of a eukaryotic chromosome.

6. List three characteristics by which cancerous cells differ from normal cells.

In Chapters 14 to 17 we address the central mystery of life. What organizes a tiny mass of material into a functioning cell capable of regulating its internal chemical composition, growing, and reproducing? What directed the single fertilized egg from which each of us originated to divide, and the resulting mass of cells to move around, grow, absorb nourishment, and take shape as a unique individual? What makes each of us distinct from other individuals but gives all of us a basic similarity as members of the human species? And what allows discerning relatives hovering over a new arrival to proclaim that it has its father's nose and its mother's crooked smile? The answer to all these questions is **genetic information**. The various different units that make up the total of an organism's genetic information (units governing characteristics such as hair color, nose size, and blood type) are called **genes**.

Each of us is a unique individual because each of us has a unique combination of genes. Genes contain information dictating what proteins our cells make, in what proportions, and how their production and interaction can be influenced by the environment in which we develop and live. Identical twins, who share identical genetic information, differ from each other only to the extent that they are exposed to different environments.

Fig. 14–1

The characteristics of living things are determined by the genes (DNA) they contain, and by the environments to which they are exposed. These seedlings all contain similar genes but they look different because each pot was grown in a different environment (different light). (Biophoto Associates)

Fig. 14–2

Sperm and eggs contain only half the number of chromosomes found in the rest of the cells in the parents' bodies; at fertilization, sperm and egg fuse, forming a zygote with the characteristic number of chromosomes found in the normal body cells of the species.

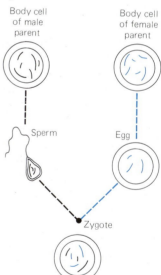

We received our genetic information from our parents: half from our mother's egg and half from our father's sperm. During the last century, biologists studied cells in the process of dividing and forming eggs and sperm. Before the cells divided, the chromosomes separated from each other in such a way that half of the chromosomes from the original cell ended up in each egg or sperm. This pattern convinced biologists that chromosomes were the bearers of genetic information. One problem remained, however; chemists found that chromosomes were composed of two substances, DNA and proteins. Which carried the genetic information?

Obviously genes must contain a variety of information. Scientists had already discovered that proteins were a diverse and complex group of molecules, but no one knew much about DNA beyond the fact that it consists of four types of nucleotide monomers, or subunits, in contrast to the 20 different amino acid monomers commonly found in proteins. DNA was assumed to be built of monotonously repeating units of the four nucleotides. Because the structure of chromosomal proteins is more complicated than that of DNA, most scientists thought the proteins carried the genetic information, and it was not until the 1940s that this was shown to be wrong.

In this chapter we shall examine the evidence that accumulated to show that the genetic material is in fact DNA, and we shall see how scientists worked out the structure of DNA. This structure is one of the simplest and at the same time the most elegant of evolutionary achievements; it dictates the duplication of DNA so that it forms exact copies that are then passed on to future generations of cells.

14–A Evidence that DNA is the Genetic Material

The first evidence to cast doubt on the belief that genes are made of proteins came from studies of **bacterial transformation**, the transfer of genetic characteristics from one bacterial cell to another. In 1928 Fred Griffith studied transfor-

LIVING NONVIRULENT CELLS.

LIVING VIRULENT CELLS

HEAT-KILLED VIRULENT CELLS

LIVING NONVIRULENT CELLS + HEAT-KILLED VIRULENT CELLS

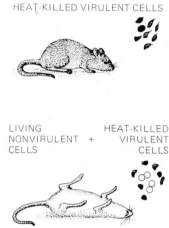

Fig. 14–3

Griffith's experiment on bacterial transformation. Virulent bacteria cause fatal disease in mice; nonvirulent bacteria or heat-killed virulent cells do not cause disease. Mice injected with both living nonvirulent cells and heat-killed virulent cells develop disease. Something from the heat-killed cells transforms the nonvirulent cells into virulent cells.

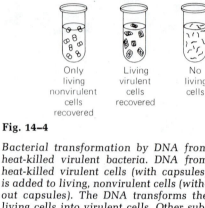

Living nonvirulent cells

DNA from heat-killed virulent cells

Only living nonvirulent cells recovered

Living virulent cells recovered

No living cells

Fig. 14–4

Bacterial transformation by DNA from heat-killed virulent bacteria. DNA from heat-killed virulent cells (with capsules) is added to living, nonvirulent cells (without capsules). The DNA transforms the living cells into virulent cells. Other substances from heat-killed virulent cells have no effect.

VEER-you-lent

mation between two strains of a pneumonia bacterium. One strain was **virulent**, that is, it killed mice into which it had been injected; the other strain was **nonvirulent** and did not cause death. Griffith showed that if dead, virulent bacteria were injected into mice along with living nonvirulent bacteria, some of the mice died. Furthermore, although these mice had received no living virulent bacteria in the injection, Griffith found living virulent cells in their corpses. He concluded that some of the genetic material from the virulent bacteria had entered the nonvirulent bacteria, transforming them to the virulent form (Fig. 14-3).

In 1944 other researchers set out to discover the chemical nature of the substance that caused this alteration. They grew virulent bacteria in the laboratory and broke them down into their chemical components: carbohydrate, lipid, protein, and DNA. They then added each of these substances to a separate culture of nonvirulent bacteria and, behold! Some cells in the culture that had received the DNA from virulent bacteria themselves became virulent, whereas the cells in cultures that had received the protein, carbohydrate, or lipid remained nonvirulent. Treatment with DNA had endowed the living bacterial cells with genetic characteristics they had not previously possessed (Fig. 14-4).

As with most revolutionary theories, the conclusion that DNA was the genetic material was not immediately accepted. Some questioned the purity of the DNA used, thinking it was contaminated with protein. More importantly, however, at that time biologists did not regard bacteria as "real" organisms. They tended to shrug off the evidence that DNA was the genetic material of bacteria as being irrelevant to an understanding of the genetic material of higher organisms.

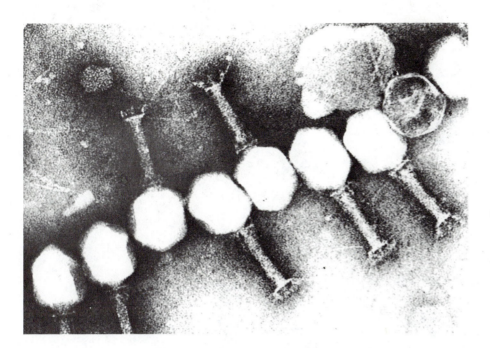

Fig. 14–5

Electron micrograph of bacteriophage T4, a virus that infects the bacterium Escherichia coli. The polygon-shaped head, which appears white in this photo, is made up of protein and contains the DNA. DNA passes out of the phage through the long, thin tail after the slender tail fibers have attached to the cell wall of an E. coli cell. (Carolina Biological Supply Company)

Another bit of evidence that genetic material is made of DNA was obtained from studies of the behavior of **bacteriophages**, known to their intimates as **phages**. These are viruses that attack bacteria. A virus consists of a molecule of DNA (or sometimes RNA) inside a protein coat. A phage attaches to the cell wall of a bacterium and injects its genetic material into the cell. The genetic material somehow "takes over" the cellular machinery of the bacterium, diverting its resources into the production of more phages instead of more bacteria.

In 1952, Alfred Hershey and Martha Chase showed that DNA is the genetic material of phages. They used radioactive forms of sulfur and phosphorus to distinguish between the protein coat and the DNA of phages. Recall (Sections 9–K and 9–L) that proteins contain sulfur but not phosphorus, and that nucleic acids contain phosphorus but not sulfur. Hershey and Chase grew phages in bacteria in a medium containing radioactive forms of these elements, so that the phage protein coats contained radioactive sulfur and the DNA contained radioactive phosphorus. They then added some of these phages to a fresh colony of bacteria and found that the radioactive phosphorus always entered the bacterial cells, whereas the sulfur remained outside (Fig. 14-7). This was another piece of evidence that genetic material consists of DNA.

This experiment also shows the advantage of using microorganisms in such situations. The DNA of viruses and bacteria is protein-free, whereas the chromosomes of eukaryotic organisms contain considerable protein. It would not have been possible to use eukaryotic chromosomes in these experiments because it is difficult to separate the DNA from its tightly bound protein without damaging the DNA itself.

back-TEER-ee-oh-fahj (the "s" is silent because the word is French)

eukaryotic: having a membrane-bound nucleus containing chromosomes (chromosomes = DNA + proteins).

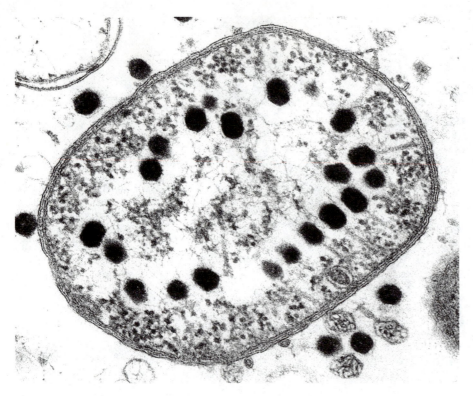

Fig. 14–6

Electron micrograph photo of a cell of the bacterium Escherichia coli infected with a virus, bacteriophage T2. The black hexagons inside the cell are new phages produced after the phage genetic material was injected into the bacterial cell; note the empty phage coats attached to the cell wall. The thin fibrous material in the cell is DNA. (From L. D. Simon, Virology 38:287, 1969).

PHAGE INJECTS DNA INTO BACTERIUM; PROTEIN COAT REMAINS OUTSIDE

Phage protein coat — Phage DNA

Bacterium

BACTERIUM PRODUCES MORE PHAGE

BACTERIUM RUPTURES, RELEASING PHAGE

Fig. 14–7

Infection of a bacterium by a bacteriophage. The virus injects its DNA into the bacterial cell, but its protein coat remains outside. Many new phages are produced inside the bacterial cell, indicating that DNA, not protein, is the hereditary material of the phage.

Circumstantial evidence that DNA is the genetic material in eukaryotes comes from the finding that almost all types of cells in an organism have the same amount of DNA, and that the amount of DNA in the eggs and sperm is only half that in the other cells. For example, Table 14-1 shows that chicken liver and kidney cells contain the same amounts of DNA, and chicken sperm contain only half this amount—this is precisely what we would expect of the genetic material. In contrast, the protein content of cells in different parts of the body varies widely and is not necessarily lower in reproductive cells. In addition, the amount of DNA in different types of organisms differs, suggesting that various species have different amounts of genetic material; again, this is what we would expect from the differences among organisms in the kinds of proteins produced and in the complexity of their body plans.

Even more convincing were the findings, in the late 1940s and early 1950s, of Erwin Chargaff and his coworkers, who discovered that the DNA of members of each species has the same chemical composition, which is different from the chemical composition of the DNA of any other species. Although all DNA

TABLE 14-1 APPROXIMATE DNA CONTENT OF SOME CELLS AND OF VIRUS PARTICLES

	DNA in pg per cell or per virus particle*
Virus particles	2.5×10^{-6}–240×10^{-6}
Cells of:	
Bacteria	6.5×10^{-4}–600×10^{-4}
Flowering plants	0.39–9.0
Vertebrates	0.39–8.0
Chicken liver (body cells)	2.5
Chicken kidney (body cells)	2.5
Chicken sperm (reproductive cells)	1.25

*Where a range of values is given, the higher and lower values are for different members of the group. pg = picograms; 1 picogram = 10^{-12} grams.

TABLE 14-2 APPROXIMATE AMOUNTS AND RATIOS OF NITROGENOUS BASES IN DIFFERENT ORGANISMS

	Percentage of nucleotides* containing each base				Base ratios†	
	A	G	C	T	A/T	G/C
Animals						
Human	30.9	19.9	19.8	29.4	1.05	1.00
Chicken	28.8	20.5	21.5	29.2	1.02	0.95
Locust	29.3	20.5	20.7	29.3	1.00	1.00
Plants						
Wheat	27.3	22.7	22.8	27.1	1.01	1.00
Fungus						
Yeast	31.3	18.7	17.1	32.9	0.95	1.09
Bacteria						
Escherichia coli	24.7	26.0	25.7	23.6	1.04	1.01
Viruses						
Phage T₇	26.0	24.0	24.0	26.0	1.00	1.00

*A = Adenine; G = Guanine; C = Cytosine; T = Thymine
†Note that each ratio is very close to 1.00

consists of the same four nucleotides (those containing the nitrogenous bases adenine, thymine, guanine, and cytosine), the nucleotides are not present in equal proportions, as was at first supposed. Instead, Chargaff and his colleagues found that the nucleotides occurred in constant percentages for each member of each species, whereas the proportions differed for members of different species. Furthermore, the number of adenine nucleotides was equal to the number of thymine nucleotides, and the number containing guanine equaled the number containing cytosine (Table 14-2). This finding eventually became a major clue in the elucidation of the structure of the DNA molecule.

AD-den-een
THIGH-mean
GWAUN-een
SITE-oh-seen

14–B The Structure of DNA

By the late 1940s most people believed that DNA indeed carries genetic information, and there was intense interest in working out the structure of the

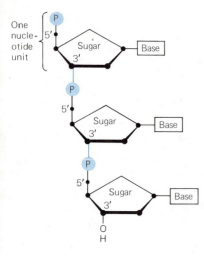

One nucleotide unit

Fig. 14–8

DNA is made up of nucleotides. The phosphate group (P) of one nucleotide is linked to the sugar of the next, forming a sugar-phosphate backbone, with the nitrogen-containing bases sticking out at the sides.

DNA molecule. Any model of DNA structure had to take several experimental findings into account:

1. DNA is made up of nucleotides. To review Section 9–L, a nucleotide is made up of a phosphate group, covalently bonded to a five-carbon sugar (deoxyribose in the case of the nucleotides that make up DNA), which in turn is covalently bonded to a nitrogen-containing base with either a single ring (thymine and cytosine) or a double ring (adenine and guanine) (Fig. 14-8).

2. Nucleotides are linked together in such a way that the phosphate group of one nucleotide attaches to the sugar of the preceding one, by covalent bonds, forming a long string of alternating sugar and phosphate groups, known as the sugar-phosphate backbone. The bases jut out at right angles to one side of the backbone (Fig. 14-8).

3. The number of nucleotides containing adenine (A) equals the number containing thymine (T), and the numbers containing guanine (G) and cytosine (C) are also equal in any DNA molecule or, in the shorthand popular with biologists, A = T, G = C.

4. The most direct evidence for the structure of DNA came from studies of the pattern of diffraction (scattering) of x-rays passed through crystals of purified DNA. In 1952, Rosalind Franklin obtained x-ray photographs indicating that DNA was arranged in a spiral, or **helix**, with the bases perpendicular to the fiber. The photographs also indicated that the sugar-phosphate backbone was on the outside of the helix, with the bases jutting into the center, and that one turn of the helix contained 10 nucleotides.

At the time Franklin made these pictures, there was a frantic race going on to solve the mystery of the structure of DNA. Linus Pauling, an American chemist, had already published a model of the structure that turned out to have an embarrassingly elementary flaw, and he was bent on retrieving his reputation. Maurice Wilkins, head of the laboratory where Franklin worked in London, was also keenly on the scent. The first people to fit all the evidence together in a workable form were James Watson, a postdoctoral researcher, and Francis Crick, an erstwhile physicist; they used a set of scale models of nucleotides to build up a structure that fitted the data collected by many others.

The structure worked out by Watson and Crick consists of two strands of DNA; the strands are arranged like a ladder, with the sides being the sugar-phosphate backbones of the two strands and the rungs being the bases. Each rung consists of a single-ring base attached to one DNA strand and a double-ring base on the other strand. A rung may consist of either an adenine opposite a thymine, or a guanine opposite a cytosine; in each rung, either base may be on either strand. The pair of bases in each rung is held together by hydrogen bonds. Two hydrogen bonds hold an adenine-thymine pair together and three bonds hold a guanine-cytosine pair together (Fig. 14-9). A and T and G and C form the most

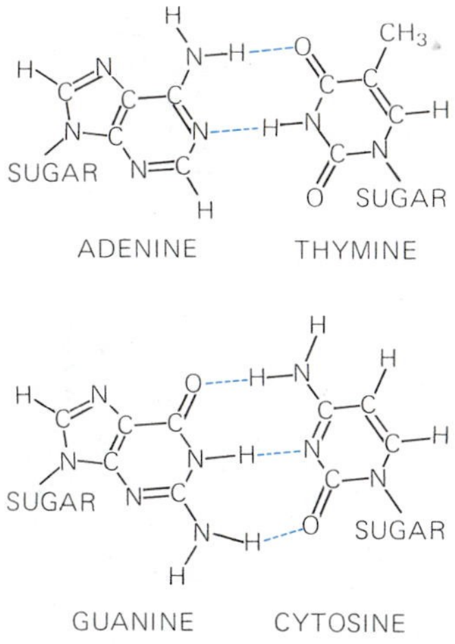

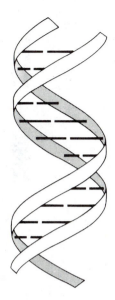

Fig. 14–9

Hydrogen bonding (color) between base pairs found in DNA. No other combination of these four bases forms such stable sets of hydrogen bonds.

ADENINE   THYMINE

GUANINE   CYTOSINE

Fig. 14–10

The double helix. The two strands are coiled around one another. Each gray strand consists of alternating sugar and phosphate groups. Color indicates the hydrogen bonds between the bases (black) that hold the two strands together.

stable combinations of hydrogen bonds, and this explains Chargaff's finding that A = T and G = C in the DNA of any species. Since each pair consists of one single-ring and one double-ring base, all the rungs of the ladder are the same width, and the backbones of the two DNA strands are always the same distance from one another. Finally, the whole ladder is twisted to form the spiral (helix) detected by the x-ray photographs; since the spiral is composed of two strands wound around each other, the DNA molecule is referred to as a double helix (Fig. 14-10).

Tremendously excited by the simple yet elegant structure they had worked out, Watson and Crick hurriedly wrote up a two-page paper for quick publication. As Watson remarked, "It was too pretty not to be true."

14–C DNA Replication

Before a cell divides, its DNA **replicates,** or duplicates itself, so that each daughter cell receives exactly the same genetic information that the parent cell had. Watson and Crick pointed out that the structure of the DNA molecule provides an easy way for the molecule to replicate. Because the two strands have complementary base pairs, the nucleotide sequence of each strand automatically supplies the information needed to produce its partner. (For example, if one strand runs A-A-T-G-C-C, then its partner must run T-T-A-C-G-G.) The publication of Watson and Crick's model of DNA structure touched off a flurry of experiments on replication.

In 1958, Matthew Meselson and Franklin Stahl published the first convincing evidence of how DNA replicates. They devised three alternate hypotheses (Fig. 14-11), any one of which would account for the observations of DNA replication at the time:

REP-lick-ates

hypothesis (pl.: hypotheses): a proposed explanation for an observed phenomenon.

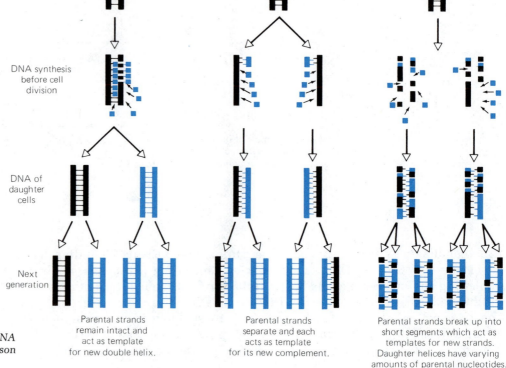

CONSERVATIVE   SEMI-CONSERVATIVE   DISPERSIVE

Parent cell's DNA

DNA synthesis before cell division

DNA of daughter cells

Next generation

Parental strands remain intact and act as template for new double helix.

Parental strands separate and each acts as template for its new complement.

Parental strands break up into short segments which act as templates for new strands. Daughter helices have varying amounts of parental nucleotides.

Fig. 14–11

Alternative hypotheses of DNA replication tested by Meselson and Stahl.

template (TEM-plate): a pattern used to guide the making of an object.

HIGH-brid

1. *Conservative replication.* In this hypothesis, the two strands of the parent DNA molecule act as templates for a completely new double-stranded molecule. The parent DNA molecule is preserved intact and goes into one daughter cell, and the new molecule goes into the other daughter cell.

2. *Semi-conservative replication.* According to this hypothesis, the two strands of the parental molecule separate, like the two sides of a zipper, and each acts as the template for formation of one new strand, which then becomes bound to it to form a complete molecule. Each daughter cell inherits a DNA molecule that is a **hybrid**, consisting of one new and one parental strand.

3. *Dispersive replication.* The parental DNA is broken up into short segments used as templates for the formation of two new double helixes, which are then somehow joined together.

In a beautifully simple series of experiments, Meselson and Stahl showed that the first and third of these hypotheses could be disproved, and provided strong support for the theory of semi-conservative replication. (Note that a hypothesis can never be proved, because we can never be certain we have taken all possible factors into account; however, it is possible to disprove a hypothesis if the data do not agree with predictions of what should happen if the hypothesis is true.)

To distinguish between "old" and "new" DNA in these experiments, Meselson and Stahl used nitrogen atoms of two different weights—^{14}N, the normal-weight atom, and ^{15}N, a heavier type. First, they grew bacteria in a nutrient medium containing ^{15}N for several generations, until the DNA of virtually all the bacteria contained ^{15}N. Next, they transferred the bacteria to a nutrient medium containing ^{14}N. They then removed cells after one, two, and three generations had passed. The DNA from the cells in each generation was purified; it was then ready for analysis of the distribution of ^{15}N and ^{14}N.

To do this, Meselson and Stahl placed each batch of isolated DNA on the surface of a salt solution in its own tube and spun all the tubes in a **centrifuge**, a machine that works on the same principle as the spin cycle of a clothes washer (Fig. 14-12). As the tubes spun, the heaviest molecules migrated downward through the salt solution the fastest; thus, the DNA containing ^{15}N migrated toward the bottom faster than ^{14}N DNA. After spinning the tubes, the investigators could actually see bands of DNA of different weights by holding the tubes up to the light. Figure 14-13 shows the patterns of bands that would be expected with each hypothesis of DNA replication.

SENT-riff-youj

Meselson and Stahl found that in the parental generation, grown in ^{15}N, all the DNA contained only ^{15}N. However, when these bacteria were grown in ^{14}N and allowed to divide to produce one new generation, the DNA of their daughter cells contained both ^{14}N and ^{15}N, and settled in a band position between that of pure ^{14}N DNA and pure ^{15}N DNA. This finding ruled out the hypothesis that replication is conservative, for in this case, two separate bands, one of ^{14}N DNA (newly synthesized) and one of ^{15}N DNA (parental) would be expected (see Fig. 14-13). Thus the first of the three hypotheses was incompatible with the experimental evidence.

Fig. 14–12

Separating DNA of different weights by centrifugation. Centrifugal force generated by spinning the tubes makes heavier DNA migrate toward the bottoms of the tubes faster. (Centrifugal force is what you feel throwing you toward the outside of the curve when your car goes around a corner too fast.)

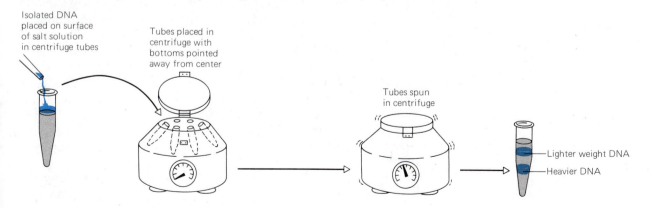

Isolated DNA placed on surface of salt solution in centrifuge tubes

Tubes placed in centrifuge with bottoms pointed away from center

Tubes spun in centrifuge

Lighter weight DNA

Heavier DNA

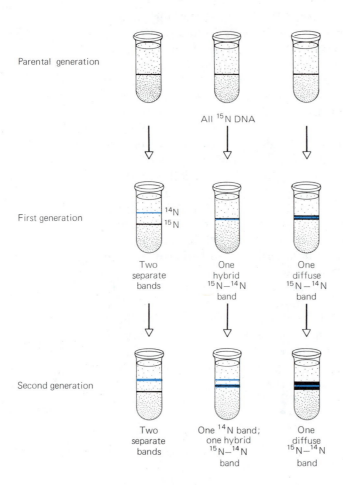

Fig. 14–13

Patterns of bands of DNA expected according to each hypothesis of DNA replication described in Figure 14–11. Meselson and Stahl found the pattern expected for semiconservative replication when they isolated DNA from successive generations of Escherichia coli.

The evidence from the third generation of bacteria permitted Meselson and Stahl to distinguish between the two remaining hypotheses. When the bacteria originally grown in ^{15}N had been grown for two generations in medium containing only ^{14}N, the DNA isolated from them formed two distinct bands. About half of the DNA was hybrid ^{14}N–^{15}N DNA, as in the first generation, and the other half contained only ^{14}N DNA. This result is incompatible with dispersive replication, which, as Figure 14-13 shows, would produce one diffuse band lying between the two actually found. However, this pattern *was* compatible with semiconservative replication. Thus Meselson and Stahl concluded that, of their three hypotheses, only semi-conservative replication was supported by the experimental evidence. Further study has shown that semi-conservative replication also occurs in eukaryotic DNA.

DNA replication begins with the separation of the two strands along the weak hydrogen bonds linking the paired bases. Nucleotides then line up opposite their pairing partners on each strand, held in place by hydrogen bonds, and an enzyme moves along the chain, hitching them up to form the sugar-phosphate backbone of a new strand.

Prokaryotic DNA: circular

14–D DNA in Cells

The DNA of a prokaryotic cell is in the form of one double helix with its ends joined to form a circle (Fig. 14-14). Eukaryotic cells (found in plants, animals, fungi, and many one-celled organisms) contain chromosomes, each composed of DNA with proteins closely attached. Biologists are still not sure whether a chromosome has one or more molecules of DNA. One experimenter has reported that the longest chromosome of the fruit fly *Drosophila* almost certainly contains a single DNA molecule stretching from one end of the chromosome to the other. This molecule is 4.0 cm long (about 1.5 inches), and since the fly itself is less than one-tenth as long, it is obvious that the DNA molecules in all its cells must be considerably coiled up. Much of the DNA is coiled around lumps of protein so that the chromosome looks like a string of beads.

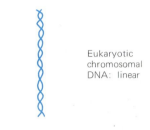

Eukaryotic chromosomal DNA: linear

Fig. 14–14

DNA is circular in prokaryotes, linear in eukaryotes.

14–E Mutation

Mutations are inheritable changes in DNA molecules. Mutation in a body cell may cause changes in the hereditary characteristics of that cell and of body parts made up of that cell's descendants; mutations in cells destined to form eggs or sperm can pass on changed hereditary traits to an organism's offspring.

mew-TAY-shuns

Mutations are brought about by **mutagenic agents**, often called **mutagens**. X-rays are mutagens that may cause breaks in the DNA molecule, possibly resulting in loss of part of a chromosome. Radioactive decay particles have similar effects; some of these are always present as background radiation that reaches us from the earth and outer space but higher levels are released by atomic weapons and by leakage from nuclear power plants following some kinds of accidents.

mew-tuh-JENN-ick
MEW-tuh-jenz

Ultraviolet radiation is also mutagenic. Ultraviolet light is used in laboratories to kill bacteria and fungus particles in the air or on surfaces where dust settles. Fortunately for us, ultraviolet rays, which are plentiful in sunlight, do not penetrate very far into living tissue, and so do not pose a threat to our offspring, because they cannot reach the human reproductive organs. However, people who spend long periods of time with their skin exposed to the sun, especially in the tropics, run the risk of developing skin cancer. Light-skinned people are especially susceptible, for they do not have much of the skin-darkening pigment melanin, which screens out ultraviolet light and thus protects underlying DNA from damage.

Certain chemicals also alter DNA. Some are similar to nucleotides and may be accidentally incorporated into DNA, where they may cause mistakes in base-pairing during future replications of the DNA. Others act by altering nucleotides in the strand; they change one of the chemical groups involved in the hydrogen bonds of base-pairing, and thus change the base-pairing properties of the nucleotide. In either case, when the strand replicates, the wrong nucleotide base-pairs to the altered member of the parental strand, again resulting in replication mistakes transmitted to all future descendants of the DNA molecule.

A small amount of mutation is perhaps beneficial, because it does change the genetic material, and through such changes improvements can occur. However, most mutations prove to be harmful to the organism. Typical mutation frequencies range from 1 in 10^6 to 25 in 10^5 for various genes in human eggs and sperm.

14–F Recombinant DNA and Genetic Engineering

The sequence of nucleotides in DNA specifies the sequence of amino acids in a corresponding protein. We now know enough about protein synthesis that scientists working in the laboratory can make segments of DNA that code for particular proteins. Our knowledge makes it possible to contemplate a new technology, **genetic engineering**, the introduction of new genetic information into cells to suit our desires. Already Har Bind Khorana has made an artificial gene and successfully introduced it into a laboratory culture of bacteria. Such work may some day lead to production of useful proteins on a commercial scale more cheaply than they can be extracted from the bodies of larger organisms, as is now done. Some scientists even envision a time when we will know enough about eukaryotic chromosomes to introduce missing genes into human eggs or early embryos, thus alleviating much suffering and dispensing with the need for lifelong treatment of genetic deficiencies. (Obviously this could be a very dangerous technology if used for other ends, and its development is strongly opposed by many people.)

One step in genetic technology that already exists is the formation of **recombinant DNA**, a mixture of DNA from two different organisms. This is produced by splicing the gene to be transplanted onto a small piece of DNA called a PLAZZ-mid **plasmid**, which readily invades a bacterium, where it is treated as part of the bacterium's DNA. By this technique, genes from any organism can be transplanted into a bacterium, which then multiplies, producing many copies of the gene. This technique can be used to produce the protein for which the gene codes or to study how the gene works.

On the other side of the coin, many people worry about what will happen if there is an accident. Suppose a strain of bacteria with a gene for a dangerous toxin were set loose into the world? Many workers feel that the chance of this happening is slight. Although the bacteria used in these experiments are *Escherichia coli,* a species universally found in the human intestine, the actual strains used in the laboratory have been out of contact with real bodies for many thousands of generations, and have evolved in such a way that they now have difficulty surviving outside their test-tube homes. The danger is further reduced by government regulation of laboratories doing recombinant DNA research.

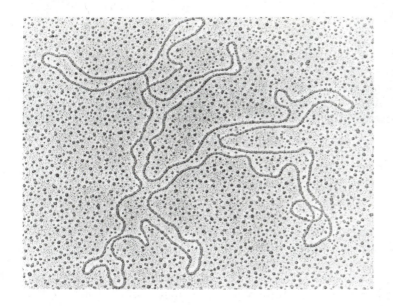

Fig. 14–15

A circle of DNA from a bacterium. (Biophoto Associates)

14–G Viruses

Viruses are particles with such simple structure that they are usually regarded as not being alive. Each consists of a single molecule of genetic material (DNA or RNA) surrounded by a mainly protein coat. Viruses cause many human diseases, and we usually think of them as nothing but a nuisance. Their reputation has been restored somewhat in the twentieth century, however, because without viruses, our understanding of the structure and replication of DNA would be negligible.

An important peculiarity of viruses is that they can reproduce only inside living cells. (This raises the interesting point that viruses probably could not have evolved until after true cells had evolved.) The viral coat, acting like a hypodermic syringe (Fig. 14-7) injects the virus's genetic material into a suitable cell. Here, the viral DNA or RNA takes over the cell's genetic machinery. The cell now produces viral RNA, which dictates that the cell make new viral coat proteins and viral genetic material as well as enzymes that destroy the cell, releasing possibly hundreds of virus particles. This happens in such viral diseases as the common cold, infectious hepatitis, polio, smallpox, influenza, and rabies. Symptoms of these diseases result when viral proteins damage the body.

The body has cellular defenses against viral damage in the form of the protein **interferon**, which kills viruses (apparently all kinds of viruses). Interferon would be a very useful medicine to treat viral diseases. During the 1970s, all interferon for medical use was laboriously extracted from human cells and so was scarce and very expensive. In 1980, drug company researchers succeeded in introducing a human interferon gene into a bacterium (genetic engineering at its most useful) and it now seems certain that interferon produced by bacteria will be available, much more cheaply, sometime in the 1980s.

in-ter-FEAR-on

14–H Cancer

Tumors are clumps of cells that grow and multiply abnormally. Some tumors are harmless, like the common wart, or fibroid cysts of the uterus; on the other hand, a tumor may become **malignant**, growing uncontrollably, destroying nearby tissues and eventually causing death. A "cancer" is a malignant tumor.

Cancerous cells differ from normal cells in three main ways: they divide more rapidly; they do not stick to each other as firmly as normal cells do; and they dedifferentiate. **Differentiation** is the process by which cells come to differ from each other as the pattern of development dictated by an individual's DNA unfolds; dedifferentiation is the loss of differentiation. In a familiar case, when you cut yourself, cells near the injury invade the wound and divide and differentiate into new skin cells, which heal the wound. Cells dedifferentiate when they become cancerous. When cells covered with cilia on the surface of the lung become cancerous, they dedifferentiate into formless cells which divide as rapidly as embryonic cells. Such cells often lose their normal ability to stick tightly to neighboring lung cells, and then they may come unstuck and **metastasize**, or travel to other parts of the body, where they may start new tumors. The main reason that we understand so little about cancer is that we understand little about how cell differentiation and division are normally controlled.

met-ASS-tuh-size

Nobody knows if there is a common cause for all cancers. Genetic changes are certainly common in cells that become cancerous, but we do not know whether or not they always occur. Many **carcinogens** (cancer-causing agents), such as chemicals and radiation, are also mutagens. On the other hand some carcinogens are not mutagens, and some mutagens are not carcinogens; this undermines the theory that mutations are always the cause of cancer.

Some forms of cancer are produced by viruses, and there are people who think that all cancers will turn out to be viral. This theory is extremely difficult to study. To produce convincing evidence, one would have to extract a virus from a cancerous cell and show that it would produce cancer in a normal cell. This is often almost impossible, first, because cancer viruses are hard to extract from the cells that contain them, and second, because cancer usually takes many years to develop, at least in humans. If a virus is introduced into a cell and cancer develops years later, it is always possible that some event in the intervening years, and not the virus itself, caused the cancer.

The viruses that cause influenza and colds are replicated and cause disease as soon as they invade a cell. Other viruses, including some that are known to cause cancer, do not immediately destroy the cells they have entered; they just sit in the cells, often for long periods, doing no damage. The genetic material of a virus becomes part of the genetic material of the cell it has invaded (Fig. 14-16); it is duplicated with the cell's own DNA and passed on to the daughter cells at cell division just as if it were one of the cell's own genes. Latent viruses of this sort are very common. We undoubtedly have dozens of them in our bodies. For instance, most adults in northern Europe and the United States contain Epstein-Barr (EB) virus. Such a latent virus may live indefinitely and be passed from parent to child. A more familiar latent virus is the herpes virus which, when stimulated by unknown factors, leaves its latent state, is reproduced and causes

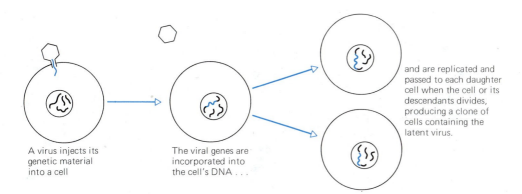

A virus injects its genetic material into a cell

The viral genes are incorporated into the cell's DNA . . .

and are replicated and passed to each daughter cell when the cell or its descendants divides, producing a clone of cells containing the latent virus.

Fig. 14–16

A latent virus in a cell.

the cell damage known as a "cold sore." (There is indirect evidence that herpes may sometimes cause cancer.)

Evidence from studies on rats and mice suggest that some carcinogens, such as asbestos, x-rays, tobacco smoke, and various bacterial and viral infections produce cancer by prodding latent viruses into action. Diseases caused by the combined action of more than one agent are already known. For instance, people must contain the EB virus before they can develop infectious mononucleosis, which is caused by another virus. There is some evidence that people who have had mononucleosis are more likely to develop Hodgkin's disease (a form of lymph cancer); so it may be that the mononucleosis virus occasionally activates the EB virus to produce Hodgkin's disease. The EB virus is invariably found in children with Burkitt's lymphoma, a lymph cancer common in parts of Africa. Burkitt's lymphoma is not common in the United States, however, which suggests that whatever factor besides the EB virus is necessary to produce the cancer is less common in the United States than in Africa.

Most human tumors are **clones**, groups of cells with identical genetic material, the progeny of one original cancerous cell. Cancer researchers think that our bodies produce malignant cells fairly frequently, but that these cells are usually destroyed by the immune system. Only occasionally does a clone grow too large for the immune system to destroy. This is why cancer therapy does not have to destroy every cancerous cell in the body to be successful. Cancer is treated by removing tumors surgically and by subjecting them to radiation and chemicals that kill them or slow their growth. If treatment can be started when there are still relatively few cancer cells, there is much more chance of reducing the number of cells to the point where the body's defenses can regain the upper hand. Unless this happens, there is usually no cure.

Cancer is the second most common cause of death in the United States and accounts for 20% of all deaths. Deaths from cancer have increased in the twentieth century. This is partly because cancer is largely a disease of later life, and people are living longer instead of dying of infectious bacterial diseases in early life. Some people, however, think that there has been an increase in cancer because carcinogens are becoming daily more common in our environment (Table 14-3).

TABLE 14-3 SOME SUBSTANCES KNOWN TO BE CARCINOGENIC

Substance	Comments
Asbestos dust Chromium compounds Some petroleum products	Workers in these industries have high risk of lung cancer
Tobacco smoke	Quit now!
Estrogen	A mammalian hormone which, in large amounts, is carcinogenic in some animals; almost certainly safe in contraceptive pills
X-rays	Many people have too many unnecessary medical x-rays
Benzene	Used to be a common solvent in chemistry labs

In recent years, treatments that seem to produce genuine cures have been found for some types of cancer, such as one form of childhood leukemia, and the death rates from other cancers have been reduced. This is partly due to new treatments and partly because people are more aware of the need to seek treatment as soon as they suspect that they may have cancer.

SUMMARY

The evidence that DNA is the genetic material in all organisms, from bacteria to oak trees, halibuts to humans, came from several lines of inquiry:

1. It is DNA that transfers hereditary characteristics from one cell to another during bacterial transformation.

2. A bacteriophage takes over the machinery of a bacterial cell by injecting only its DNA into the cell.

3. All the body cells of individuals of the same species contain the same amount of DNA; reproductive cells contain half this amount of DNA. Members of different species contain different amounts of DNA.

4. The DNA of all members of any species has the same ratio of nitrogenous bases.

The DNA molecule is built as a double helix, with two strands of a sugar-phosphate backbone forming the sides of a twisted ladder. The strands are connected by perpendicular "rungs" consisting of the base-pairs adenine and thymine or guanine and cytosine, with each base hydrogen-bonded to its complement on the opposite strand. A prokaryotic cell's genetic material consists of one circular, double-helical DNA molecule; eukaryotic nuclei contain linear chromosomes consisting of DNA and proteins.

DNA replicates semiconservatively; the double helix unwinds, with the disruption of the weak hydrogen bonds between base pairs, and each base attracts a nucleotide containing its complementary base. Enzymes link the nucleotides so lined up, forming a new complementary strand for each original strand of the helix.

Mutations are inheritable changes in the DNA. X-rays, ultraviolet radiation, and various chemicals are among the mutagens that may cause loss of parts of the DNA or changes in the sequence of nucleotides, which are passed on in future replications of the DNA.

Recombinant DNA experiments introduce foreign genes into bacterial cells, which then produce the protein coded by the gene. Scientists can now introduce new genes into existing organisms. The use of such techniques poses new moral problems for society.

Cancer is a collection of diseases in which the normal genetic mechanisms that control cell division do not function. Cancer cells also lose their normal ability to stick to other cells, and they lose their differentiated state. Latent viruses, which are duplicated when the host's cells duplicate, are probably necessary to the formation of many types of cancer. Development of such a cancer is often triggered, not by the virus itself, but by another infection or by an environmental carcinogen.

SELF-QUIZ ——

Choose the *one best answer* for these multiple choice questions.

1. One reason DNA is believed to be the genetic material is that:
   a. all the body cells of an individual seem to have identical amounts and compositions of DNA, while reproductive cells have half the amount of DNA found in body cells.
   b. the proteins are the same from cell to cell in an individual, but the DNA differs: thus the DNA must be the material that makes different tissues different.
   c. DNA is the largest type of macromolecule found in living organisms.
   d. DNA is found in the nucleus of eukaryotic organisms.

2. A nucleotide consists of:
   a. A, G, T, and C
   b. nitrogenous bases
   c. a sugar, a phosphate group, and a nitrogen-containing ring compound
   d. a sugar-phosphate backbone

3. In a DNA molecule:
   a. the nitrogenous bases bond covalently to the phosphate groups

   b. the sugars bond ionically to the nitrogenous bases
   c. the sugars bond to the nitrogenous bases by hydrogen bonds
   d. the nitrogenous bases bond to each other by hydrogen bonds
   e. the sugars bond to the phosphate groups by covalent bonds and to the nitrogenous bases by hydrogen bonds.

4. Indicate which bases are represented by 1, 2, and 3 in this diagram of part of a DNA molecule:

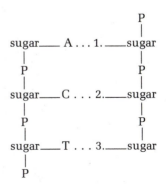

5. If replication were conservative, which of the following banding patterns would Meselson and Stahl have found in the *fourth* generation of cells? (Note: generation 0 = generation where all the DNA is labeled ^{15}N; concentration refers to concentration of DNA.)

    a. a high concentration band containing only ^{15}N DNA and a high concentration band containing only ^{14}N DNA.

    b. a low concentration hybrid band containing both ^{14}N and ^{15}N DNA and a high concentration band containing only ^{14}N DNA.

    c. a high concentration band containing only ^{14}N DNA.

    d. a high concentration band containing only ^{14}N DNA and a low concentration band containing only ^{15}N DNA.

QUESTIONS FOR DISCUSSION

1. Why is it necessary for eggs and sperm to contain only half the amount of genetic material found in the other cells of the body?

2. Why is the constancy of DNA content from cell to cell in an organism considered to be evidence that DNA is the genetic material? Is it necessary for cells of an organism to contain identical genetic information? Is it possible for an organism to have different genetic information in different cells of the body?

3. What is the biological importance of the fact that the sugar-phosphate backbones of the DNA double helix are held together by covalent bonds, and that the cross bridges between the two strands are held together by hydrogen bonds?

4. In the bacteria used by Meselson and Stahl, a new generation is produced every 20 minutes when the cells grow under ideal conditions. After placing the cells in the fresh ^{14}N nutrient medium, how long should Meselson and Stahl have waited before taking their samples of first-, second-, and third-generation cells?

5. In the first generation of Meselson and Stahl's experiment, the results showed a hybrid band of DNA containing both ^{14}N and ^{15}N. Which of the following is the best interpretation of these results?

    a. the results are consistent with semiconservative replication.

    b. the results support both semiconservative and dispersive replication.

    c. the results are not consistent with conservative replication.

    d. neither dispersive nor conservative replication can take place.

6. Why is it necessary to limit the amount of x-rays a person is exposed to over a given period of time? Which organs must be especially well shielded from x-ray exposure?

PROTEIN SYNTHESIS 15

OBJECTIVES

When you have studied this chapter, you should be able to:

1. Describe the genetic code and explain why it must be a triplet code.

2. Given a DNA coding strand and a table of codons, determine the complementary messenger RNA strand, the codons which would be involved in peptide formation from that messenger RNA sequence, and the amino acid sequence which would be translated.

3. Describe the role of DNA, messenger RNA, transfer RNA, ribosomes, and amino acids in protein synthesis.

4. List the steps involved in protein synthesis at the ribosome level.

5. Describe how transcription of DNA to RNA is controlled in a prokaryote.

I n Chapter 14, we saw that the genetic information of an organism is contained in its DNA. This information is a set of instructions specifying the order in which amino acids are to be joined to form a polypeptide or protein. But just how is the information borne in DNA translated into a functioning protein?

Since both DNA molecules and polypeptides are long and unbranched, it seemed reasonable to assume that the sequence of nucleotides in the DNA determines the sequence of amino acids in a protein. By the mid-1960s, biochemists had worked out how amino acids are fitted into a protein, according to the instructions carried in the DNA. They discovered that the other type of nucleic acid, RNA, acts as a go-between, carrying genetic information from DNA to the protein-synthesizing machinery of the cell (Fig. 15-1).

The next question was raised by the observation that cells produce different proteins, or different quantities of the same proteins, at different times in their lives. Why? And how is a cell's protein production geared to its current needs? In this chapter we will study these two questions, the central issues of genetics on the molecular level.

15–A RNA and DNA

In the 1940s, studies of the pancreas and other organs that synthesize a lot of protein made it clear that RNA, as well as DNA, is necessary for protein synthesis. In fact, because DNA remained in the nuclei of these eukaryotic cells, and

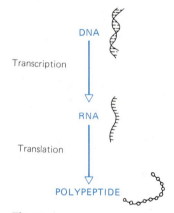

Fig. 15–1

Information is stored as the sequence of nucleotides in DNA, rewritten into the sequence of nucleotides in RNA and finally expressed as the sequence of amino acids in a polypeptide (protein).

pancreas (PAN-kree-uhs): an organ near the stomach that manufactures many digestive enzymes and protein hormones.

263

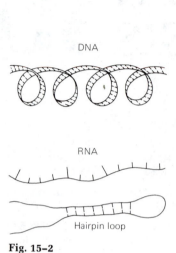

DNA

RNA

Hairpin loop

Fig. 15–2

*Most DNA is double-stranded
and wound in a helix. RNA is
always single-stranded although
it may bend to form hairpin
loops.*

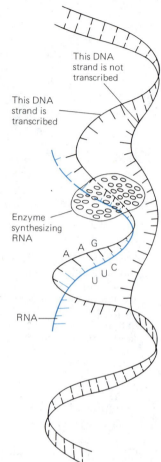

This DNA
strand is not
transcribed

This DNA
strand is
transcribed

Enzyme
synthesizing
RNA

A   A   G

U   U   C

RNA

Fig. 15–3

Transcription of RNA from DNA.

because RNA was found in the cytoplasm (where protein synthesis takes place),
it seemed probable that RNA played a direct role in protein manufacture.

Both DNA and RNA are made up of nucleotides, but there are several
differences between the two types of nucleic acids. First, the two contain slightly
different sugars: ribose in RNA, deoxyribose in DNA, reflected in the names
ribonucleic acid (RNA) and deoxyribonucleic acid (DNA). Second, RNA contains
nucleotides with the nitrogenous base uracil instead of the very similar thymine
found in corresponding DNA nucleotides; the other three bases—guanine,
adenine, and cytosine—are found in both DNA and RNA. A third difference is
that RNA is single-stranded, whereas DNA is double-stranded, although the
RNA molecule may double back on itself, forming hairpin loops that look like
double-stranded portions (Fig. 15-2).

RNA is synthesized on a DNA template by a process known as **transcription**
(= written across). Part of the DNA double helix uncoils and an enzyme moves
down one DNA strand, matching RNA nucleotides to their base-pair comple-

day-OX-ee-RYE-bose

YOUR-uh-sill

ments on the DNA strand and joining them to form a long strand of RNA (Fig. 15-3). The base-pairing rules for DNA hold true: guanine and cytosine pair, and thymine on the DNA is still paired to adenine in the RNA; but uracil in RNA is paired to adenine in DNA.

We shall discuss three kinds of RNA: messenger, transfer, and ribosomal. **Messenger RNA (mRNA)** carries the genetic information from DNA to the cell's protein-making machinery, the **ribosomes. Transfer RNA (tRNA)** carries amino acids to the ribosomes, where the amino acids are joined together to form a polypeptide chain. Each kind of amino acid is carried by its own kind of tRNA. **Ribosomal RNA (rRNA)** is a major component of ribosomes.

RYE-boh-soams

15–B The Genetic Code

Since it was produced by transcription of a gene, an mRNA molecule contains the same directions for polypeptide synthesis as does the gene. Conversion of an mRNA's genetic information into the structure of a polypeptide is known as **translation** (because RNA structure is translated into polypeptide structure). Before we can understand how the various types of RNA interact during translation, we must know how genetic information is coded in the DNA molecule that determines RNA structure.

The structure of an organism's proteins determines the biochemical reactions it can perform and, indeed, everything about the organism. Biochemists reasoned that the genetic information must, directly or indirectly, determine the sequence of amino acids in proteins and so dictate their structure. Since both DNA and polypeptides are linear, unbranched molecules, it seemed possible that the order of nucleotides in DNA designated the order of amino acids in polypeptides. (We speak of polypeptides, rather than proteins, in discussing protein synthesis because some proteins are made up of more than one polypeptide chain, each of which is synthesized separately. Hemoglobin, for instance, is a protein made up of four polypeptide chains which are assembled into the final protein after they are synthesized.)

But how is the amino acid sequence of a polypeptide coded in the structure of a DNA molecule? We can make a shrewd guess that, since there are 4 kinds of nucleotides in DNA, the genetic code must have a 4-letter "alphabet." Since there are 20 common amino acids, it is impossible that each letter of the "alphabet" stands for an amino acid; that would leave us with 16 amino acids that could never be placed into proteins. It is also impossible that the "words" of the genetic language are 2 letters long: 4 letters arranged in all possible pairs give a total of only 16 combinations—still too few (Fig. 15-4). But if we lengthen the "words" to 3 letters, we have an ample number of combinations—64. Therefore, the smallest possible "word" length in "genetic language" is three nucleotides.

By 1960, there was considerable evidence that the genetic code was, in fact, a triplet code; however, there remained the problem of determining which triplets code for which amino acids. Biochemists devised a way to prepare artificial RNAs with known nucleotide sequences; when these were put into solutions containing ribosomes, amino acids, transfer RNAs, and various other

THE FOUR CODE LETTERS

A
C
G
U

SIXTEEN DOUBLETS FROM THE FOUR CODE LETTERS

AA	AC	AG	AU
CA	CC	CG	CU
GA	GC	GG	GU
UA	UC	UG	UU

Fig. 15–4

Possible length of "words" in the genetic code and the number of combinations.

The abbreviations for the amino acids are:

Ala	Alanine
Arg	Arginine
Asn	Asparagine
Asp	Aspartic acid
Cys	Cysteine
Gln	Glutamine
Glu	Glutamic acid
Gly	Glycine
His	Histidine
Ile	Isoleucine
Leu	Leucine
Lys	Lysine
Met	Methionine
Phe	Phenylalanine
Pro	Proline
Ser	Serine
Thr	Threonine
Trp	Tryptophan
Tyr	Tyrosine
Val	Valine

TABLE 15-1   CODONS FOUND IN MESSENGER RNA

(To find the amino acid specified by a particular codon, find the row marked with the first base of the codon [at the left] and go across this row until you reach the column headed by the second base. Then find the third base marked at the far right of the table. The three *Stop* codons signal the position where the ribosome stops reading and terminates the polypeptide chain. The codon AUG initiates synthesis of a polypeptide.)

First Base		Second Base — U		C		A		G		Third Base
U		UUU	Phe	UCU	Ser	UAU	Tyr	UGU	Cys	U
		UUC	Phe	UCC	Ser	UAC	Tyr	UGC	Cys	C
		UUA	Leu	UCA	Ser	UAA	*Stop*	UGA	*Stop*	A
		UUG	Leu	UCG	Ser	UAG	*Stop*	UGG	Trp	G
C		CUU	Leu	CCU	Pro	CAU	His	CGU	Arg	U
		CUC	Leu	CCC	Pro	CAC	His	CGC	Arg	C
		CUA	Leu	CCA	Pro	CAA	Gln	CGA	Arg	A
		CUG	Leu	CCG	Pro	CAG	Gln	CGG	Arg	G
A		AUU	Ile	ACU	Thr	AAU	Asn	AGU	Ser	U
		AUC	Ile	ACC	Thr	AAC	Asn	AGC	Ser	C
		AUA	Ile	ACA	Thr	AAA	Lys	AGA	Arg	A
		AUG	Met	ACG	Thr	AAG	Lys	AGG	Arg	G
G		GUU	Val	GCU	Ala	GAU	Asp	GGU	Gly	U
		GUC	Val	GCC	Ala	GAC	Asp	GGC	Gly	C
		GUA	Val	GCA	Ala	GAA	Glu	GGA	Gly	A
		GUG	Val	GCG	Ala	GAG	Glu	GGG	Gly	G

substances needed for protein synthesis, they directed the synthesis of polypeptides.

In 1961 scientists found that an artificial RNA containing only uracil nucleotides produced a peptide chain containing only the amino acid phenylalanine; they deduced that the RNA code word UUU stands for phenylalanine. (The DNA code would be the complement of this, AAA.) Finding the amino acids coded by words containing different letters required more ingenuity, but by 1965 the complete genetic code had been worked out. The code words, or **codons**, carried by messenger RNA are shown in Table 15-1. Note that 3 of the 64 triplets do not code for any amino acids: UAG, UAA, and UGA are "Stop" signals that terminate the synthesis of a polypeptide chain. The other 61 combinations are codons for amino acids. Because many amino acids have more than one codon, the code is said to be **degenerate**.

We can go along a messenger RNA molecule and mark off its nucleotides by threes to obtain the amino acid sequence coded for; for this reason, it was believed that the genetic code was nonoverlapping. Indeed, it would be very restrictive if the second word had to begin with the second or the third letter of

FEE-nill-AL-uh-neen

CODE-onz

the first word. We now know that although the code for a particular protein chain is read without overlapping the words, some viruses, and probably some bacteria too, have genes that share lengths of DNA. In some cases, the end of one gene overlaps the beginning of another; in others, a gene contains a second gene within it. Still another example of overlapping genes is found in a virus where both genes begin at the same place; the first gene ends at a Stop codon, but the protein-synthesis machinery sometimes skips this Stop sign and continues to form protein along the messenger RNA until it reaches two Stop signals in a row. Both the longer and the shorter protein formed are necessary to the virus.

In eukaryotes, genes for many polypeptides are interrupted by segments of DNA that are not part of the code for the polypeptide. We do not yet know why this rather surprising arrangement has evolved. When mRNA is transcribed, the sections of non-coding DNA are transcribed too, but they are removed from the mRNA molecule by enzymes before the mRNA is translated into a polypeptide.

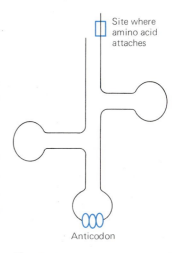

Site where amino acid attaches

Anticodon

Fig. 15–5

A transfer RNA molecule.

15–C Protein Synthesis

Protein synthesis requires many different things, including ribosomes (made of protein and ribosomal RNA), transfer RNAs (Fig. 15-5), messenger RNA, amino acids and various enzymes.

It is still unclear just how the translation of mRNA into a polypeptide starts in a eukaryote. In prokaryotes, protein synthesis begins when an AUG codon of an mRNA molecule binds to a small ribosome subunit. (Ribosomes come in two pieces.) A large ribosome subunit attaches to this complex, forming a complete ribosome bound to mRNA.

Each amino acid is brought to the ribosome by its own tRNA, which holds the amino acid in place until it has been joined to the growing polypeptide chain. Transfer RNA has one part that attaches to the particular amino acid, and another part, the **anticodon**, that base-pairs to the mRNA codon specifying its amino acid. The anticodon is made up of the three bases complementary to the three that make up the codon. For instance, the tRNA that carries the amino acid methionine bears the anticodon UAC, which will base-pair with the mRNA codon AUG. After the first tRNA with its amino acid has attached to the mRNA-ribosome complex, a tRNA with an anticodon complementary to the second codon of the mRNA binds next door. Now an enzyme links the two amino acids carried by the two bound tRNAs. The tRNA for methionine now leaves the ribosome and will eventually pick up another molecule of methionine. Meanwhile, the other tRNA and the mRNA move along the ribosome; this brings the next codon onto the ribosome, where the appropriate tRNA soon attaches by its anticodon, and so on, until the ribosome reaches a Stop codon on the mRNA.

A single mRNA molecule may be translated into protein more than once. Each molecule typically has several to over a hundred ribosomes attached to it and transcribing its message as they move along. Eventually the mRNA molecule is broken up into its component nucleotides; the average lifetime of an mRNA molecule in a bacterium is about 2 minutes.

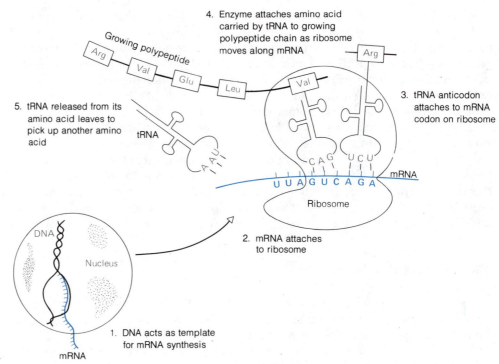

Fig. 15–6

The events of protein synthesis.

15–D Antibiotics

Several **antibiotics** (antibiological agents) help cure diseases by interfering with protein synthesis in the disease-causing organism. Because the ribosomes of bacteria are slightly different from those of eukaryotic cells, some antibiotics prevent protein synthesis in bacteria but not in their hosts. Others block protein synthesis in all cells, but can be used to combat bacterial or fungal infections because these parasites normally carry out protein synthesis faster than their hosts; interfering with protein synthesis deprives the parasites of the advantage that they normally have over the body defenses of their hosts. **Tetracyclines** block the binding of transfer RNAs to bacterial ribosomes. **Puromycin** also binds to the ribosome and becomes linked to the growing polypeptide chain; since puromycin cannot move on the ribosome, it blocks further construction of the protein. **Cycloheximide**, which blocks protein synthesis only on ribosomes of non-bacterial cells, is used to treat fungal infections.

TET-ruh-SIGH-cleans

PURE-oh-MY-sin

SIGH-cloe-HEX-ih-mide

15–E Control of Protein Synthesis

At any one time a cell is making only some of the proteins specified by its total genetic makeup. A colony of bacteria in the laboratory may be able to grow when placed in a nutrient medium containing only inorganic salts and a single source of organic carbon, for example, glucose. On this diet the cells must make all the enzymes necessary to convert glucose to all the other organic molecules that they need. Cells in enriched media (containing inorganic salts plus a variety of sugars, amino acids, and other small organic molecules) grow much faster, since they need not spend so much energy on making enzymes to produce foodstuffs available ready-made in their environment. If these cells are transferred to a less nutritious medium, their growth and reproduction slow down as they switch their protein-synthesis machinery to making all the enzymes needed to produce a wide variety of vital organic molecules.

We saw before that the average messenger RNA molecule lasts only about 2 minutes in bacterial cells; it is then broken down and its nucleotides are used to produce yet other messenger RNA molecules. In this way the cell can control fairly precisely just what proteins are made, and how much of each. But what determines which proteins a cell makes?

Our understanding of how protein synthesis is controlled in bacterial cells is based on the Nobel prize–winning work of François Jacob and Jacques Monod, published in 1961. According to their findings, bacterial cells contain a number of regulatory genes that code for small protein molecules, which are produced constantly in small quantities in the cell. Some of these regulatory proteins are **repressors**; they block protein synthesis by binding with a particular site on the DNA and preventing the enzyme that makes RNA from attaching at that site and transcribing a nearby section of DNA into messenger RNA. The structure of a repressor is such that it will change shape and leave the DNA, unblocking the nearby genes, if it binds with a type of molecule whose presence in the cell calls for production of the protein coded by the gene.

For example, Jacob and Monod found that the bacterium *Escherichia coli* (normally found in the human intestine, and also a standard experimental subject) has three genes lying in a row on the cells' DNA: one codes for a membrane protein that brings the sugar lactose into the cell, one codes for an enzyme that breaks lactose down to form glucose and galactose, and a third codes for an enzyme that probably has a related function. If the bacterium is given glucose or sucrose as a source of carbon, this set of genes is turned off by a repressor protein which prevents RNA synthesis; there is no need for the proteins coded by the genes. However, if the bacterium is transferred to a medium with lactose as the sole carbon source, lactose molecules diffuse into the cell and combine with the repressor protein in such a way that it comes unstuck and no longer blocks RNA synthesis. An RNA-synthesizing enzyme binds to the DNA and transcribes the three genes to mRNA. The ribosomes then make the proteins coded by the mRNA; these are the proteins that permit the cell to take in lactose more quickly and to use it once it is inside the cell.

So, in this example, the presence of a particular molecule removes a repressor protein and permits manufacture of enzymes needed to deal with that molecule. This mechanism provides a way to "turn on" (transcribe), at need, a gene which is normally "turned off." Prokaryotic cells also have ways to turn genes off when their products are not needed. There are still, however, unanswered questions about the control of protein synthesis in prokaryotes.

Control of protein synthesis in eukaryotes is even more complex and less well understood. It is clear that regulation may occur at many steps in the sequence from DNA to protein. Certain proteins enhance the transcription of some genes into RNA, offsetting the effects of other proteins, which tend to keep the DNA tightly coiled and thus unavailable for RNA synthesis. Some hormones also act by binding to particular sites on the chromosomes and speeding transcription of certain genes. Genes in liver and intestinal cells can be turned on as a result of the arrival of certain new food molecules, but this takes longer—possibly hours or days—than the turning on of genes in bacterial cells.

Observations on the control of protein synthesis in eukaryotes that have not yet been adequately explained include the following:

1. Many genes and their corresponding mRNA molecules contain segments that do not code for parts of polypeptide chains. These segments are cut out of the mRNA before it is translated.

2. Many mRNA molecules never reach the cytoplasm or participate in protein synthesis. They are formed, and rapidly destroyed, in the nucleus.

3. Some mRNAs are destroyed more rapidly than others.

4. Some ribosomes cannot translate some mRNAs.

SUMMARY

Like DNA, RNA is composed of nucleotides; the two differ in that RNA contains ribose instead of deoxyribose and uracil instead of thymine; also, RNA is single-stranded, whereas DNA is double-stranded. All cellular RNA is synthesized on a DNA template. Ribosomal RNA (rRNA) makes up part of the ribosomes; transfer RNA (tRNA) carries amino acids to the ribosomes and holds them in place while they are joined into the protein chain; and messenger RNA (mRNA) carries instructions for which amino acids to put where in the chain.

The genetic code carried by mRNA consists of codons, each three nucleotides long, specifying the various amino acids to be placed into a polypeptide chain. The code is degenerate, with most amino acids being specified by more than one codon. Three of the 64 codons are signals for protein synthesis to stop.

A number of antibiotics work by interfering with protein synthesis in disease-producing organisms.

A cell does not make all of the proteins encoded by its genes, but only those it needs. The amounts and kinds of protein the cell produces can be controlled at several different stages in protein synthesis. In bacterial cells the binding of regulatory proteins to DNA may prevent transcription of DNA to RNA, and the regulatory protein may become bound or unbound as it combines with other molecules related to the function of the protein encoded by the gene. In eukaryotes some proteins keep parts of the DNA coiled up and unavailable for transcription; this can be counteracted by other proteins or by some hormones. The mRNA released from the nucleus is also controlled. We are still far from understanding the control of protein synthesis in eukaryotes.

SELF-QUIZ

1. Which of the following is *not* true of the genetic code?
    a. There are three nucleotides per codon
    b. Each codon codes for only one kind of amino acid
    c. The codons are carried on messenger RNA
    d. There is just one codon for each type of amino acid
    e. There are no spacers to designate where one codon leaves off and the next begins

2. The following represent nucleotide sequences in DNA. Write out the messenger RNA sequences that would be transcribed from them; then write out the amino acid sequences that would be produced by translation of the mRNA:
    a. T-A-C-A-A-G-T-A-C-T-T-G-T-T-T-C-T-T
    b. T-A-C-G-T-T-G-C-T-G-C-C-T-G-C-C-G-G

3. According to current ideas concerning protein synthesis:
    a. transfer RNA molecules specific for particular amino acids are synthesized along a messenger RNA template in the cytoplasm
    b. transfer RNA molecules transport messenger RNA from the nucleus to the ribosomes
    c. messenger RNA, synthesized on a DNA template in the nucleus, provides information that deter-mines the sequence in which amino acids will be linked durin; translation
    d. the ribosome can start to "read" the instructions for making a protein at any point on an RNA molecule
    e. transfer RNA attaches to a complementary section of messenger RNA and attracts the proper amino acid to the spot

4. Given the three types of RNA and ribosomes, amino acids, and the necessary enzymes, list the steps that occur in protein synthesis.

5. Under normal conditions, the transcription of DNA to RNA in a prokaryote is controlled by:
    a. regulatory proteins that detach from or attach to DNA depending on the presence of smaller molecules in the cell
    b. regulating the amount of RNA-synthesizing enzyme in a cell
    c. transferring cells to nutrient media with different small molecules
    d. changes in the lifespan of RNA molecules before they are destroyed
    e. antibiotics that attach to ribosomes and block protein synthesis

QUESTIONS FOR DISCUSSION

1. Each type of transfer RNA has its own type of enzyme to bind it to an amino acid. Why is this important?

2. Suppose a bacterial cell contained a mutation that changed one of the nucleotides in an anticodon of transfer RNA. How might this mutation affect protein synthesis?

3. We have seen why the genetic code could not consist of codons with fewer than three nucleotides each. What factors might have selected against codons of more than three nucleotides?

Essay:
PROTEINS AS EVOLUTIONARY PUZZLE PIECES

The discovery of the genetic code and the invention of machines to help find the sequence of amino acids in a protein opened up an exciting new field, the use of proteins as living biochemical fossils. Traditionally, evolutionary "trees" have been constructed by comparing living and fossil organisms, and by determining the age of rocks in which the fossils are found. Analysis of protein structure and comparison of proteins from various modern organisms provide another line of evidence that can be used to confirm evolutionary trees constructed from fossil evidence, and can sometimes shed light on points where the fossil record is inadequate. The sequence of amino acids in its proteins can be used, just like the structure of its teeth or bones, to trace the probable ancestry of a living organism.

To construct an evolutionary tree that includes most living organisms, we must study a protein that is common to a variety of organisms. Some biochemists have chosen to work with cytochrome *c*, a vital part of the electron transport chain in cellular respiration. All aerobic organisms synthesize cytochrome *c*, and the variations of this molecule from one species to another can be compared. The relatively small size of this protein—from 103 to 112 amino acids in various organisms studied—is also an advantage.

The first step in constructing an evolutionary tree for protein structure is to isolate the protein from a number of different species and to determine the amino acid sequence in each. Next, a computer is used to compare the proteins with one another, and to tabulate their similarities and differences. The more similar two species' amino acid sequences, the more closely related the species are likely to be.

If the computer is also supplied with the genetic code, it can construct the "missing links" between related species; that is, it can indicate the probable structure of the ancestral cytochrome *c* possessed by a species which was the common ancestor of two related species. For example, suppose one species has the amino acid isoleucine in a particular position in its cytochrome *c*, and another species has proline in that position. If we assume that one of these amino acids was present in the ancestral protein, then there must have been mutations in two adjacent nucleotides to convert the DNA of one species to that of the other. If, however, we assume that both species came from a common ancestor whose cytochrome *c* had leucine at that position, then each species could have arrived at its present amino acid by mutation of only one nucleotide (Fig. 15-7). The latter case is more probable.

One interesting finding in comparative protein studies is that some positions in a protein molecule are occupied by the same amino acid in every species for which that protein has been analyzed. Is this coincidence? It seems more likely that these invariant amino acids cannot be replaced without destroying the protein's folding pattern, and hence its function. An individual with such a protein would be at a disadvantage compared to its

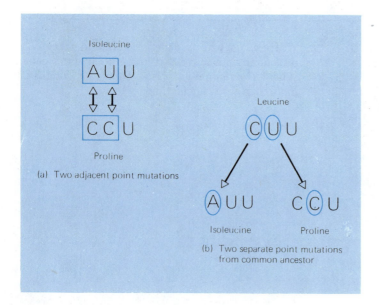

Fig. 15–7

Two ways of imagining the evolutionary relationships between proteins having the amino acids isoleucine and proline at a particular position. (The RNA molecules coding for the proteins are shown.) (a) One amino acid could have been substituted for the other as a result of two adjacent mutations (A to C and U to C). (b) Each amino acid could have been substituted for leucine as a result of two separate mutations (C to A or U to C).

normal neighbors, and so variations in the invariant amino acids are selected against. Variations in some of the other amino acids do not seem to make so much difference, although they may well be responsible for the "fine tuning" that makes a protein function optimally in a species' particular environment.

The relationships among vertebrates have been studied by examining the structure of hemoglobin, a protein that is found in all of them. **Hemoglobin** is the blood's oxygen-carrying molecule, and it imparts its red color to the blood. Hemoglobin consists of four polypeptides: two alpha chains and two beta chains. Analysis of the beta chains has shown that humans and chimpanzees have identical beta chains in their adult hemoglobin; gorilla and human beta chains differ by only one amino acid; pigs differ from humans by about 17 amino acids; and horses and humans differ by 26.

The alpha chains of hemoglobin are very similar to the beta chains and, in fact, humans and other vertebrates have several molecules that are all very similar to the beta chains of hemoglobin. One of these is **myoglobin,** a muscle protein that takes up oxygen from hemoglobin in the blood and holds it ready for the muscle to use. There are also different types of hemoglobin chains made only before birth; beta chains are not made until a late stage in fetal development. (Alpha chains are produced throughout fetal and adult life.) The fetal hemoglobins have a greater affinity for molecular oxygen than the hemoglobin of adults does; this permits the fetus to obtain oxygen from the mother's circulatory system. Hence, fetal hemoglobin is an adaptation permitting the fetus to obtain an adequate oxygen supply without direct access to air.

How did so many different types of molecules with such similar structures come into being? Most likely, at some point in the evolutionary past an ancestral gene for an oxygen-carrying protein was replicated in more than one copy. This idea is supported by the fact that there is more than one copy of some genes in the nuclei of modern organisms. Suppose that some copies of the gene underwent mutations resulting in proteins with slightly different, but advantageous, functions: individuals with two or more proteins that were similar to each other, but that were each adapted to performing slightly different functions, would be selected for over individuals with just one protein to serve all these functions by itself. In other words, multiple copies of genes could undergo adaptive radiation, changing from one to several different proteins, each adapted to its particular function, and all together conferring selective advantage on the individual that possessed them over the individuals that had only one unspecialized protein to do all the jobs.

16 MENDELIAN GENETICS

OBJECTIVES When you have studied this chapter, you should be able to:

1. Use the following terms: parental (P_1) generation, first filial (F_1) generation, second filial (F_2) generation, dominant, recessive, homozygous, heterozygous, segregation, independent assortment, incomplete dominance, homologous chromosomes, crossing over.

2. Define and compare the terms phenotype and genotype and their relationships to the terms dominant and recessive.

3. Use a Punnett square to illustrate a one- or two-character cross, and work out the genotypic and phenotypic ratios expected from such crosses.

4. Correlate the pattern of inheritance of genetic characteristics in breeding experiments with the behavior of the chromosomes during meiosis and fertilization.

5. In your own words, state the rules of inheritance, segregation, and independent assortment which were Mendel's most important contribution to genetics.

Since before the dawn of history, human beings have understood that offspring of plants and animals inherited some characteristics from their parents. Prehistoric people doubtless recognized a child's resemblance to its parents, bred calves from the cows that gave the most milk, and saved the most productive corn for seed, just as we do today.

Until about 1900, however, people assumed that the characters of parents blended in their offspring. Charles Darwin himself believed that this was the case. The idea of blending inheritance is expressed when we say that a person or an animal has "mixed blood."

In 1900, biologists rediscovered the work of Gregor Mendel. More than 30 years earlier, Mendel had shown that inherited characters do not really blend. Instead they exist and are inherited in the form of discrete entities which we now know to be segments of DNA molecules (Chapter 14); these segments are called **genes**.

An organism produced by sexual reproduction receives half of its genes from its mother and half from its father. Inherited characters appear to blend because all the genes in the new combination interact with each other and slightly change the final outcome in the individual's appearance, body chemistry, and behavior.

In this chapter we shall examine the patterns of inheritance that emerge as the genes pass from one generation to the next.

16–A Mendel's Genetic Experiments

Mendel was a monk, and later the abbot, at the monastery of Brünn, in what is now Czechoslovakia. For his experiments he used ordinary garden peas, which he grew in the monastery garden. Peas have one great advantage for genetic experiments: they can be cross-fertilized or self-fertilized, as the experimenter wishes. The shape of a pea flower is such that the pollen of one flower usually fertilizes the egg of the same flower; however, by removing the male flower parts and transferring pollen by hand from one plant to another, any given pea plant may be used to fertilize any other plant.

Mendel identified several strains of pea plants that bred true for a particular trait, which means that every plant showed the trait, generation after generation. He chose seven different traits (characters) to study; for each character, he had available pure-breeding lines of plants that showed contrasting forms of the character (Fig. 16-1). Let us follow one of his single-character experiments, a study of the inheritance of flower color.

In this experiment, Mendel crossed a pure-breeding strain of red-flowered pea plants with a pure-breeding strain that produced white flowers. A cross between genetically unlike strains of organisms produces genetically mixed offspring known as **hybrids**. When the hybrid offspring of the red and white

HIGH-bridd

Fig. 16–1

Mendel studied seven different traits of pea plants, each of which appeared in two different forms. In each pair, the dominant form (discussed in Section 16–B) is the one on the left.

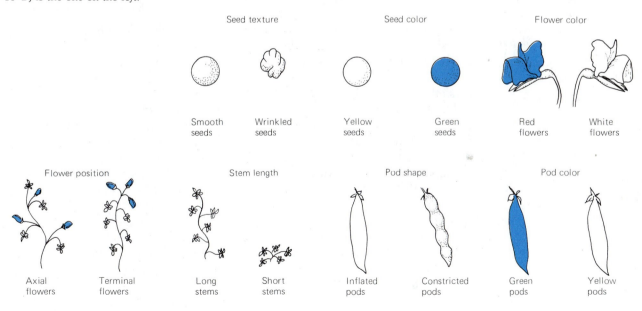

Seed texture — Smooth seeds / Wrinkled seeds

Seed color — Yellow seeds / Green seeds

Flower color — Red flowers / White flowers

Flower position — Axial flowers / Terminal flowers

Stem length — Long stems / Short stems

Pod shape — Inflated pods / Constricted pods

Pod color — Green pods / Yellow pods

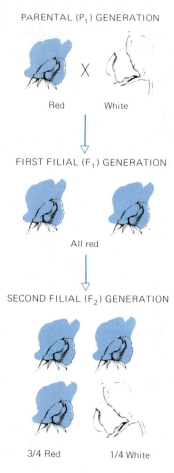

PARENTAL (P₁) GENERATION

Red   White

FIRST FILIAL (F₁) GENERATION

All red

SECOND FILIAL (F₂) GENERATION

3/4 Red   1/4 White

Fig. 16–2

In a cross between pure-breeding red-flowered and pure-breeding white-flowered pea plants, all the offspring were red-flowered. Self-pollination of these offspring yielded about 3/4 red-flowered and 1/4 white-flowered plants.

hoe-mow-ZIE-gus
HET-er-oh-ZIE-gus

flower cross matured, they all produced red flowers (Fig. 16-2). This was somewhat unexpected; what had happened to the gene for white flowers that should have been passed to the offspring from the white-flowered parental strain?

Mendel continued this experiment by allowing the hybrids to self-pollinate, and he collected more than 900 seeds from them. When they were planted, most of these seeds grew into red-flowered plants, but about a quarter of them produced white-flowered plants. It is clear that the white-flower gene had not been lost, although it had not been expressed in the first generation of hybrids.

Mendel realized that these results could be explained by assuming that a genetic trait is governed by a pair of genes. In the pure-breeding red-flowered stock, he thought, each plant carried two genes for red flowers. Furthermore, he reasoned, each reproductive cell must receive only one member of this pair of genes and so each new offspring also had a pair of red-flower genes, one from the egg and one from the pollen. Similarly, all the plants in the white-flowered stock bore, and could pass on to their offspring, only white-flower genes. When he crossed the two stocks to produce hybrids, each white-flowered parent contributed one white-flower gene and each red-flowered plant contributed one red-flower gene to the hybrid offspring. Each member of the hybrid generation contained one red- and one white-flower gene, even though its flowers were red.

When these hybrid plants reproduced, some reproductive cells received a red- and some a white-flower gene, and each kind could combine with another cell containing either a red- or a white-flower gene at fertilization; thus, some white-flowered plants appeared in the next generation.

The theory that genes could combine in an individual but then separate so that only one or the other went into the reproductive cell to produce the next generation became known as Mendel's **law of segregation**.

16–B Dominance and Recessiveness

The disappearance and subsequent reappearance of the white flower color, and similar results with the other traits he studied, led Mendel to formulate his **law of dominance**: when contrasting genes for a trait are present in hybrid individuals, one of them may appear as an observable trait of the individual and mask the presence of its partner. The gene that appears, or expresses itself, is said to be **dominant**; the masked gene is said to be **recessive**. The presence of the recessive gene can be detected only if the individual is **homozygous recessive**, that is, contains two identical recessive genes, in which case the recessive trait is expressed in the individual. **Homozygous dominant** individuals (that is, individuals with two identical dominant genes) and **heterozygous** individuals (those with two different genes, in this case one dominant and one recessive) both express the dominant gene and cannot be distinguished by casual inspection.

Geneticists often use a shorthand method of designating genes, using a capital letter to designate the dominant gene of a pair and the lower case of the same letter to designate the recessive: for example, the red- and white-flower genes studied by Mendel can be written R for the red gene and r for the white.

Because of dominance, it is hard to tell whether an individual that expresses a dominant gene is homozygous or heterozygous for that gene. Such an individual is said to have a dominant **phenotype**, that is, an observable appearance that expresses the dominant gene. Its **genotype**, that is, its true genetic makeup, may be either *RR* (two doses of the red-flower gene) or *Rr* (one red-flower and one white-flower gene). An individual with a recessive phenotype (white flowers in our example) must have the genotype *rr* (two white-flower genes). An individual's genotype is fixed at the time of fertilization; its phenotype, however, results from the interaction of all of its genes with one another and with factors in the environment. For example, the phenotype "dwarf" in a plant might arise because the plant possesses "dwarf" genes, or because the plant is poorly nourished, even though it has "tall" genes.

FEE-no-type

GEE-no-type

16–C Diagrams of Genetic Crosses

Figure 16-3 shows a diagram of Mendel's red- × white-flower crosses. In the **parental (P₁) generation**, the parents had two different genotypes: the pure-breeding red-flowered plants had the genotype *RR*, and the pure-breeding white-flowered plants were *rr*.

Sexual reproductive cells are called **gametes**; in plants the gametes are the **sperm**, which is produced following growth of the pollen grain, and the **egg**. Before gametes form, the two genes of a pair segregate from one another, and so each gamete contains only one flower color gene—R in the case of gametes from the red-flowering plant, r from the white-flowered plant. When the gametes

GAM-eats

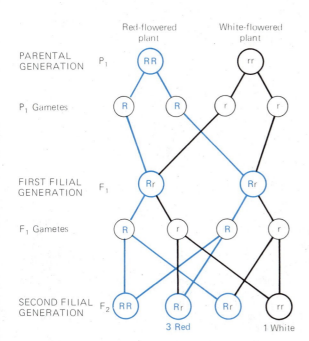

Fig. 16–3

Genotypes of reproductive cells (gametes) and offspring when a plant homozygous for red flower color is crossed with one homozygous for white flower color, and the hybrid offspring then self-pollinate. This is the same cross shown in Figure 16-2.

FILL-ee-al combine at fertilization each member of the hybrid (known as **first filial**, or **F₁**) generation receives one R and one r gene, giving it a pair of flower color genes.

In the gametes produced by the F₁ plants, the R and r genes are in different gametes. Since an egg has an equal chance of receiving either an R or an r gene, and since it is equally likely to be fertilized by a sperm nucleus bearing either R or r, four combinations of gametes are possible to produce the **second filial**, or **F₂, generation:** (1) an R egg nucleus and R sperm nucleus (RR); (2) an R egg and r sperm nucleus (Rr); (3) an r egg with an R sperm nucleus (rR); and (4) an r egg and r sperm nucleus (rr). Once fertilization is complete, (2) and (3) are genetically indistinguishable. Hence the possible genotypes in the F₂ generation are RR, Rr, and rr, and we expect to find them in a ratio of $1RR{:}2Rr{:}1rr$, since there are two ways to obtain the Rr combination and only one way to obtain each of the other combinations. The ratio of phenotypes is three red-flowered plants to one white-flowered plant, since the RR plants and the Rr plants all have red flowers. This 3:1 (or ¾:¼) ratio is typical of a cross involving a single dominant/recessive gene pair.

The line type of diagram shown in Figure 16-3 may be difficult to follow; as

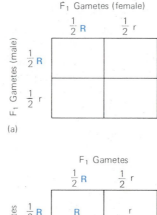

(a)

(b)

(c)

Fig. 16–4

Constructing a Punnett square. (a) Gametes from each parent on Punnett square. It does not matter whether the female gametes are the ones across the top (as shown) or down the side, as long as the gametes from each parent all appear on the same side. (b) Punnett square with the gametes written across the top now also placed in each square beneath them. (c) Gametes from the side added into each box. Each two-gamete combination represents a possible combination of a gene pair at fertilization, when two gametes join.

an alternative, we may use a **Punnett Square** (named after the geneticist Reginald Crundall Punnett) to keep the crosses straight. To construct a Punnett square, we first write down the genotypes of the parents and determine what genetic combinations will be present in their gametes. Next, we draw a set of boxes like those in Figure 16-4, and write the gametes from one parent above the top row of boxes and the gametes from the other parent to the left of the boxes, as shown. To obtain the possible offspring of the cross between the two parents, we write the gametes in the boxes, as shown. When all the gametes have been so entered, we have in each box a combination of two gametes, that is, a fertilization, producing a member of the next generation. This gives us the expected genotypes of the offspring, and we can obtain the phenotypes if we know which genes are dominant and which recessive.

16–D Independent Assortment

In addition to his crosses with one character, Mendel carried out crosses with plants which showed two pairs of contrasting genes. In one of these, Mendel crossed plants homozygous for seeds that were both smooth and yellow with plants homozygous for wrinkled, green seeds. All the F_1 offspring were smooth and yellow, showing that smooth and yellow were dominant. Self-fertilization of the F_1 plants produced an F_2 generation with the following characteristics:

315 smooth and yellow
101 wrinkled and yellow
108 smooth and green
 32 wrinkled and green

The ratio of these phenotypes in the F_2 generation is about 9:3:3:1. Mendel proposed that these data were best explained by assuming that the genes governing seed color and seed texture were inherited independently, a phenomenon now known as Mendel's **law of independent assortment**. For example, when we consider only yellow vs. green seeds, we find $315 + 101 = 416$ yellow and $108 + 32 = 140$ green, a ratio of 2.97:1, quite close to the 3:1 ratio of a one-character cross, which is what it is. Similar results are found in considering smooth vs. wrinkled seeds.

In this example, let S = smooth, s = wrinkled, Y = yellow and y = green. The P_1 plants must have had genotypes $SSYY$ (smooth, yellow) and $ssyy$ (wrinkled, green); they produced gametes SY and sy respectively. All members of the F_1 generation have the genotype $SsYy$. Self-fertilization of the F_1 plants, however, produces a more complex situation. According to Mendel's law of independent assortment, the members of each gene pair are inherited independently of the members of other gene pairs; thus, each F_1 plant produces four kinds of gametes—SY, sy, Sy and sY—in equal proportions (Figure 16-5). Note that each gamete receives *one member of each pair of genes*. To find all possible genetic combinations in the offspring, these gametes can be written along the sides of a Punnett square. Since any female gamete can be fertilized by any male gamete,

TABLE 16-1 PHENOTYPES AND GENOTYPES, AND THEIR RATIOS, RESULTING FROM A SELF-CROSS OF *SsYy* INDIVIDUALS

Phenotype	Ratio	Genotype
smooth, yellow	9/16	1/16 SSYY:2/16 SSYy:2/16 SsYY:4/16 SsYy
smooth, green	3/16	1/16 SSyy:2/16 Ssyy
wrinkled, yellow	3/16	2/16 ssYy:1/16 ssYY
wrinkled, green	1/16	1/16 ssyy

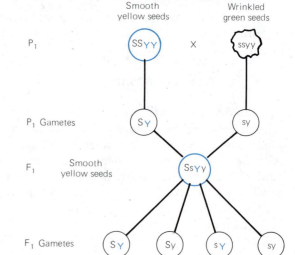

Fig. 16–5

Independent assortment of members of two heterozygous gene pairs during gamete formation. The F₁ individual produces four different types of gametes in equal proportions. Some of the gametes have the same combinations of genes as those received from the parents (SY and sy), but two new combinations are also produced (Sy and sY).

Fig. 16–6

Punnett square for fertilizations of the gametes in Figure 16-5. Gametes produced by the F₁ individuals are arranged outside the boxes; combinations in boxes represent possible genotypes (letters) and phenotypes (circles around letters) in the F₂ generation. Phenotypes are denoted by colored circles for yellow seeds, black circles for green, smooth circles for smooth seeds, and irregular circles for wrinkled seeds.

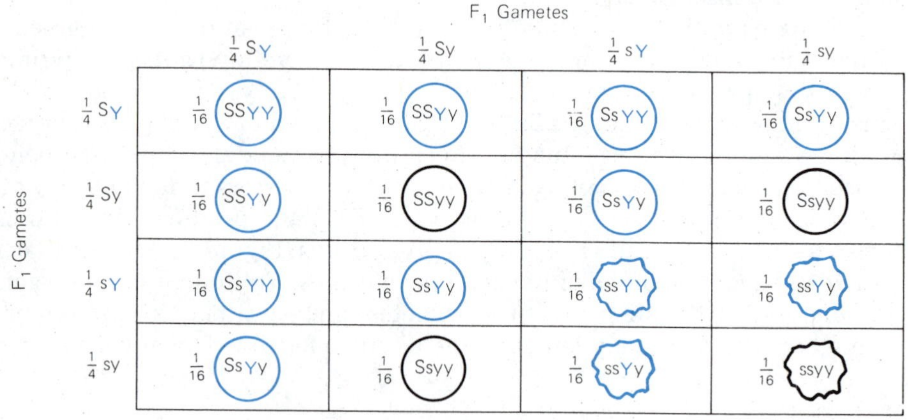

F₁ Gametes

	¼ SY	¼ Sy	¼ sY	¼ sy
¼ SY	1/16 SSYY	1/16 SSYy	1/16 SsYY	1/16 SsYy
¼ Sy	1/16 SSYy	1/16 SSyy	1/16 SsYy	1/16 Ssyy
¼ sY	1/16 SsYY	1/16 SsYy	1/16 ssYY	1/16 ssYy
¼ sy	1/16 SsYy	1/16 Ssyy	1/16 ssYy	1/16 ssyy

there is a total of nine possible genotypes falling into four phenotypes in the F_2 generation (Fig. 16-6).

Table 16-1 shows the genotype and phenotype ratios for this **dihybrid** (two-character) cross. A 9:3:3:1 phenotypic ratio in the F_2 generation is characteristic of a dihybrid cross. The genotypic ratio is 1:2:2:1:4:1:2:2:1.

16–E Incomplete Dominance

The pairs of genes studied by Mendel all exhibited a dominant:recessive relationship; to be sure, Mendel chose his pairs because they behaved as distinct alternatives. Many gene pairs are now known to exhibit **incomplete dominance**, or **codominance**, in which the heterozygote has a different phenotype (as well as a different genotype) from either homozygote.

For example, in snapdragons, flower color is controlled by genes that show incomplete dominance. Homozygous plants have either red or white flowers; heterozygotes have pink flowers. After a red-flowered plant is crossed with a white-flowered plant, the F_1 plants are all pink, and they are all heterozygous for red and white flower color. Half their gametes will contain the gene for red flowers, and half will contain the gene for white flowers. The F_2 phenotype and genotype ratios will both be 1:2:1, just the same as the F_2 genotype ratios for any monohybrid cross (Fig. 16-7). Since the heterozygote produces pink flowers, the expected phenotype ratio for the F_2 plants is 1 red-flowered:2 pink-flowered:1 white-flowered.

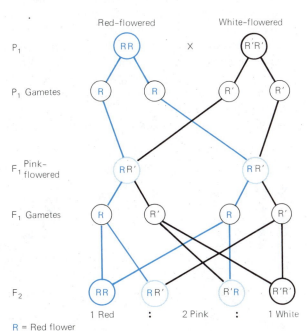

Fig. 16–7

Diagram of a cross involving incomplete dominance. In snapdragon plants, the red- and white-flower genes show incomplete dominance: heterozygotes have pink flowers.

16–F Meiosis

Mendel explained the results of his breeding experiments by proposing that each plant contained two genes for flower color in its body cells but only one flower color gene in each gamete. We now know that genes are parts of chromosomes, and every chromosome bears many genes. Each body cell of a pea plant contains chromosomes in pairs, called **homologous** (similar) **chromosomes** (homologues); each chromosome in the pair carries one of the two genes that make up a gene pair (Fig. 16-8). Moreover, each gamete contains only one member of each pair of homologous chromosomes; this bears out Mendel's theory. The body cells are said to be **diploid** (two sets of chromosomes) and the gametes **haploid** (one set of chromosomes). This also holds true for most other plants and for most animals, including humans.

This arrangement is vital for organisms that reproduce sexually if the number of chromosomes (and genes) is not to double in each generation. If each gamete contained the diploid number of chromosomes, then the fertilized egg, and the new individual which developed from it, would contain twice as many genes as each parent. In sexually reproducing organisms, this does not happen. A special form of nuclear division, known as **meiosis,** reduces the diploid chromosome content of a nucleus to the haploid number. Meiosis ensures that each haploid gamete contains one member of each pair of chromosomes found in a diploid cell.

Before meiosis begins, each chromosome is replicated (duplicated) (Section 14-C) and becomes a set of **sister chromatids**, joined at a structure called the **centromere** (see Fig. 16-9). Thus, each nucleus starting meiosis has the equivalent of four sets of homologous chromosomes; it requires two nuclear divisions, halving the chromosome number at each division, to produce gamete nuclei with one set of chromosomes each.

In the first stage of meiosis, two nuclei are formed by separating homologous chromosomes; sister chromatids stay together, still joined at the centromere. How do the chromosomes separate so precisely that each new nucleus receives one member of each pair of homologous chromosomes? Before the centromeres attach to the meiotic spindle, each pair of homologues lines up together (Fig. 16-10). The homologous sets then attach to the spindle fibers, and members of

DIP-loid
HAP-loid

my-OH-sis

SENT-roe-meer

Fig. 16–8

Two pairs of homologous chromosomes. Homologues are usually of the same length, have their centromeres (clear circles) at the same positions, and bear members of the same gene pairs in the same order. For example, the long pair of homologous chromosomes shown here bears the gene pair A and a, showing that the individual is heterozygous for this trait, while the individual is homozygous recessive for the next gene pair, b and b, and so forth.

Fig. 16–9

A chromosome consists of DNA, proteins, and RNA. When DNA replicates, its two strands separate, and each serves as the template for the formation of another strand. The two identical double-stranded DNA molecules so formed are joined at the centromere and are called sister chromatids; they become chromosomes when they separate during nuclear division (meiosis, discussed here, or mitosis, discussed in Section 10-J).

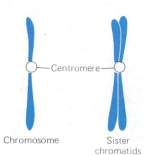

Chromosome   Sister chromatids

PROPHASE I

Chromosomes have been replicated, forming sets of sister chromatids attached at centromeres; each set pairs with its homologue

PROPHASE II

Each group prepares to enter another division.

METAPHASE I

Spindle forms, and chromosomes move to equator. Centromeres of homologous sister chromotids attach to spindle fibers on opposite sides of equator. Nuclear envelope begins to disappear

METAPHASE II

Spindles form; chromatids move to the equator, where centromeres attach to spindle fibers. Centromeres divide.

ANAPHASE I

Each set of sister chromatids moves away from its homologue; this gives rise to segregation of genes carried on homologous chromosomes

ANAPHASE II

Sister chromatids separate

TELOPHASE I

The chromosomes have formed distinct groups.

TELOPHASE II

Four haploid nuclei are formed, each has one member of each pair of chromosomes from original cell

Fig. 16–10

Basic events of meiosis; details vary in different organisms. Formation of female gametes involves unequal division of cytoplasm following nuclear division.

each pair move along the fibers toward opposite poles of the cell. Thus, when the chromosomes form two groups, each group contains one member of each pair of homologous chromosomes.

In the second division of meiosis, the sets of sister chromatids attach to the spindle, and the centromeres finally divide, releasing the sister chromatids as independent chromosomes, ready to move to opposite poles of the cell. When this is done, each group of chromosomes forms a haploid nucleus, containing one member of each original pair of homologous chromosomes.

We can see, now, how the inheritance patterns for the gene pairs observed by Mendel came about. Segregation of members of a gene pair occurs when homologous chromosomes separate in the first division of meiosis (Fig. 16-11). Independent assortment of gene pairs on nonhomologous chromosomes comes about because the different chromosome pairs can line up several different ways before the first meiotic division (Fig. 16-12).

16–G Linkage, Crossing Over, and Mapping

Mendel's work languished in obscurity until 1900, when it was rediscovered almost simultaneously by three biologists in different parts of Europe. They realized that Mendel's results dovetailed elegantly with the observations of chromosomal movements during meiosis, which had been discovered in the 34 years since publication of Mendel's results. It quickly became accepted that the genes Mendel studied were carried by the chromosomes.

As geneticists began to test Mendel's ideas, applying them to many different organisms, they began to find pairs of genes that were not inherited in the ratios predicted by Mendel. If an individual is heterozygous for two pairs of genes (for example, $AaBb$), four types of gametes—AB, Ab, aB, and ab—should occur with equal frequency. However, with many pairs of genes this did not occur; offspring were receiving too many of two of the combinations, and too few of the other two.

This can be explained by assuming that each chromosome is occupied by many genes. Chromosomes, in fact, carry hundreds or thousands of genes apiece, and each member of a pair of homologous chromosomes is occupied by the same kinds of genes, in the same order. A particular position on a chromosome may be occupied by any one of two or more genes that exist for that trait (see Fig. 16-8). For example, in pea plants, one chromosome may have the R gene at its flower-color position, and its homologue might have either R or r at the same position.

HOMOLOGOUS CHROMOSOMES
HETEROZYGOUS AT THE "A"
LOCUS

PROPHASE I

METAPHASE I

Fig. 16–11

The law of segregation reflects the events of meiosis. Homologous chromosomes are separated so that each member of the pair ends up in a separate gamete. The two paired genes borne on a pair of homologous chromosomes must end up in different gametes.

TELOPHASE I

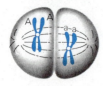

METAPHASE II

TELOPHASE II

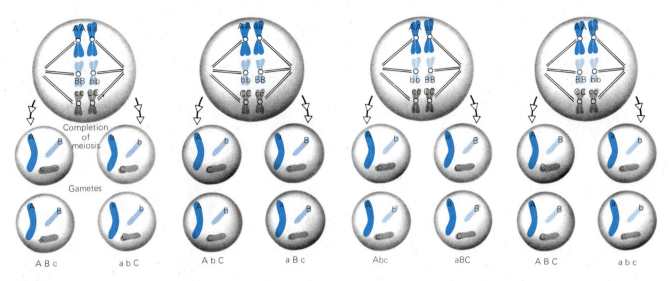

Fig. 16–12

Independent assortment occurs because sets of homologous chromosomes can line up in any combination before the homologues separate during the first division of meiosis. Here, a cell heterozygous for gene pairs Aa, Bb, and Cc is shown with the three homologous chromosome pairs lining up four different ways, resulting in eight possible gene combinations in the gametes that finally form.

Because chromosomes separate from their homologues during meiosis, we would expect that any two genes borne on the same chromosome would end up in the same gamete; in other words, they will act as though they are **linked** together. In fact, Mendel was either lucky or clever that the seven pairs of traits he studied were located on six of the seven different chromosome pairs of pea plants.

Let us consider an example of genetic linkage. Suppose that genes A and B are located on the same pair of homologous chromosomes. We cross an individual homozygous for the dominant forms of both traits *(AABB)* with one that is homozygous recessive for both *(aabb)*. The hybrid offspring of this cross will all be *AaBb*, having received the gamete *AB* from one parent and the gamete *ab* from the other. If we cross these hybrids among themselves, we come out with the following results (phenotypes):

¾ showing the traits A and B

¼ showing the traits a and b

This is quite different from the 9:3:3:1 ratio we observed in the two-character cross in Section 16-D; in fact, it is a 3:1 ratio, like those of the one-character cross in Section 16-C. If we look at the gametes formed by the F_2 hybrid individuals, we see that, since A and B are linked, and a and b are linked, these pairs of genes cannot assort independently; A and B are carried together into one gamete by the chromosome that bears them both, while a and b are carried into the opposite gamete by the homologous chromosome (Fig. 16-13). Hence we can identify

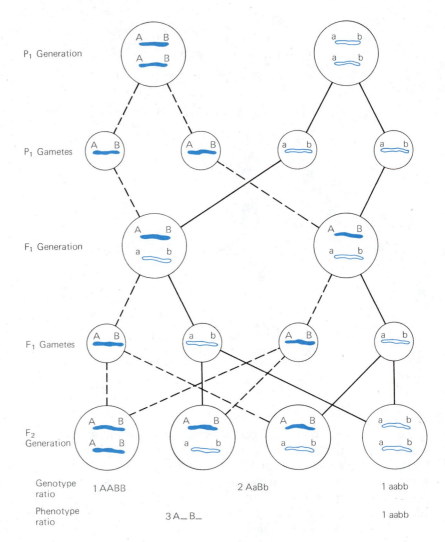

Fig. 16–13

Cross involving two gene pairs linked on the same chromosome. The genes on each chromosome move as a unit rather than assorting independently (compare with the A and B pairs in Figure 16-12). Thus the phenotype and genotype ratios in the F$_2$ generation are the same as for a cross involving one gene pair rather than a cross with two unlinked gene pairs.

linked genes by studying the ratios of offspring obtained in the F$_2$ generation. If the ratio is the Mendelian ratio of 9:3:3:1, for a two-character cross, the genes are assorting independently, and they are probably located on different chromosomes. However, if we find instead a 3:1 ratio, we are looking at linked genes.

During meiosis, homologous chromosomes line up together before they segregate into different nuclei. At this time parts of the chromosomes may cross, break off, and exchange pieces of DNA with their homologues, a phenomenon known as **crossing over**. If the point of crossover is between two genes that we are studying, this crossing over can rearrange the genes that were previously linked so that they are now on opposite chromosomes (Fig. 16-14). The closer together two genes are on a chromosome, the less likely it is that a crossover will occur between them. If they are close together, the two genes will stay on the same chromosome, and be inherited together, more often than not. As we study genes that are further and further apart on the same chromosome, however, it is more

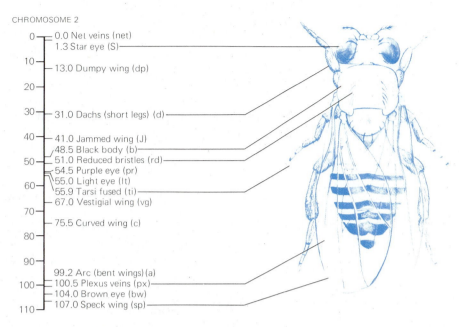

CHROMOSOME 2

0 — 0.0 Net veins (net)
— 1.3 Star eye (S)
10 —
— 13.0 Dumpy wing (dp)
20 —
30 — 31.0 Dachs (short legs) (d)
40 — 41.0 Jammed wing (J)
— 48.5 Black body (b)
50 — 51.0 Reduced bristles (rd)
— 54.5 Purple eye (pr)
— 55.0 Light eye (lt)
60 — 55.9 Tarsi fused (ti)
— 67.0 Vestigial wing (vg)
70 —
— 75.5 Curved wing (c)
80 —
90 —
— 99.2 Arc (bent wings) (a)
100 — 100.5 Plexus veins (px)
— 104.0 Brown eye (bw)
110 — 107.0 Speck wing (sp)

Fig. 16–15

A map of some genes known to occur on one of the chromosomes of the fruit fly Drosophila melanogaster. The map shows mutant genes that have been identified at each position. Note that two of the mutations, star eye and jammed wing, are dominant, but that most mutations are recessive.

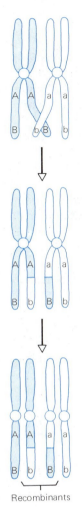

Recombinants

Fig. 16–14

Crossing over occurs when homologous sets of sister chromatids are lined up at the beginning of meiosis. Chromatids may cross, break off, and rejoin onto another chromatid. Eventually this produces four chromosomes with different gene combinations; there are some new gene combinations not seen in the parental chromosomes.

and more likely that crossing over will occur between them. The distance between genes can be estimated by performing a large number of crosses, examining the phenotypes of the offspring, and calculating the percentage of offspring showing recombination of the genes for the traits in question (in our example above, by determining the percentage of offspring that received *Ab* or *aB* chromosomes in gametes from their parents, rather than the chromosomes *AB* and *ab* that the parents contained).

By accumulating the data from a large number of such crosses, using different combinations of genes, geneticists have been able to construct **chromosome maps** for several organisms; that is, they have been able to determine which genes are together on which chromosomes, and approximately how far apart the genes must be (Fig. 16-15). We now know a great deal about the chromosome maps of such organisms as the fruit fly *Drosophila*, laboratory mice, corn, pink bread mold *(Neurospora)*, and some bacteria and viruses, all of which are often used for genetic experiments.

dross-OFF-ill-uh

SUMMARY

The modern science of genetics—the study of how genes governing hereditary characteristics are inherited—is based on Gregor Mendel's work. Mendel applied mathematical analysis to construct a theoretical model which would explain the results he obtained through breeding hundreds of pea plants. Mendel's conclusions may be summarized in modern terms in this way:

1. Genetically based traits are determined by discrete units, or genes, which are passed from parent to offspring during reproduction.

2. A plant or animal contains pairs of genes that determine its genetic characteristics.

3. Genes for a trait may occur in different forms, and one form of a gene (dominant) may hide the presence of another form (recessive) with which it is paired in a heterozygous individual (law of dominance).

4. During meiosis, the two members of each gene pair separate from one another and pass into different cells (law of segregation).

5. At fertilization, each offspring receives a pair of genes for each characteristic, one member of each pair from the gamete of each parent.

6. The genes from each parent remain distinct in the offspring and may reappear in the phenotype of later generations even if they are masked by the phenomenon of dominance in some individuals in intervening generations.

7. During meiosis, the genes of one pair assort independently of genes of other pairs, so long as they are located on different chromosomes (law of independent assortment).

The behavior of genetically determined traits in breeding experiments is paralleled by the behavior of the chromosomes during meiosis. This parallelism provides part of the evidence that genes are borne on chromosomes. Genes located on the same chromosome are linked and are inherited together except when they are separated by crossing over during meiosis.

SELF-QUIZ

The following problems will test your understanding of the ideas in this chapter.

1. Mendel found that the gene for tallness in peas (*T*) is dominant to the gene for shortness (*t*). What offspring phenotypes would be expected from the following crosses, and in what ratio?
   a. heterozygote self-fertilized
   b. homozygous tall × heterozygote
   c. heterozygote × homozygous short

2. Two *Drosophila* (fruit flies) with normal wings are crossed. Among 123 progeny, 88 have normal wings and 35 have "dumpy" wings.
   a. What inheritance pattern is shown by the normal and dumpy genes?
   b. What were the genotypes of the two parents?

3. If a dumpy-winged female (from Problem 2) is crossed with her father, how many normal-winged flies will be expected among 80 offspring?

4. A number of plant species have a recessive gene for albinism: homozygous albino (white) individuals are unable to synthesize chlorophyll. If a tobacco plant heterozygous for albinism is allowed to self-pollinate, and 500 of its seeds germinate, then:
   a. how many of these offspring will be expected to have the same genotype as the parent plant?
   b. how many seedlings will be expected to be white?

5. Sniffles, a male mouse with a colored coat, was mated with Esmeralda, an alluring albino. The resulting litter of six all had colored fur. The next time around, Esmeralda was mated with Whiskers, who was the same color as Sniffles. Some of Esmeralda's next litter were white.
   a. What are the probable genotypes of Sniffles, Whiskers, and Esmeralda?
   b. If a male of the first litter were mated with a colored female of the second litter, what phenotypic ratio might be expected among the offspring?
   c. What would the expected results be if a male from the first litter were mated with an albino female from the second litter?

6. A kennel owner has a magnificent Irish setter which he wants to hire out for stud. He knows that one of its ancestors was Erin-go-braugh, who carried a recessive gene for atrophy of the retina. In its homozygous state, this gene produces blindness. Before he can charge a stud fee, he must check to make sure his dog does not carry this gene. How can he do this?

7. In *Drosophila*, the gene for short-legged *(d)* is recessive to the gene for normal leg length *(D)*, and the gene for hairy body *(h)* is recessive to the gene for normal body *(H)*. Make a Punnett square for each of the following crosses:
   a. *DdHh* × *Ddhh*
   b. *DDHh* × *Ddhh*
   c. *DdHh* × *ddhh*
   d. What proportion of the offspring from the cross shown in (b) would be expected to show the normal phenotype for both traits?

8. In tomato plants, the gene for purple stems *(A)* is dominant to the gene for green stems *(a)*, and the gene for red fruit *(R)* is dominant to the gene for yellow fruit *(r)*. If two tomato plants heterozygous for both traits are crossed, what proportion of the offspring are expected to have:
   a. purple stems and yellow fruits?
   b. green stems and red fruits?
   c. purple stems and red fruits?

9. If 640 seeds resulting from the cross in Question 8 are collected and planted, how many are expected to grow into plants with:
   a. red fruit?
   b. green stems?
   c. green stems and yellow fruits?

10. If one of the parents from Question 8 is crossed with a green-stemmed plant heterozygous for red fruits, what proportion of the offspring are expected to have:
    a. purple stems and yellow fruits?
    b. green stems and yellow fruits?
    c. green stems and red fruits?

11. Pooh had a colony of tiggers whose stripes went across the body. His American pen-pal, Yogi, sent him a tigger whose stripes ran lengthwise. When Pooh crossed it with one of his own animals, he obtained plaid tiggers. Interbreeding among the plaid tiggers produced litters with a majority of plaid members, but some crosswise- and lengthwise-striped animals were also produced. Draw the crosses made by Pooh, showing the genotypes of the tiggers which account for the coat patterns observed.

12. In cattle, the gene for straight coat *(S)* is dominant to the gene for curly coat *(s)*. The gene pairs for red *(RR)* and white *(R'R')* coat color show an absence of dominance; heterozygotes have a roan coat *(RR')* (red lightened by intermixed white hairs). A farmer has three groups of cows: white ones in the clover patch, red in the alfalfa field, and roan in the cornfield. He has a roan bull, Ferdinand, who services the cows in all three fields.
    a. What color calves should he expect in each field, and in what proportions?
    b. Ferdinand dies from a bee sting and the farmer decides to make his herd of cows exclusively roan in memory of his beloved bull. He sells all the red and white cows, and vows to sell any red or white calves born subsequently. What color bull should he buy to replace Ferdinand, if he wants to sell as many calves as possible?

13. The gene for pea comb *(P)* in chickens is dominant to the gene for single comb *(p)*, but the gene for black *(B)* and white *(B')* feather color show incomplete dominance, *BB'* individuals having "blue" feathers. If birds heterozygous for both pairs of genes are mated, what proportion of the offspring should be:
    a. single-combed?
    b. blue-feathered?
    c. white-feathered?
    d. white-feathered and pea-combed?
    e. blue-feathered and single-combed?

For the crosses in Problems 14 and 15, state whether the genes are linked or unlinked; if linked, tell which genes occur on the same chromosome in the female parent.

14. A female *Drosophila* heterozygous for the recessive genes sable body and miniature wing was mated with a sable-bodied, miniature-winged male, and the following progeny were obtained:
    249 sable body, normal wings
     20 normal body, normal wings
     15 sable body, miniature wings
    216 normal body, miniature wings

15. In *Drosophila*, the gene for red eye is dominant to the gene for purple eye and the gene for long wings is dominant to the gene for dumpy wings. A female fly heterozygous for both traits is crossed with a male which has purple eyes and dumpy wings. The F_1 are:
    609 red eyes, normal wings
    614 red eyes, dumpy wings
    622 purple eyes, normal wings
    616 purple eyes, dumpy wings

QUESTIONS FOR DISCUSSION ——————————————————————

1. In a "test cross," an individual of dominant phenotype but unknown genotype is bred to a homozygous recessive individual. This procedure can be used to determine the unknown genotype. If an *rr* individual is bred to an *R_* individual (where _ could be either *R* or *r*), what is the expected outcome if the unknown genotype is *RR*? If it is *Rr*? If no *rr* offspring are obtained, how sure are we that the unknown genotype is *RR*?

2. Mendel worked with two pairs of genes that were on the same chromosome but gave results not much different from the 9:3:3:1 ratio expected of unlinked genes. How frequently must such genes cross over in order for their linkage to go undetected in experiments like Mendel's?

3. How could you tell that genes such as those in discussion question #2 were really linked after all?

4. Evaluate the saying "alike as two peas in a pod" in light of your study of this chapter.

Essay:
SECRETS OF MENDEL'S SUCCESS

At the time of Mendel's famous experiments, many prominent botanists were trying to discover how plants pass genetic information to their offspring. Their inquiries were singularly fruitless, whereas Mendel's work put forth a simple yet clear theory that has withstood the test of time. How did Mendel, working alone, see things that his contemporaries in the midst of the scientific community did not?

One reason for Mendel's success was his choice of an experimental subject. The popular choices were several species of hawkweeds, which, unbeknown to botanists in the nineteenth century, do not always follow the rules of sexual reproduction. They sometimes produce seeds without benefit of pollen, in which case the offspring have no male parent. It would obviously be extremely confusing to try to sort out patterns of heredity when you think you know the parents but really don't!

Mendel avoided this pitfall. He spent a great deal of time deciding what characteristics the ideal subject should have, and he concluded that he required a plant whose reproduction could be readily controlled. He also needed a plant with contrasting characters that could be reproduced dependably for generation after generation. Peas were available in many such varieties, each breeding true and invariably showing its distinct characteristics.

It is also easy to control the reproduction of peas because of their flower structure. Most familiar types of flowers have male parts (**stamens**) and female parts (**carpels**) exposed to the air, where the carpels can receive pollen blown or rubbed off neighboring plants or carried on the bodies of insects. Peas and their relatives, however, have a peculiar petal structure that completely

———————————————————

STAY-min
CAR-pull

encloses the reproductive parts in such a way that the pollen from other flowers cannot enter, and each flower normally pollinates itself (Fig. 16-16). In order to perform a cross between different strains, Mendel opened the flower that was to donate pollen and plucked off a stamen. He then opened the flowers of the other parent and dusted pollen onto their carpels. To be certain that none of the carpels was fertilized by pollen from its own flower, Mendel amputated the stamens before their pollen was mature. In order to obtain large numbers of offspring, he hand-pollinated many dozens of flowers. This was a tedious process, but it ensured that Mendel knew the parentage of every pea he collected.

Mendel had studied mathematics and probability theory, and so he knew the importance of obtaining a large number of offspring from each cross to minimize the effects of "sampling error" that results from looking at a small number of cases. He also realized that he should study just one characteristic at a time, and he followed each characteristic through many generations. In this way he was able to discern the patterns of dominance and recessiveness, the paired nature of genes, and gene segregation. When he came to study two traits together, Mendel's mathematical ability quickly helped him to grasp the fact that he was working essentially with two one-character crosses at once, and to formulate the law of independent assortment.

Perhaps Mendel's most important contribution, however, was his recognition of genes as discrete particles. This is far from obvious. Many hereditary characteristics—human height, intelligence, and skin color, for example—are continuous over a broad range, and most biologists of Mendel's day believed in a blending theory of inheritance. This blending theory did not fit in well with the theory of evolution by natural selection, which required that inherited variations be maintained from generation to generation, rather than merging into some great average. Charles Darwin was among the evolutionists who were extremely uncomfortable with the blending theory of inheritance, but Darwin never knew that Mendel's work would have removed one of the chief objections to the theory of evolution by natural selection.

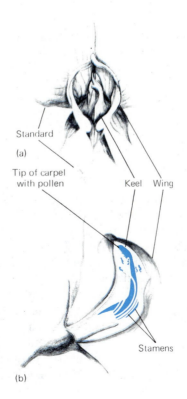

Standard

(a)

Tip of carpel
with pollen

Keel   Wing

Stamens

(b)

Fig. 16–16

Flower of the garden pea, the subject chosen by Mendel for his genetic experiments. (a) External view, showing the modified petals: standards, wings, and keel. (b) Side view of flower cut open to show relationship of reproductive parts to the petals. Since the reproductive parts (stamens and carpel) are enclosed by the keel, the pea flower normally self-fertilizes.

INHERITANCE PATTERNS AND GENE EXPRESSION

17

OBJECTIVES

When you have studied this chapter, you should be able to:

1. Explain how mutations may affect the protein encoded by a gene, and how this is related to phenotypic expression of mutant genes.

2. Given data from an appropriate breeding experiment, recognize the 1:2:1 and 2:1 ratios characteristic of lethal genes, and demonstrate knowledge of the inheritance patterns expected from parents carrying lethal genes by working out crosses correctly.

3. State the possible genotypes of people with blood types A, B, AB, and O, and use your knowledge of these genotypes to solve problems.

4. Explain the difference between genes with multiple forms and polygenic characters, and give an example of each.

5. State the pattern of sex determination ("sex chromosome" complement of each sex) and inheritance of sex-linked genes for mammals, and use this information in working out sex-linkage problems.

6. Explain the difference between sex-linked and sex-influenced characteristics and give examples of each.

7. List at least five factors that may affect the expression of a particular gene in an organism.

8. Describe the inheritance pattern found in the human genetic disorders hemophilia, red-green color blindness, sickle-cell anemia, Tay-Sachs disease, and phenylketonuria.

Genes, which carry hereditary information, are lengths of chromosomal DNA coding for the sequence of amino acids in polypeptides or proteins. In Chapter 9 we saw that proteins must fold up properly to work, and that their folding depends on having the correct amino acids in the correct order. A mutation, or change of structure, in the DNA of a gene may prevent the formation of a protein altogether, or it may cause the cell to make an altered or inactive protein. The seriousness of a mutation depends on how much it affects the protein and on how essential the protein is to the organism.

In the twentieth century, it has been realized that some inherited abnormalities are the expression of mutant genes in the phenotype of an organism. Much research has been devoted to finding out how such genes are inherited and expressed, and how they may be affected by factors in the organism's environment.

phenotype: The actual characteristics shown by an organism, determined by the interplay of its genes and the environment.

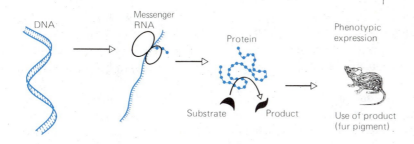

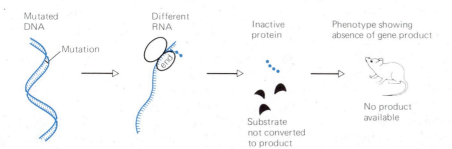

Fig. 17–1

Example of how a mutation results in a recessive gene. The DNA codes for an enzyme (protein) needed for pigment in the fur. The mutant makes an incomplete protein, and so the animal does not produce the pigment. An albino animal results.

In this chapter we shall look at some inheritance patterns not seen by Mendel, concentrating on human examples. We shall also see how these patterns relate to the fact that genes express themselves as their products, the proteins coded for by their DNA.

17–A Lethal Genes

In Chapter 16 we saw several examples of dominant/recessive gene pairs. Why is one gene dominant and one recessive? Often this is a difficult question to answer, but in some cases the reason is clear. For instance, mutant DNA that codes for no protein, or for an inactive protein, will not be expressed and so will be a recessive gene. In heterozygous individuals, the normal gene will produce its protein, while no functional protein will be made by the recessive gene. The phenotype of the individual will appear normal and the normal gene will be dominant. In an individual homozygous for the recessive gene, no protein will be made to produce the normal expression of the trait; the recessive phenotype will be the absence of the normal trait. An albino plant will result from "absence of chlorophyll" or a dwarf plant from "absence of growth hormones."

If a protein is essential to life, an organism that fails to produce an active form of that protein will die, and the defective gene will be known as a **lethal gene**. Dominant lethal genes are possible, but most are rapidly eliminated because they cause the death of the organism that carries them. Recessive lethal genes, however, may cause no harm to a heterozygous individual, and so they may spread to future generations and become quite common in the population. It

dominant gene: gene that appears in an organism's phenotype when paired with a gene for another form of the same trait.

recessive gene: gene that is masked when paired with a gene for another form of the same trait (e.g., white flower color in peas is recessive to red flower color, which is dominant).

heterozygous: having genes for two different forms of a trait in the same gene pair (e.g., genes for both red and white flower color).

homozygous: having two genes for the same form of a trait (e.g., two white- or two red-flower genes).

has been calculated that each human being carries about 30 lethal recessive genes in the heterozygous condition. This is higher than the figure for many other organisms and is at least part of the reason that marriages between close relatives produce a higher proportion of abnormal offspring in humans than in most other species.

Tay-Sachs disease, a disorder resulting in deterioration of the brain and death by about the age of four, is the result of a lethal recessive gene. The homozygous recessive individual lacks an enzyme needed to clear a certain lipid out of the nerve cells of the brain. One in 30 people of East European Jewish extraction is heterozygous for this disorder, which so far is incurable.

If just one copy of a "normal" gene does not produce enough of its protein for normal body functioning, the gene will show incomplete dominance. In these cases the heterozygote has a different phenotype from either homozygote. For example, in humans there is a lethal gene that causes **brachyphalangy**, or shortening of the middle bone in the fingers, in heterozygotes; this makes the fingers appear to have only two bones instead of three. In homozygotes, this gene results in abnormal skeletal development. Homozygous babies lack fingers and show other skeletal defects which cause death in infancy.

BRACK-ee-fal-AN-jee

In a cross between two brachyphalangic people, one out of every four children would be expected to be homozygous for the lethal gene and die during infancy; half would be expected to be heterozygous and show brachyphalangy; and one-fourth would be expected to be normal (Fig. 17-2). This 1:2:1 ratio is typical of lethal genes in which the normal gene is incompletely dominant.

Some lethal genes are mutations of genes that code for proteins so essential that without them the embryo does not develop normally. In animals that produce several offspring at a time, embryos that die early may be resorbed back into the uterus and a 2:1 ratio may be observed when the remaining offspring are born: two-thirds heterozygotes to one-third homozygous normal offspring (Fig. 17-3). For example, the short-tail gene *(t)* in mice causes early embryonic death in the homozygote. The embryo is then resorbed. If such embryos are taken from the uterus early in pregnancy, they are seen to have no backbone and none of the tissue that later forms the muscles, kidneys, and many other important organs. Heterozygotes *(Tt)* have shorter tails than normal mice *(TT)*.

A famous human gene that is frequently lethal in the homozygous condition is the one responsible for sickle-cell anemia. The gene for sickle-cell anemia

Fig. 17–2

A lethal gene in humans. Note the characteristic genotype and phenotype ratios for crosses of (heterozygous × heterozygous) and (normal × heterozygous).

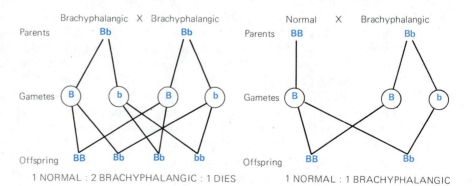

Brachyphalangic X Brachyphalangic
Parents   Bb   Bb

Gametes   B   b   B   b

Offspring   BB   Bb   Bb   bb
1 NORMAL : 2 BRACHYPHALANGIC : 1 DIES

Normal   X   Brachyphalangic
Parents   BB   Bb

Gametes   B   B   b

Offspring   BB   Bb
1 NORMAL : 1 BRACHYPHALANGIC

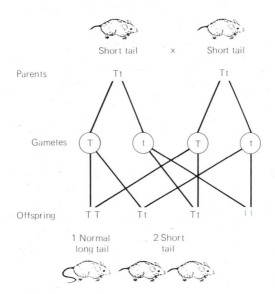

Fig. 17–3

A lethal gene producing a 2:1 ratio among the offspring. The short-tail gene in mice causes death of the homozygous recessive embryo early in development (color).

causes substitution of one amino acid for another in one of the polypeptide chains (the beta chain) of hemoglobin, the protein that makes blood red. As a result of this change, red blood cells exposed to abnormally low oxygen levels assume a sickle shape (or "sickle"), due to formation of stacks of hemoglobin molecules inside the cell. The sickled cells become stuck in the smaller blood vessels and impede circulation to the organs supplied by these vessels. Symptoms of this poor circulation include tiredness, muscle cramps, and irritability.

BAIT-uh; HE-mow-glow-bin

The sickle gene shows incomplete dominance, with both normal and sickle beta chains produced in heterozygous people. Without special blood tests, these people may not know that they possess the sickle gene. People homozygous for the sickle gene are more severely affected because all of their beta chains are defective. About 90% of these people die at an early age.

We might expect that such a lethal gene would quickly be eliminated from the population by natural selection, as those homozygous for the sickle gene died too young to pass it on to their offspring. Yet in large areas of tropical Africa 20 to 40% of people are heterozygous for the gene. In the 1950s it was noticed that these people live in precisely the areas most afflicted with a virulent form of malaria. Possessing at least one copy of the sickle gene lowers a person's chances of developing this disease. Cells containing sickle hemoglobin sickle much faster when they are infected with the parasite that causes the disease. When a cell sickles, it is eaten by one of the body's scavenger white blood cells, which clean up damaged and dead cells. Since the infected sickle cells are cleared quickly from the body, the parasites do not usually develop to the stage where they can cause disease. In malaria-infested regions, therefore, it is advantageous to be heterozygous for the sickle gene, which protects against a common deadly disease, even though the sickle gene is usually lethal in the homozygous state.

An individual heterozygous for a genetic condition is referred to as a **carrier**, whereas an individual homozygous for the condition is called an **affected indi-**

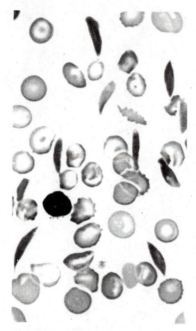

Fig. 17–4

Red blood cells from a person with sickle-cell anemia. Normal cells are discs. Sickle cells become sickle-shaped when the oxygen level in the blood is low. (Carolina Biological Supply Company)

Fig. 17–5

The metabolic blocks in phenylketonuria and albinism.

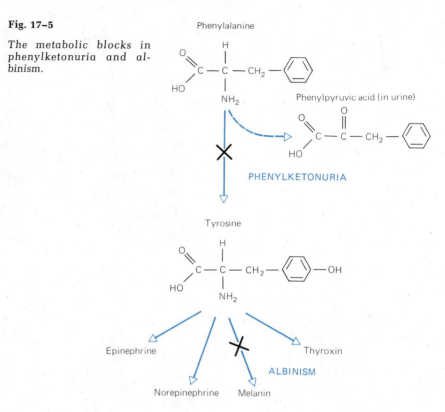

vidual. People heterozygous for the sickle gene are sometimes referred to as "having" sickle-cell anemia. This is unfortunate, since it suggests that the carrier is less fit than the normal homozygote, which is not usually the case.

17–B Inborn Errors of Metabolism

met-TAB-boll-ism

The sum of all the body's biochemical reactions is called its **metabolism**. Many genes code for proteins that are enzymes for one step or another in the body's metabolism. A gene that codes for an inactive or inefficient enzyme can give rise to an **inborn error of metabolism**; many of these are known.

FEN-nill-key-tone-YOUR-ee-uh

Albinism and **phenylketonuria** (**PKU**) are two human hereditary disorders that happen to be on the same metabolic pathway (Fig. 17-5). Victims of PKU are homozygous recessives who lack the enzyme that normally converts the amino acid phenylalanine to another amino acid, tyrosine; phenylalanine is converted instead to phenylpyruvic acid, which builds up in the bloodstream until it reaches toxic levels, damaging the brain and causing mental retardation if left untreated. Phenylpyruvic acid in the urine of these people gives a characteristic odor. PKU can now be controlled via a special low-phenylalanine diet during childhood. When brain development is complete, the PKU victim can adopt a normal diet; however, a woman with this condition must return to the low-

FEN-ill-AL-an-een
TIE-roe-seen

FEN-ill-pie-RUE-vic

phenylalanine regimen during pregnancy in order to prevent abnormal brain development in her fetus. Many states now require that newborns receive a blood test for PKU and several other metabolic disorders.

People homozygous for the albinism gene lack an enzyme that normally converts tyrosine to melanin, the pigment that makes eyes, hair, and skin brown or black. True albinos have white hair and very light skin and eyes. You may wonder whether victims of PKU are also albino, since they cannot make the tyrosine that is eventually converted to melanin. The answer is no, because tyrosine can be obtained in the diet as well as from conversion of phenylalanine. However, people with PKU usually have light coloring. A person could, of course, be both albino and PKU if he or she were homozygous for both genes.

MELL-uh-nin

17–C Genes with Multiple Forms

Up to this point, we have considered only genes with two distinct forms. Most genes consist of hundreds of nucleotides, and so mutations can occur in many different sites on a gene, producing many different forms of the gene. Of course, any one individual can contain no more than two different forms of any gene, since genes occur in pairs on homologous chromosomes.

Possibly the most familiar gene with multiple forms is found in the human ABO blood groups. There are three main gene forms in this system: I^A, I^B, and i. The I^A and I^B forms code for two different enzymes, each attaching a particular sugar to a protein on the surface of red blood cells. The i form does not code for an enzyme. Both enzymes are produced in a person having both the I^A and the I^B gene; i is recessive to I^A and I^B. Table 17-1 shows the possible genotypes and phenotypes in this blood group system.

In addition to the ABO blood group system, there are many other blood group systems, dictated by genes on various chromosomes; the Rh blood group is probably the most familiar.

Blood groups have an interesting application in deciding questions of parentage, such as in paternity lawsuits or in cases where someone suspects that babies may have been mixed up in hospital nurseries. Only a few drops of blood are needed to determine the blood types of the child and its supposed parents, and this genetic evidence can be used to decide whether a particular person or

TABLE 17-1 **POSSIBLE GENOTYPES FOR DIFFERENT BLOOD GROUPS (PHENOTYPES) IN THE HUMAN ABO BLOOD GROUP SYSTEM**

Blood group (phenotype)	Genotype	Frequency in U.S. population
O	ii	45%
A	$I^A I^A$ or $I^A i$	41%
B	$I^B I^B$ or $I^B i$	10%
AB	$I^A I^B$	4%

couple could have had a child of a particular blood type. Such evidence can show, for example, that a certain man could not have fathered a particular child, but it can never prove that he did. Since there are many men in the world with each blood type, having a blood type that makes paternity possible does not prove that a man actually is the father.

17–D Polygenic Characters

In contrast to multiple forms of a single gene, **polygenic characters** are traits governed by many genes at many different chromosome positions, possibly even on different pairs of chromosomes. Familiar examples in humans include height, intelligence, body build, and hair and skin color. Human skin color is a much-studied polygenic trait, and there is some debate over how many different genes are involved in determining this trait. A very dark-skinned person has genes which code for production of the dark pigment melanin at all the skin-color–gene positions, whereas a light-skinned person has many non–melanin-producing genes at these positions. The genes are thought to be additive: the more melanin-producing genes a person has, the darker the skin.

Polygenic characters are difficult to study since it is hard to disentangle the effects of each gene on a particular phenotypic character. Environment may further muddy the waters; for instance, in our example, people of any skin color can become somewhat lighter or darker depending on exposure to the sun.

Polygenic characters are so common that it is no wonder that people in the nineteenth century believed in blending inheritance; the phenotypes of various individuals in each species show a wide range, and offspring of parents at the extreme ends of the range generally come out about midway on the scale (Fig. 17-6).

17–E Sex Determination

In many species, the most obvious difference in phenotype between individuals is their sex. It is also one of the most far-reaching differences, for sex affects many organs that are not directly involved in sexual reproduction, and the sex hormones are influential in the phenotypic expression of many other genes. In most familiar animals, sex is genetically determined.

The simplest cross to produce half males and half females (a 1:1 ratio in the offspring) is one between a homozygote and a heterozygote. In humans and most other mammals, and in birds, this is what happens: one sex is heterozygous and one homozygous for an entire pair of chromosomes, called the **sex chromosomes** (Fig. 17-7). In most mammals, males are heterozygous for the pair of sex chromosomes, having one X and one Y chromosome; females have two X chromosomes. Birds are the other way around: females are heterozygous, with one Z and one W chromosome, while males are ZZ.

Available evidence indicates that all individuals have the genes needed to develop into members of either sex. For example, embryonic birds and mammals

PARENTAL
GENERATION

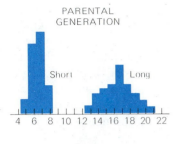

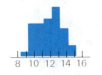

F₁ GENERATION

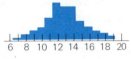

F₂ GENERATION

Fig. 17–6

Ear length in corn, an example of a polygenic trait. Crosses between a short-eared strain and a long-eared strain yielded F₁ plants with ears of medium length; the F₂ generation showed a wider spread in ear length, although few ears were as long or as short as those found in the original parental strains. Numbers = length of ears, in cm; height of bars = percent of individuals with each ear length in each generation.

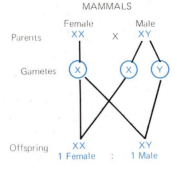

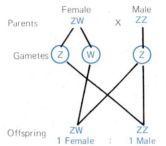

Fig. 17–7

How sex is determined by inheritance of two different chromosomes in mammals and birds.

have been induced to grow into members of the genetically "wrong" sex by hormone treatment. Under normal circumstances, hormones from the ovary or testis maintain the correct sex of each individual.

In humans, the Y chromosome seems to be somehow important in determining maleness. Most people with a Y chromosome are phenotypically male, while those with only X chromosomes are phenotypically female (Table 17-2). The abnormal sex chromosome numbers shown in Table 17-2 result from **nondisjunction** of the sex chromosomes during meiosis; the sex chromosomes fail to segregate normally during meiosis, and gametes form with one too many or one

segregate: separate into different nuclei during meiosis.

TABLE 17-2   PHENOTYPES FOR VARIOUS SEX CHROMOSOME COMPLEMENTS IN HUMANS

Sex chromosomes	Phenotype*
XX	Normal female
XY	Normal male
—	—
XXX	Female; fertile or sterile
X (Turner's syndrome)	Female; sterile, ovaries rudimentary or absent
XXY (Klinefelter's syndrome)	Male; possible mental retardation
XXXY	Male
XYY	Male; tall, acne-prone

*Defects in various genes involved in hormone production can alter the phenotype normally exhibited by a particular sex chromosome combination.

too few chromosomes. (Another instance of nondisjunction is **Down's syndrome**, in which chromosome number 21 or 22 (see Fig. 17-8) fails to segregate normally.)

Possession of a particular set of sex chromosomes causes an individual to develop into a male or a female via a process known as **sex differentiation**. Surprisingly, all of the genes involved in sex differentiation that have been studied thus far lie on chromosomes other than the sex chromosomes. As a result, we still do not know how the sex chromosomes influence sex!

17-F Sex Linkage

Although we have yet to discover any sex-determining genes located on the sex chromosomes, we do know of a variety of other genes carried by the X chromosome, and of a very few that may be on the Y chromosome. In mammals, parts of the X chromosome are homologous with parts of the Y chromosome, and the genes on these portions act like any other genes. A portion of the X chromosome has no corresponding bit on the Y chromosome, however, and genes located on this area are said to be **sex-linked** (Fig. 17-9). In male mammals, any recessive gene on the nonhomologous part of the X chromosome will be expressed in the phenotype, since there is no gene on the Y chromosome to mask it.

Fig. 17-8

The chromosomes of a normal human male. (a) The 46 sets of sister chromatids, stained and separated from a single cell about to undergo division. (b) A karyotype, made by cutting out the sets of chromosomes in part (a) and pairing them with their homologues in a conventional order; this is done to determine whether a person has a normal set of chromosomes (he does). (Carolina Biological Supply Company)

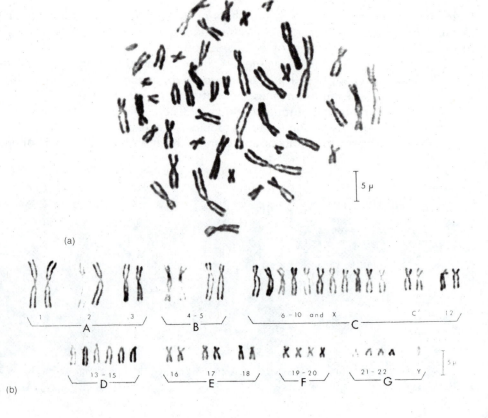

A female must have two copies of such a gene before it shows in her phenotype. Recessive sex-linked phenotypes, therefore, are more common in male mammals than in females. Since many recessive genes are deleterious, this is one of the reasons that more male than female mammals of any age die.

Red-green **color blindness** and hemophilia are two well-known recessive sex-linked traits in humans. Let us consider a cross between a woman homozygous for normal color vision and a color-blind man. All the children of this marriage will have normal color vision, since all receive an X chromosome bearing a normal gene from their mother (Fig. 17-10). Imagine that the daughters in this family marry men with normal color vision, while the sons marry women who are carriers of the gene for color blindness (this is equivalent to a brother-sister marriage with respect to color blindness). In the next (F_2) generation, we would find the ratio of 3 normal vision to 1 color-blind, as expected from a cross involving one gene pair. Our results have this added twist, however: all the color-blind children are male! The appearance of the recessive trait in one sex much more often than in the other is the hallmark of a sex-linked trait.

Hemophilia is caused by a recessive gene that occurs only on the X chromosome. If a woman has the gene she usually also has a dominant, normal gene at the hemophilia position on her other X chromosome and so she does not have

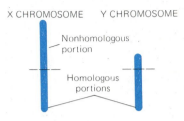

Fig. 17–9

The X and Y chromosomes of male mammals are homologous for part of their lengths, but a large part of the X chromosome has no corresponding gene positions on the Y chromosome.

HEE-moe-FEEL-ee-uh

Fig. 17–10

Inheritance of human red-green color blindness, controlled by genes on the X chromosome. In a marriage of a woman homozygous for normal color vision and a color-blind man, all the children have normal color vision. If the children marry people with the same genotypes as their siblings, half of the male offspring are expected to be color-blind, inheriting the mother's X chromosome bearing the color-blindness gene. Girls that inherit this maternal X chromosome receive an X chromosome with a normal gene from their fathers, and so they are not color-blind.

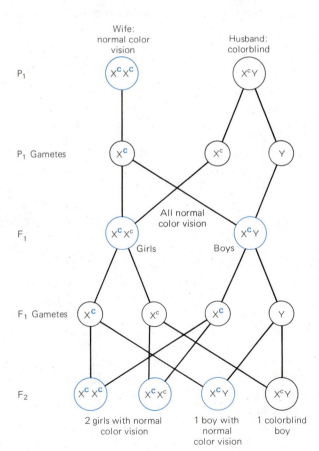

hemophilia. A man with the gene, on the other hand, has no second X chromosome bearing a normal gene and so his hemophilia gene will be expressed. A person with hemophilia produces very little of a protein needed for blood to clot and so may bleed to death after even a slight cut. Hemophilia can now be controlled (but not cured) by injection of clotting factor extracted from normal blood. As a result, some men with hemophilia now live to grow up and reproduce. If a man with hemophilia marries a woman who is heterozygous for the gene, a daughter with hemophilia may be born. (A daughter would have to inherit X chromosomes bearing the gene from *both* of her parents to have hemophilia; in the past, men with hemophilia seldom lived to become fathers, and so there are no records of females with hemophilia.)

Queen Victoria was the world's most famous hemophilia carrier. Her hemophiliac son, Leopold, Duke of Albany, and her two carrier daughters, Princesses Alice and Beatrice, spread the gene through the royal houses of Europe, including those of Russia, Prussia, and Spain. For a time hemophilia was called the "royal disease," but fortunately no modern monarchs have inherited the gene.

17–G Some Factors Influencing Gene Expression

All the genes possessed by an individual determine its genetic potential: what might be. What actually happens, however, is another matter. Embryonic development is influenced by the interaction of all the genes, as they express themselves by producing or failing to produce polypeptides and proteins. Environmental factors also play an important role in development; in the last decade or two there have been several true "horror stories" in which drugs taken by pregnant women have caused improper development of the fetus, or cancers (loss of control of cell division) later in the baby's life.

Although their main role is influencing the organs involved in sexual reproduction and related functions, sex hormones may also influence other characters. **Sex-influenced genes** express themselves to a greater or lesser degree as a result of the level of sex hormones. For example, a bull may have genes for high milk production, but he will not produce milk because he has only low levels of female hormones. However, because of these genes, he would make an excellent sire for a dairy herd. Similarly, males and females both have the genetic potential to produce the organs characteristic of the opposite sex, but they develop organs typical of their own sex because they have more of the appropriate hormones (both males and females have hormones characteristic of the opposite sex, but at much lower levels).

In humans the gene for male pattern baldness, while not located on the sex chromosomes, is nevertheless much influenced by the presence of male hormones. In men, the gene acts as a dominant because of the presence of male sex hormones; in women, it acts as a recessive, so that a female must have two doses of the gene before she loses her hair.

Gout is another trait whose expression is influenced by sex. In gout, painful deposits of uric-acid salts build up in the tissues, especially in the joints of the

Fig. 17–11

Male pattern baldness is a sex-influenced trait; its expression is enhanced by the presence of male sex hormones.

big toes. The gene for gout is expressed much more in the presence of male than of female sex hormones. In Victorian literature, gout figured largely as a reason for the temper tantrums of crotchety old men. Avoiding red wine and rich and spicy foods was supposed to alleviate the condition, but this treatment tried the temper of its victims still further. Fortunately, gout can now be treated.

Sex hormones are only one of many factors that can influence the phenotypic expression of a gene. Many traits are controlled mainly by one gene pair, but are influenced to some extent by the products of other genes, called **modifier genes**. It was long believed that eye color in humans was controlled by a single pair of genes, with brown eyes dominant to blue. It is now known that there are also at least two pairs of modifier genes involved, and it is possible, though extremely uncommon, for blue-eyed parents to have brown-eyed children.

Gene expression is also influenced by factors in the external environment. A Himalayan rabbit grows black fur only on those parts of the body which remain cool (Fig. 17-12); a human being becomes darker (or redder!) when exposed to bright sunlight for a time. A person with a poor diet does not reach the height made possible by his or her genes; in many countries, the modern generation towers above its parents as a result of improved nutrition. Different hormones may be produced at different times of life and so age also plays a part in gene expression; consider the many changes accompanying puberty, such as voice change and growth of the testes in males; breast enlargement and a characteristic pattern of body fat deposition, giving the rounded contours of the female figure; and growth of hair in the armpits and pubic area in both sexes.

SUMMARY

Genes express themselves by coding for polypeptides or proteins. The severity of a mutation depends on how much it affects the protein encoded by the gene. Some mutations result in lethal genes, which cause death if not counteracted by a normal copy of the gene. Most familiar lethal genes are recessive, and cause death only in the homozygous condition.

Changes in less vital proteins may cause metabolic disorders, as exemplified by albinism and phenylketonuria.

Mutations at different places in a gene may result in the existence of several different forms of a gene. Such multiple forms are found in the human ABO blood group. Many characters are determined by several different gene pairs at a time; these polygenic characters show a wide range of phenotypes.

Sex is determined by the presence of like or unlike sex chromosomes; in humans and most other mammals, females have the sex chromosome combination XX and males are XY. Traits carried on the sex chromosomes are said to be sex-linked, whereas those on other chromosomes, but depending on sex hormones for their expression, are sex-influenced.

An individual's phenotype depends on what mix of genes it has, how these genes are influenced by the proteins (or molecules such as hormones or pigments produced by these proteins) encoded by other genes, and what factors of the external environment impinge on it. External factors influencing gene expression include temperature, light, and diet.

Fig. 17–12

Expression of coat color genes in the Himalayan rabbit is related to skin temperature. Black fur grows on parts of the body with skin temperature below 33°C. If fur is shaved from a warmer part of the body, and an ice pack applied while the fur grows back, the new fur will also be black.

SELF-QUIZ

1. Review the information on brachyphalangy, Section 17–A.
   a. If two brachyphalangic people marry, what are their chances of having a child with normal fingers?
   b. If a brachyphalangic person marries a normal person, what phenotypic ratios can be expected among their offspring?

2. A geneticist studying the various genes that govern coat color in mice is trying to develop true-breeding strains of each possible coat color. He carries out several generations of matings among mice with yellow coats and always obtains some offspring of other colors.
   a. What does this indicate about the genotype of yellow mice?
   b. The geneticist tallies up his results over several generations and finds that he has obtained a total of 184 yellow mice and 95 of other colors. What does this suggest about the nature of the yellow gene?
   c. Why did the geneticist never obtain a homozygous yellow mouse?
   d. How could he prove what became of the homozygous yellow offspring?

3. A pedigree of ABO blood groups for several generations of humans is shown below. Circles represent females; squares represent males. Marriages are represented by horizontal lines directly connecting two people, and children are connected to their parents by a vertical line down from the marriage line. For example, (b) and (c) are married to each other, and (d) is one of their two sons. Give the possible genotype(s) for each individual marked with a letter.

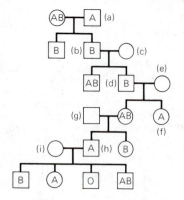

4. Boom-Boom Bustley, popular chorus dancer at the "Tie Your Garter" Club in Las Vegas, is suing Sonny Rae, stand-up comedian at the club across the street, for support of her child, which she claims was fathered by Rae. Blood tests show that Ms. Bustley's blood type is B, Mr. Rae's is O, and the baby's type is also O. What should the jury decide?

5. What would be the expected genotypic ratios among the children of a woman whose father had hemophilia, and whose husband is normal?

6. Under what circumstances is it possible for both a father and his son to have hemophilia?

7. Red-green color blindness in humans is a sex-linked recessive trait.
   a. In a large family in which all the daughters have normal vision and all the sons are color-blind, what are the probable genotypes of the parents?
   b. If a normal-sighted woman whose father was color-blind marries a color-blind man, what is the probability that their son will be color-blind?
   c. What is the probability that the couple in (b) will have a color-blind daughter?

8. If a species of mammal has some members which carry a sex-linked lethal trait that causes early death and reabsorption of the embryo, what sex ratio would be expected among the offspring of a female carrier and a normal male?

9. It is often said that men inherit baldness from their maternal grandfathers via their mothers. In light of what you have learned about this trait, is this a valid statement? Explain.

QUESTIONS FOR DISCUSSION

1. One problem with genetic counseling is that people who learn they are carriers for genetic diseases such as hemophilia, Tay-Sachs, sickle-cell anemia, or phenylketonuria may consider this a terrible stigma. Men have even been known to deny paternity of their children and divorce their wives for infidelity when told that the child had inherited a deleterious recessive gene from each parent. What kinds of arguments and counseling would you use, if you were a genetics counselor, in an attempt to induce a healthier, more productive response to such a discovery?

2. People who are carriers for sickle-cell anemia have blood of lower oxygen capacity than most people, so they are probably exposed to greater than usual risks if they become divers, jet pilots, or mountaineers. Otherwise they have no physical handicaps; nevertheless, they have frequently been denied access to various professions as a result of the common ignorance and prejudice against genetic disorders. Since this is the case, a proposed nationwide screening for sickle-cell carriers may well do more social harm than good. Is it not better for carriers to remain in ignorance of genetic conditions about which nothing can be done? If not, why not?

3. Until the advent of modern technology, hemophiliac men usually died before they reached reproductive age. Nowadays they can be provided with "clotting factor," a blood extract that permits them to lead normal lives and live to have children. The treatment costs about $25,000 a year per person. Can society or should society insist that such men be sterilized so that they cannot perpetuate their disease if taxpayers have to pay the bill for their medication?

4. Table 17-2 shows that individuals with a single X chromosome are known to occur, but not individuals with only a Y chromosome. Why do you think this is?

5. Every so often the Ann Landers column has a letter from a mother whose husband or in-laws have been chiding her for having daughters instead of sons. Is this censure justified? Why?

6. Name some factors besides those mentioned in the chapter that may influence gene expression.

SUGGESTED READINGS

Bridges, C. B. "Sex in relation to chromosomes and genes." *American Naturalist,* **59**:12, 1925. The classic treatise on sex determination.

Harrison, D. *Problems in Genetics with Notes and Examples.* Reading, MA: Addison-Wesley, 1970. A useful review and practice book.

Mendel, G. J. *Experiments in Plant Hybridisation.* Edinburgh: Oliver and Boyd, 1965. An English translation of Mendel's original paper, together with comments and a biography of Mendel by others.

Sayers, Dorothy L. *Have His Carcass.* New York: Harcourt, Brace, Jovanovich, 1932. A mystery novel about a human genetic trait.

Stern, C. *Principles of Human Genetics,* 3rd ed. San Francisco: W. H. Freeman, 1973. An excellent human genetics text.

Watson, J. D. *Molecular Biology of the Gene,* 3rd ed. Menlo Park, CA: W. A. Benjamin, 1976. A very clearly written molecular biology textbook.

Winchester, A. M. *Heredity, Evolution and Humankind.* St. Paul, MN: West, 1976. Discusses genes, their evolution and expression; contains a good section on human genetics.

Part Four

PLANTS

18 PLANTS: THE INSIDE STORY

OBJECTIVES

When you have studied this chapter, you should be able to:

1. List the functions of the stems and leaves.

2. Sketch or describe the structure of a leaf, and explain how this structure is related to the leaf's role in photosynthesis.

3. List four functions of roots, and compare the advantages and disadvantages of fibrous versus taproots in carrying out these functions.

4. List the components of a good soil, and explain the importance of each to the plants growing in it.

5. Name the two types of vascular tissue in plants, describe or sketch their location in the plant, and state their functions.

6. State, and briefly explain, two mechanisms that may move sap upward in plants.

7. Explain how materials move in the phloem of a plant, and state what determines the direction of movement.

Biophoto Associates

Fig. 18–1

Sunlight shining through a tree canopy. The leaves grow so that they intercept a maximum amount of light. (Biophoto Associates)

Imagine spending your entire life with the lower half of your body buried in damp earth, and the upper half basking in the warm sun by day, chilled by the dew at night, and bowing as the breezes pass. Nowhere to go. Nothing to do. It is not a life that appeals to even the most indolent among us, but for most plants it is the only way to survive.

Plants, of course, are photosynthetic organisms, using solar energy to make their own food from simple inorganic materials. Given enough light and water, minerals from the soil, and carbon dioxide from the air, a plant can make all the complex chemicals it needs to sustain life. In this chapter we shall examine how the body of a flowering plant is built and how its architecture allows it to carry out its life processes. Much of what we say is also true of non-flowering plants, such as club mosses, ferns, and gymnosperms (Section 2–H).

18–A The Plant Body

The leafy canopy.   In most plants, the task of capturing sunlight falls largely to the leaves. If we look attentively at a plant, we notice that the leaves are arranged in such a way that they maximize the plant's ability to intercept light (Fig. 18-1). Leaves in general are broad and flattened, spreading a great deal of surface area to the sun, while at the same time minimizing the weight that must be supported. The green, photosynthetic tissue of a leaf is stretched out amid a supporting network of **veins** (Fig. 18-2). The veins are composed of **vascular tissue** (tissue specialized for transport), which conducts water to the leaf and takes manufactured food back to the rest of the plant, in addition to providing support.

Although its place in the sun assures a leaf of the energy it needs for photosynthesis, it also carries the threat of desiccation: the heat of the sun readily vaporizes water, which may then leave the leaf and drift away into the air. The **epidermis** (the outer layer of leaf cells) secretes a layer of waxy **cuticle** that

desiccation   (dess-ih-KAY-shun): drying out.

epp-ih-DERM-iss
CUTE-ih-cull

(a)

(b)

Fig. 18–2

Photosynthetic structures. (a) Cells containing green, photosynthetic chloroplasts. (b) Veins support a leaf's soft photosynthetic tissue and provide a transport network. (Biophoto Associates)

impermeable: not allowing the passage of the named substance from one side to the other.

stow-MAH-tah (sing.: stoma; STOW-mah)

MEE-so-fill

retards evaporation and thus helps the plant to maintain a favorable water balance (Fig. 18-3). An especially thick cuticle imparts the characteristic shine to the leaves of many favorite house plants.

The cuticle, however, presents another problem: it is impermeable not only to water vapor, but also to carbon dioxide and oxygen, which the plant must obtain from the atmosphere. If we were to peel the epidermis off the underside of a leaf, however, and look at it through the microscope, we would see that it is perforated with tiny pores called **stomata** (= "mouths") (Fig. 18-4). Each of these is surrounded by two liplike **guard cells**, which can open or close the pore depending on the plant's need for more carbon dioxide versus its need to conserve water. The stomata open into air spaces inside the leaf, where carbon dioxide can circulate to the adjacent photosynthetic cells. Cells in the middle two layers, the **mesophyll**, contain most of the chloroplasts, the organelles where photosynthesis occurs. With the vascular tissue to bring it water, the stomata and air spaces to take in carbon dioxide, the broad surface area to absorb sunlight, the cuticle to retard water loss, and, most importantly, the chloroplast-packed mesophyll, leaves are well adapted to carry out photosynthesis.

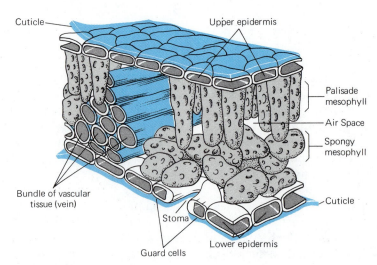

Fig. 18–3

Cross section through a leaf.

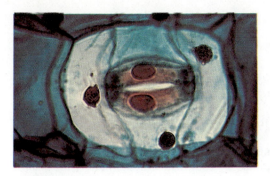

Fig. 18–4

A stoma ("mouth") between a pair of lip-like guard cells on the underside of a Tradescantia leaf. The guard cells are pink. Cell nuclei are red. (Carolina Biological Supply Company)

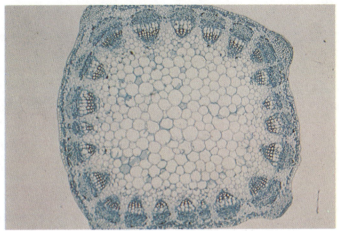

Fig. 18–5

Cross section of a stem of alfalfa. The vascular bundles are arranged in a ring near the outside of the stem. They transport materials up and down the stem between the roots and leaves, and their thick cell walls help to support the plant. (Carolina Biological Supply Company)

Stems.   The stems of a plant serve a variety of functions. First, they support the leaves, holding them up to the sun. The sturdy vascular tissue of stems is important for this function (Fig. 18-5). Wood is composed entirely of vascular tissue. In soft, **herbaceous** stems, the water content of other tissues in the stem is also very important for support; think of how little support a flower has when its stem loses water and wilts.

Stems also transport substances to and from the leaves, again in vascular tissue. The vascular tissue of stems is continuous with that in roots and leaves, and it forms the central part of the pathway from root tip to shoot tip, transporting water and food from one to the other, and to points between.

Stems may serve other functions: the stems of many herbaceous plants are green and carry out photosynthesis; the stems of cacti store water and carry on photosynthesis as well; and underground stems, such as the familiar potato, hide stored food from hungry herbivores.

Like the leaves, young stems are protected from the dry air by a cuticle perforated with stomata; in older stems, the epidermis bursts as the stem grows and is replaced with thick, protective **bark**.

The plant underground.   There is more to a plant than meets the eye. While the leaves spread out in the light and air, the roots weave through the soil, carrying on their own tasks. Chief among these are gathering water and minerals needed by the photosynthesizing leaves and anchoring the plant body.

A single root grows out from a sprouting seed, and in some plants this remains the largest and most prominent part of the root system, with small, thin side branches. Such a root is called a **taproot** (Fig. 18-9a). A taproot is an admirable anchor, and since food coming down from the top of the plant passes first into the taproot, it is also ideally situated to be a food depot. Many familiar vegetables are taproots of this sort—carrots, beets, turnips, and radishes, to name a few. In most plants, however, the first root gives off branches that rapidly overtake it in size, or roots form at the base of the stem, forming a fibrous root

Essay:
WHEN DOES A PLANT EAT WITH ITS LEAVES?

Many plants can take up nutrients through the leaf surfaces. This ability is carried to fascinating lengths by carnivorous plants, which not only absorb food through the leaves, but first use the leaves to capture and digest animal prey.

Carnivorous plants inhabit acid bogs. Acidity retards the growth of bacteria that release nutrients from dead organic matter, and also makes the nutrients that are released very soluble, so that they leach away easily. Thus, many plants in acid bogs cannot obtain adequate supplies of nutrients through their roots. Indeed, the roots of carnivorous plants are very small, serving mainly to anchor the plant and absorb water. The animal prey caught by the leaves of carnivorous plants is rich in protein, a good source of nitrogen, and also contains other minerals that plants need. Like other green plants, carnivorous plants produce sugars by photosynthesis, and indeed they find little competition for sunlight because few non-carnivorous plants can survive in their habitats.

There are three basic types of carnivorous plants: active traps (Venus flytrap), pitfall traps (pitcher plants), and flypaper traps (sundews). All three have certain adaptations in common. The trap is a modified leaf, supplied with nectar glands that exude substances attractive to prey; in most other plants, nectar glands are confined to the flowers (which also consist of modified leaves). The trapping leaves also have glands to produce enzymes that digest trapped insects. A third feature in common is the modification of leaf hairs in ways that aid in capturing prey.

(a)

(b)

Fig. 18–6

Venus's flytrap, an endangered carnivorous plant native only to the Carolinas. (a) An entire plant; the bilobed leaves have a red lining attractive to insects. Sensitive hairs on the inner surface of the leaves respond to an insect's jostling by starting an electrical impulse that causes rapid changes in the water content of certain leaf cells; the leaf folds up quickly and imprisons the prey. Hairs along the edges of the leaf form a cage around the prey. (Carolina Biological Supply Company) (b) Part of a fly protruding from a trap. Digestive glands on the interior surface of the trap secrete enzymes that digest away the insect's soft parts. The rest of the fly blows away when the trap re-opens in readiness for another victim. (Biophoto Associates)

(a)   (b)

Fig. 18–7

A pitcher plant, a pitfall type of carnivorous plant. (a) The leaves of pitcher plants are tubular; a hood over the opening prevents entry of rain. Insects are attracted by nectar-secreting glands on the lip of the pitcher. Just beyond these glands is a slick area where unwary insects slip and plunge to a pool of digestive juices below. Downward-pointing hairs inside the pitcher cause the prey to skid into the pitcher and prevent them from crawling back out. (Biophoto Associates) (b) A leaf cut in half to show trapped insects. (Carolina Biological Supply Company)

Fig. 18–8

Sundew, a flypaper-type carnivorous plant. (a) Each leaf is covered with hairs that secrete glistening, sticky droplets attractive to insects. (Biophoto Associates) (b) Once an insect becomes entangled in the sticky secretion, nearby hairs grow towards it and hold it more firmly. Again, digestive enzymes are secreted and the digestion products absorbed through the leaf surface. (Carolina Biological Supply Company)

(a)   (b)

Fig. 18–9

Root systems. (a) Taproot system of an oak seedling, with one long taproot and many smaller branch roots. (b) Fibrous root system with many roots of about equal length and thickness growing from the spreading underground stem of a violet plant.

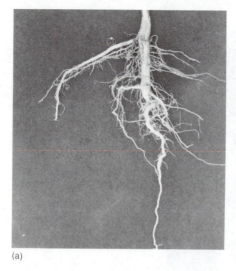

(a)

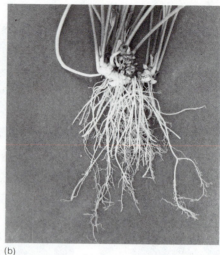

(b)

system, with no distinct main member (Fig. 18-9b). The spread of a fibrous root system enables it to absorb and use the water and minerals from a large volume of soil. In woody plants, spreading roots also help to support the top of the plant. Many tree seedlings start life with taproots but eventually produce well-branched systems; the notion of a pine or oak tree as a giant carrot is incorrect.

Cells in the outer area of the newer roots take in water and minerals and move them inward toward the center of the root. Here lies the root's vascular tissue, which is continuous with the vascular tissue of the stems and leaves (Fig. 18-10). Because the absorptive areas tend to be near the ends of the root system, it

Fig. 18–10

Cross section of a buttercup root. (a) The small core of vascular tissue in the center of the root is surrounded by root cortex whose cells contain many small, round starch-storing bodies, stained purple in this preparation. (b) Closeup of the central cylinder of vascular tissue. The "star" of large, thick-walled purple-stained cells is the xylem, the vascular tissue that transports water. Between the arms of the star lies the phloem, the vascular tissue that transports food. (Carolina Biological Supply Company)

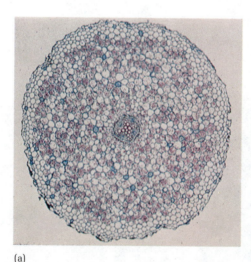

(a)

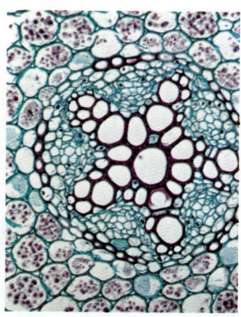

(b)

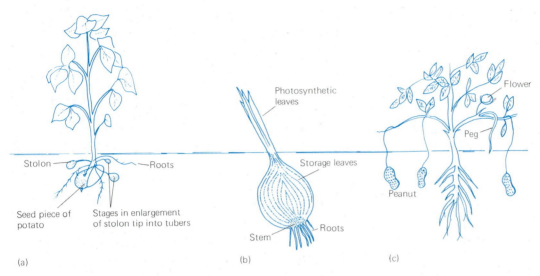

Fig. 18–11

Underground, but not roots. (a) A potato is an underground storage stem (tuber). Potato plants are grown from pieces of potatoes, each containing an "eye" (bud) that sprouts to form the stem and leaves. As food is produced, it is stored in modified underground stems, the stolons, whose tips swell and form tubers. (b) An onion is composed of modified, fleshy food-storing leaves formed underground at the bases of the green photosynthetic leaves. (c) The peanut plant bears flowers above the ground, but after the petals fall the female part of the flower elongates, forming a "peg" which is driven into the loose, sandy soil. Here it enlarges as the peanut shell, enclosing the nutritious seeds, the peanuts.

is important to move a large ball of earth when transplanting a shrub or tree. The larger roots close to the base of the stem (or tree trunk) have such thick coverings that they cannot absorb water, but only transport that taken in by the smaller roots.

Roots are not the only plant structures found underground. As mentioned before, underground stems may store food (potatoes and irises); they may also run beneath the soil and send up new clusters of leaves at intervals (as do many kinds of weeds). Onions and similar bulbs are clusters of buried, food-storing leaves. The peanut plant even buries its young fruits after the flowers wither, and the seed pods (peanut shells) develop under the surface of the sandy soil (Fig. 18-11).

18–B Roots and Soil

The health of a plant is intimately linked to the condition of the soil around its roots. Soil is mostly rock particles formed as the underlying rock breaks down. The type of rock determines what minerals are present in the soil.

The soil's water content, of course, is crucial to the plant's well-being. Soil can hold water on the surface of each particle and in narrow spaces between

Clay

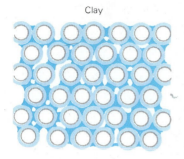

Sand

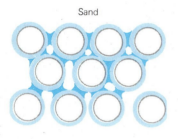

Fig. 18–12

Comparison of water held by clay and sandy soils. Light color represents water held as a film on the surface of each particle; darker color is water held by capillarity (Section 9-E) in the tiny crevices between particles. Narrower spaces (in clay) hold water more tightly than wider spaces.

particles. The smaller the particles in a soil, the more water that soil can hold, since smaller particles have a greater collective surface area and more small-sized spaces between particles than an equal volume of larger particles. In fact, clay soils, with the smallest particle size, tend to hold water so tightly that plants cannot withdraw much of it. The large particles of a sandy soil, on the other hand, often let so much water drain out that the soil is too dry for most plants. A soil with a mixture of particle sizes is often best. Rainfall exceeding the soil's water-holding capacity drains away, carrying valuable minerals with it; this is called **leaching**.

Organic matter is a valuable component of soil. Manure, dead leaves, and bits of wood act like sponges, soaking up water and swelling when it rains, and releasing water slowly later. The alternate shrinking and swelling of organic particles keeps rock particles loosened so that roots can grow through the soil easily. Organic matter on top of the soil (mulch) acts as a sunshield, retarding evaporation of water from the soil to the air.

Does organic matter make better fertilizer than inorganic products? Plants can take up nutrients only in certain chemical forms, and inorganic fertilizers generally provide these forms, bypassing the steps of microbial digestion needed to make organic fertilizers ready for plants. There is probably no valid basis for the claim of superior nutrition from organic fertilizers, but they do provide soil conditioning as well as minerals. In addition, the slow decomposition of organic matter releases minerals into the soil gradually, allowing plants a sustained source of nutrients. By contrast, some nutrients in inorganic fertilizers may leach rapidly out of the soil after each treatment.

Soil organisms are an important component of the soil because they digest organic matter and release its minerals. Legumes—the pea family, including beans, clover, alfalfa, vetch, peanuts, and some trees—develop root nodules that house **nitrogen-fixing bacteria** (Fig. 18-13). The bacteria convert atmospheric nitrogen (N_2), which plants cannot use, into ammonium (NH_4^+), a form plants can use to make proteins. Because legumes increase the soil's nitrogen content, they are planted as part of crop rotation programs.

Fig. 18–13

Root nodules on a cowpea plant. (Carolina Biological Supply Company)

Fig. 18–14

Sunflower seedlings grown in various nutrient solutions. Left to right: nutritionally complete medium, followed by media lacking only sulfur, nitrogen, phosphorus, and potassium, respectively.

Many fungi in the soil grow close around, or even into, the roots of plants, forming mutually beneficial associations. The fungi have superior ability to absorb minerals from the soil solution, and they pass these on to the roots; the roots release organic molecules that the fungi can use as food. Most familiar forest trees form these associations with soil fungi; indeed, pine seedlings imported into Puerto Rico and Australia grew poorly until supplied with fungus-containing soil from thriving North American pine forests.

Oxygen in the soil is important because most living things, including plant roots, need oxygen to carry out respiration. There is little oxygen in a water-logged soil. The burrowing of earthworms and insects loosens the soil and allows oxygen to penetrate more easily.

Minerals dissolved in the soil water are carried to the root cells' membranes. Like all cell membranes, those of root cells are selectively permeable, admitting some substances more readily than others. There are special transport mechanisms for taking in larger amounts of some substances, such as nitrogen, potassium, phosphorus, and sulfur (Fig. 18-14). Nitrogen, phosphorus, and potassium are the soil minerals that plants require in the greatest quantities; commercial fertilizers are rated by the percentage of each; "5-10-5," for example, contains 5% nitrogen, 10% phosphorus, and 5% potassium, by weight. Plants also require substantial amounts of calcium and magnesium, which are plentiful in most soils; they need much smaller quantities of iron and such **micronutrients** as zinc and copper.

Minerals in the soil are not necessarily available to the roots of plants. Both minerals and soil particles carry electrical charges; often the soil particles with negative charges bind positively charged minerals, such as iron, so strongly that they will not dissolve and disperse in the soil water. The acidity of the soil also affects the solubility of minerals and in this way affects the nutrition of plants; some plants require acid soil, but others grow better in a limey (alkaline) soil.

Although a plant is selective about the substances it admits, it cannot totally prevent the entry of substances that are unnecessary, or even toxic, to living organisms. For example, experimenters have tried to recycle nutrient-rich wastes from sewage treatment plants as fertilizers and have found that the plants take up pollutants in addition to nutrients. Wastes from industrial towns often contain such high levels of toxic substances—such as selenium, antimony, cadmium, and tin—that plants fertilized with the wastes contained more toxic substances than is considered safe for human consumption.

18–C Internal Transport

A plant's roots absorb water and minerals but cannot feed themselves in the darkness of the soil. The leaves rely on water and minerals from the roots, which they in turn supply with food. The plant can carry out this complementary division of labor between relatively distant organs because of its vascular tissue, a transport system linking its parts and moving substances between them.

ZIE-lum
FLOW-um

Mal-PIG-ee

The vascular tissue is of two types, called **xylem** and **phloem**, which lie close together. The wood of a tree is xylem; the tree's phloem lies just inside the bark, and will come off attached to bark peeled away from the underlying wood. In 1679, the Italian scientist Marcello Malpighi performed an experiment which demonstrated the functions of the two tissues. Malpighi girdled a tree, a procedure that involves peeling off the bark in a complete ring around the tree. This disconnects the phloem above the girdle from that below, but leaves the xylem intact. After the girdling treatment, a swelling appeared in the bark just above the stripped area; the fluid exuded from the swelling was sweet (we now know that it contains sugar). The leaves seemed to be unaffected for many days. Eventually, however, the leaves wilted and then died, and the entire tree was soon dead (Fig. 18-15).

From these observations, Malpighi concluded that phloem transports food, such as the sugar in the fluid exuded from the bark, to the roots (we now know that it also transports food from leaves or food storage areas to growing leaves, stems, flowers, or fruits at the top of the plant as well). With their food supply cut off, the roots died when they had used up their stored food. The leaves remained

Fig. 18–15

A girdling experiment performed by Malphighi. (Left) The bark, including the phloem, was removed in a complete ring around the tree. (Center) A sugary fluid leaked out of the phloem above the girdle, but the leaves remained green for a time, supplied with water through the still-intact xylem. (Right) The wilting and death of the leaves showed that the roots had finally died of starvation, cut off from the food normally supplied by the phloem.

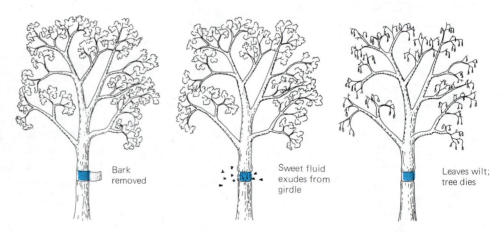

Bark removed

Sweet fluid exudes from girdle

Leaves wilt; tree dies

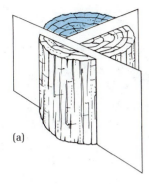

(a)

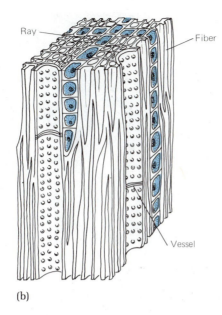

Ray

Fiber

Vessel

(b)

Fig. 18–16

Structure of wood in an angiosperm (flowering) tree. (a) The wood is cut along the planes shown here. (b) A block of wood corresponding to the colored section of part (a). The vessels conduct sap, and the fibers are nonconducting cells that give extra strength. Rays are made up of living cells specialized for conduction sideways (for instance, to the living cells in the bark area); they may also store small amounts of materials.

healthy for a time, and Malpighi concluded that xylem transports water to the leaves, because leaves picked off a plant and left without water die within hours.

We now know that xylem transports a mixture of water and minerals, called **sap**. But how does the xylem move the sap to the leaves? The answer is especially hard to imagine for tall trees, which may reach heights of almost 100 metres; the observation of Malpighi and other microscopists that the cells of the xylem are just so many tiny dead and empty pipes stacked from one end of the tree to the other makes the problem even more perplexing (Fig. 18-17).

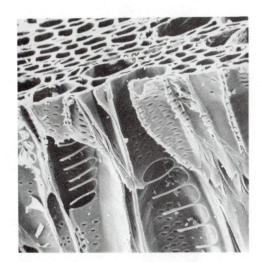

Fig. 18–17

Xylem tubes in a piece of wood as they appear with the scanning electron microscope. (Biophoto Associates)

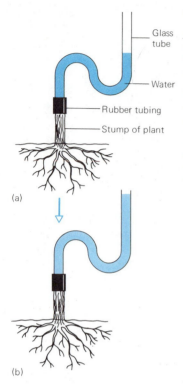

(a)

(b)

Fig. 18–18

Demonstration of root pressure. (a) A curved glass tube is sealed to a freshly decapitated plant and is then filled with water. (b) Sap rising into the tube from the roots raises the water level in the tube until the hydrostatic pressure, exerted by the weight of the water held up in the tube, equals the root pressure; the pressure can be calculated from the final height of the water in the tube.

Fig. 18–19

Root pressure in short plants may give rise to guttation, the exuding of sap droplets from the ends of the leaves. (Biophoto Associates)

Fig. 18–20

Stephen Hales found that sap movement depends on leaves by comparing the rate of water absorption to the surface area of the leaves on different branches.

How a little sap goes a long way. One way that sap moves in some plants is by a push from below, known as **root pressure**. We can demonstrate this by cutting a plant off at its base and fixing a glass tube over the stump (Fig. 18-18). The roots will push sap up into the tube, sometimes to a height of a foot or more. To do this, the roots must be well supplied with oxygen, for they are using energy to take up minerals by active transport, so fast that water follows by osmosis and becomes trapped in the xylem. More water coming in behind pushes from below, so that the sap has nowhere to go but up (Fig. 18-19).

However, not all plants can produce root pressure, and many plants are much taller than the heights attained by sap under root pressure. We must look for another mechanism to account for the rise of sap to the tops of tall trees.

In 1727, the English clergyman Stephen Hales demonstrated that **transpiration**, the evaporation of water from leaves, can pull sap up through a plant. Hales cut similar leafy branches from a tree. He removed varying quantities of leaves from the branches and set each branch in a container with a measured amount of water. Hales found that branches with leaves drew much more water out of their containers than did leafless boughs; furthermore, the amount of water removed from the container was roughly proportional to the area of leaf surface on the branch (Fig. 18-20). He decided that it was some activity of the leaves that caused sap to rise in the plant.

In another experiment, Hales found that this movement occurred only when the leaves were dry and exposed to air (Fig. 18-21). A branch with its leaves underwater could not "perspire" (or transpire, as we would say today), and thus little water moved through the branch. When the leaves were dry, however, they "perspired" freely into the air and pulled water up into the tube.

In yet another experiment the energetic Hales dug a deep hole next to a pear tree, exposing a root. He attached tubes of water and mercury to the root in series

Fig. 18–21   CHAPTER 18 PLANTS: THE INSIDE STORY ———————— 321

In this experiment, Hales showed that water moves through a plant only when its leaves are dry. A glass tube 7 feet long was attached to a branch and filled with water. (a) No water moved through the branch when the leaves were immersed in water, even though the water in the tube was being pulled down by the force of gravity. (b) However, when the leaves were dry the branch would pull water up the tube against the pull of gravity.

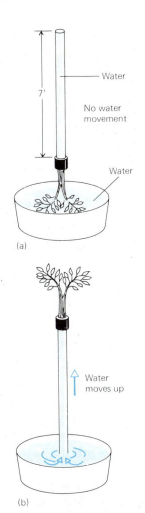

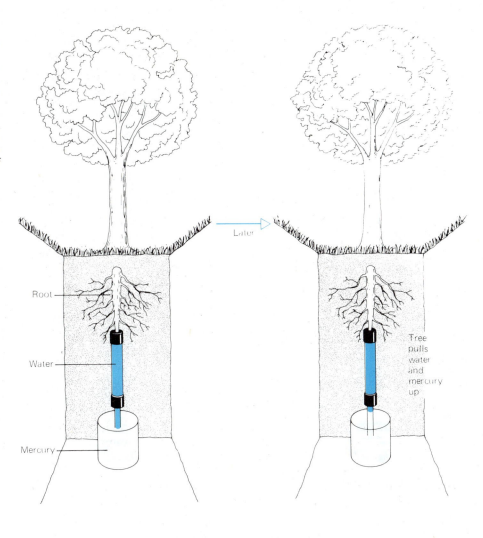

Fig. 18–22

Hales measured the pull exerted by a transpiring tree by attaching a water-filled tube to the cut root of a pear tree. He then measured how far mercury in contact with the water was pulled up into the tubes as the tree withdrew water.

so that he could measure the pull exerted by the root (Fig. 18-22). The root pulled more strongly on sunny, dry days than on cloudy or damp days, and the pull slackened at night. These were exactly the results to expect if evaporation of water from the leaves were indeed the driving force that pulled water up through the tree.

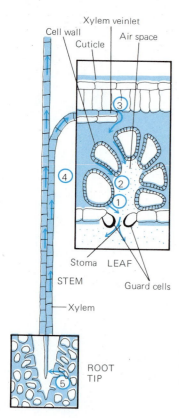

Xylem veinlet
Cell wall
Cuticle
Air space
3
4
2
1
Stoma LEAF
STEM
Guard cells
Xylem
ROOT TIP
5

Fig. 18–23

Movement of the "transpiration stream" through a plant depends on the cohesiveness of water and its natural tendency to move into a water-deficient area. Movement begins as water evaporates into the air spaces of the leaves (1) and progresses toward the root tips (5), as described in the text. (After Weier, Stocking, and Barbour, Botany 4th ed., p. 229. New York: John Wiley & Sons.)

Modern studies of the structure of plants and the properties of water have shown that sap moves up the tree by the following chain of events (Fig. 18-23):

1. Transpiration occurs: water evaporates from the walls of leaf cells into the air spaces inside the leaves and then moves out of the leaf into the atmosphere through the stomata.

2. The loss of water leaves a water deficit in the cell wall, which is quickly made good by movement of water in from the walls of neighboring cells; these obtain replacement water from their neighbors, and so on, until the cell donating the water is a conducting cell in the xylem at the tip of a veinlet in the leaf.

3. Since water molecules attract one another strongly, they stick together, or cohere; pulling water out of the top of a xylem column is like pulling on a rope of water that extends all the way down the column, out through the root, and into the soil; all the water in the xylem moves up a bit, and the sequence continues.

Because transpiration sets this process in motion and the cohesion of water molecules allows it to continue, the mechanism is called the **transpiration-pull water-cohesion** mechanism of xylem transport.

Flow in the phloem. The other vascular tissue, phloem, carries food and other organic substances. Most of the food moves in the form of sucrose (table sugar) or other sugars. Some sugars arriving near the root tips are combined with nitrogen taken in from the soil, forming amino acids, which may then travel to other parts of the plant through the xylem or phloem. Phloem may also contain minerals taken back into the plant body from dying leaves.

The conducting cells in the phloem of a flowering plant are **sieve tube members**, stacked atop one another to form long **sieve tubes**. Perforated **sieve plates** at the end of each cell allow materials to pass from one cell to the next. Unlike the conducting cells of the xylem, sieve-tube members contain cell membranes and living cytoplasm, but there is no nucleus or other organelles that might impede the flow.

Botanists are not entirely convinced that they understand how phloem transport works. The best model to date was proposed in 1926 by Ernst Munch (Fig. 18-24). This model uses the osmotic properties of water. Water tends to flow across selectively permeable membranes, such as cell membranes, toward the side with a higher total concentration of **solutes** (dissolved particles) (see Section 10–F). According to this theory sieve-tube members obtain solutes, especially sucrose, by active transport. Water follows the solutes by osmosis, building pressure inside the sieve-tube member. Soon the pressure pushes the solution into the next sieve-tube member, and so on. Eventually the solution reaches an

SUE-crose

amino acids: the molecules that are linked together to form proteins.

active transport: moving molecules of a substance to the side of a membrane where that substance is already more concentrated, by expending cellular energy (such as ATP).

Fig. 18–24

Munch's model system for mass flow, the proposed mechanism for movement of substances in the phloem. The selectively permeable membrane is permeable to water but not to sucrose. Colored dots represent sucrose; colored arrows represent movement of water or solution. At the beginning of the experiment, a sucrose solution is placed in the left-hand chamber, and plain water in the right; the tube is then immersed with the two ends in a water bath. Water enters the sucrose solution by osmosis across the membrane, and the solution rises into the neck of the tube and flows across to the right-hand chamber, where water is eventually forced out through the membrane. This system stops when the sucrose concentration becomes equal in the two chambers, but in a living plant one area constantly produces sucrose as others remove and use sucrose, and movement would continue.

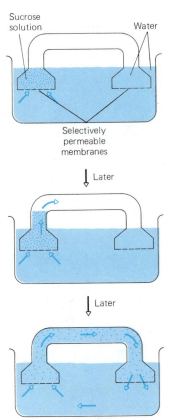

area where tissues around the phloem are removing the solutes; as water follows the solutes out of the sieve tube, the pressure decreases. Because sucrose is the chief solute involved, this occurs mainly in a tissue that needs energy from sugar for growth, or in an organ actively storing food, such as a root, seed, or fruit.

So, in a living plant, some areas are solute "sources," continually making sugar (in leaves) or releasing it (storage roots or stems); other areas are solute "sinks," continually removing solutes. Movement always goes from a high-solute, high-pressure area to a low-solute, lower-pressure area. This mechanism is called the **mass flow** theory.

The mass flow theory is consistent with the presence of cell membranes, cytoplasm, and high internal pressures observed in sieve-tube members. It can also explain another interesting finding: the direction of movement in the phloem may reverse from time to time. To do this, the model requires only that one area of the plant lower its solute concentration below that in the area originally receiving the flow. Suppose, for example, that a leaf was producing food and sugar that was being transported to a storage area in the root: the phloem contents would be moving downward in the plant. But then suppose the plant began to produce flowers, which would need food to grow; food made in the leaves might now be moved upward as the developing flowers removed sugar and minerals and lowered the concentration of solutes at the top of the stem.

Practical applications.   Knowing how plants handle various substances can be helpful in managing plants and plant pests. For example, leaves may absorb some substances sprayed onto their surfaces and move them into the phloem, which transports the substance throughout the plant. This is the basis of **foliar feeding,** or fertilizing with a nutrient spray on the leaves. Other substances must be worked into the soil because only the roots will absorb them.

Chemicals that are not absorbed may also be sprayed onto the outside of a plant; **topical pesticides** protect the outside of leaves by killing leaf-chewing insects or invading fungi, but will not damage the living cells inside the leaves or contaminate fruits produced after the spraying, or roots used as food. **Systemic pesticides,** watered into the soil or sprayed on the leaves, are absorbed and distributed throughout the plant. Systemic pesticides combat internal pests, such as leaf miners, insect larvae which live entirely inside leaves and never come into contact with sprays applied externally. These pesticides are useful to protect ornamental plants but cannot be used on food crops because they are toxic to humans as well as to insects.

SUMMARY

The basic plant body plan includes leaves, stems, and roots. Leaves are usually broad and flattened; they grow in patterns that allow maximum use of sunlight. The combination of cuticle and stomata allows leaves to exchange gases while reducing water loss. Stems serve to support the leaves and to transport substances to and from leaves; in addition, they often carry out photosynthesis or store food and water. Roots anchor the plant, absorb and transport substances from the soil, and store food.

Roots take up most of the plant's water and mineral needs from the soil. The quality of the soil is determined by rock particle type and size, rainfall, organic matter, soil organisms, and oxygen.

The vascular tissues transport substances throughout the plant body. Xylem carries sap (water and minerals) from the roots to the leaves; some movement in the xylem is due to root pressure, but most can be attributed to transpiration pull acting on highly cohesive water in the xylem column. Phloem carries food and minerals between areas of manufacture (leaves or stems), storage (roots or stems), and use (stems, roots, growing leaves, flowers, or fruits), moving from areas of higher pressure to areas of lower pressure.

Knowing how roots and soil interact, and how plants transport food and other substances internally, can help us manage plants.

SELF-QUIZ _____

1. Which of the following is *least* apt to be a function of stems?
   a. support of leaves
   b. absorption of minerals from soil
   c. photosynthesis
   d. storage of food
   e. transport of food

2. Name the structures in a leaf responsible for each of the following functions.
   a. admission of air from the atmosphere into the leaf
   b. transport of water into the leaf from the stem
   c. transport of food out of the leaf to other parts of the plant
   d. closing the leaf off from the air when water loss is too great
   e. covering the leaf and retarding loss of water

3. A fibrous root system is apt to perform which function better than a taproot system?
   a. absorption   c. food storage
   b. anchorage   d. transport

4. For each soil component listed below, state, in six or fewer words, its importance to plants:
   a. rock particles   c. soil bacteria
   b. oxygen   d. organic matter

5. Xylem is *not* found in:
   a. leaves
   b. stems
   c. roots
   d. all of the above
   e. none of the above

6. Which vascular tissue transports food from a potato to a potato flower?

7. For movement of sap by root pressure, the plant must have:
   a. enough minerals in the soil
   b. enough water in the soil
   c. enough oxygen in the soil
   d. living root cells
   e. all of the above

QUESTIONS FOR DISCUSSION

1. In a study of maple trees in areas with varying degrees of air pollution, a researcher found that leaves from trees in more polluted areas had fewer stomata and more surface hairs than leaves from unpolluted areas. How might you account for these differences?

2. Leaf structure varies from one species to another, and it is often closely correlated with the plants' habitat. In what types of habitat might you expect to find leaves with:
    a. extra thick cuticle
    b. little or no cuticle, little or no xylem, and no stomata

3. Potatoes are storage tubers (underground stems). Why must potato farmers be concerned to rid their crops of insect pests that eat large quantities of potato leaves?

4. Modern chemical analysis can tell very accurately just what elements are found in plants, and in what quantities. Why isn't this a good method for determining the nutritional requirements of plants?

5. Why do green plants need oxygen?

6. Would the rate of transpiration increase or decrease under the following conditions?
    a. higher humidity
    b. closing of the stomata by the guard cells
    c. increasing sunlight
    d. increased wind

7. Refer to the photographs below. These three cuttings were taken from the same plant at the same time. Cutting #1 was placed in water immediately. Cutting #2 was left lying on the table for half an hour; a 2-inch length was then cut off the bottom end, and it was placed in a container of water. Cutting #3 was also left lying on the table for half an hour; it was then put in water without further cutting. The photographs were taken 2½ hours after the cuttings were removed from the plant. How can you account for the differences in the three cuttings?

8. Why is it best to cut flowers in the early morning or in the evening? Why should flowers be cut with longer stems than will be needed in the planned arrangement?

9. Why should evergreen trees and shrubs be watered thoroughly before the ground freezes for the winter?

10. What does wilting indicate about the movement of water in a plant? Does wilting signal that the soil should be watered?

11. The virus known as "beet yellow" is transmitted from plant to plant by aphids, small insects which feed by inserting the sucking mouthparts into cells in the phloem. Why does the disease spread through the plant rapidly?

GROWTH AND REPRODUCTION OF FLOWERING PLANTS

1999

OBJECTIVES

When you have studied this chapter, you should be able to:

1. Describe the sequence of events as a root tip or shoot tip grows in length, using the names of the following important areas: root cap, apical meristem, leaf primordium, axillary meristem, zones of elongation and maturation.

2. Identify the following parts of a bean seed: seed coat, cotyledons, embryonic axis, first foliage leaves.

3. Identify the following on a plant or on a photograph or drawing of a plant: areas where apical meristems would be found, axillary meristems, annual rings, sapwood, heartwood, terminal bud, bud scales, bud scale scars, leaf scars, one year's growth in a cross section of a woody stem, one year's growth in length of a twig.

4. List or recognize four differences between monocots and dicots.

5. Name the parts of a flower and state the role of each in reproduction of the plant.

6. Describe the sequence of events as pollination, fertilization, and development of the seed take place in a flower.

7. Briefly define, and give examples of, the following hormone-mediated phenomena in flowering plants: tropisms, apical dominance, photoperiodic response, short-day flowering, long-day flowering.

8. Explain the advantage of asexual, or vegetative, reproduction to a plant, and list some means of vegetative propagation; explain why humans use vegetative propagation of plants, and name some methods that we use for vegetative propagation.

A human child grows according to a well-defined program; growth stops well before we reach adulthood, and during most of our lives no further growth occurs. This **determinate** mode of growth is in marked contrast to the **indeterminate** growth of plants; a plant may keep growing, adding new leaves, branches, and roots, throughout its life. In fact, its new parts enable the plant to obtain more sunlight, water, and minerals, which help it continue to grow. In addition, if a plant loses a leaf, stem, or root, it can compensate by producing replacement parts.

Most of a plant's body consists of cells that are specialized for particular functions; these cells will never divide again. A plant's ability to grow is due to its possession of **meristems**, groups of cells that retain the capacity to divide and produce new cells. Some of these new cells mature as parts of specialized tissues and organs; others remain meristematic; that is, they are able to continue the cycle of division. Eventually, some meristems produce flowers, a plant's reproductive organs, and the flowers produce seeds of the next generation.

MARE-ih-stems

MARE-ih-stem-AT-ic

All phases of a plant's growth are controlled by chemical messengers, or **hormones**, which travel from one part of the plant to another and coordinate growth.

In this chapter we shall follow the ongoing saga of the meristems and the hormones that influence them in the growth and reproduction of flowering plants.

19–A The Life of a Bean

The bean seed. We begin with an ordinary bean seed, a reproductive package containing a plant embryo and its food supply (Fig. 19-1). The "skin" of the bean is a protective **seed coat** wrapped tightly around the embryo. If we peel the seed coat off, we can separate the two "halves" of the bean embryo; these are really the two extremely fat "seed leaves" of the embryo, enfolding the tiny, easily overlooked main **axis** of the plant-to-be. Because they have two seed leaves, or **cotyledons**, beans belong to the group of flowering plants called **dicotyledons**. The cotyledons of a bean are enlarged because they contain the

cot-uh-LEE-duns
DIE-cot-uh-LEE-duns

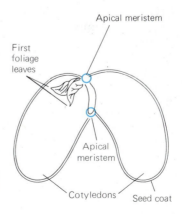

Fig. 19–1

A bean seed split open. The embryo consists of the two large cotyledons and a tiny plant axis with the two first foliage leaves already partly developed. The apical meristems will become the growing tips of the root and shoot.

embryo's food supply; many other kinds of plants (such as corn, Fig. 19-12), store their seed food separately, outside the embryo itself.

The dried beans that we buy at the grocery store are among the most dehydrated of living entities—in fact, in this dormant state the beans look dead. However, under the right conditions the bean can break dormancy and **germinate**, or begin its growth into a photosynthetic bean plant.

Germination.   A seed needs a particular set of environmental conditions for successful germination. Its first requirement is water; you may already know that you can speed the germination of beans in your garden if you first soak them in water overnight. The seeds also need a favorable temperature, and they require oxygen to respire as they use the energy of their stored food to build new body structures.

Root growth.   As the bean seed germinates, the first structure to enlarge is the seedling's first root. At the tip of the root is the **root cap**, a thimble of cells that covers the growing root tip and protects it from the rough soil particles (Fig. 19-2). The outer cells of the root cap are scraped off as the root pushes its way through the soil.

AY-pick-al

Within the root cap lies the **quiescent zone**, an area of cells that are usually not actively dividing. Around and above the quiescent zone is the **apical meristem** (apical = of the tip), a region of actively dividing cells, producing cells for both the root cap and the root itself. In the **zone of elongation**, just above the apical meristem, newly produced cells lengthen and push the root cap and apical meristem through the soil; this is the actual force of growth in the root. Above the zone of elongation, cells in the **zone of maturation** have reached full size and are becoming specialized in structure and function. The outer cells in the zone of maturation produce slender extensions, called **root hairs**, which anchor the root among the soil particles.

Once this first, or **primary root**, of the bean seedling is established it begins to produce new branches, which rapidly increase in size, so that by the time the plant is a few weeks old it has a spreading, fibrous root system.

Growth of the shoot system.   With its root system established, the seedling is ready to produce its shoot system. The apical meristem of the shoot lies between the two tiny **first foliage leaves** in the bean embryo. The pattern of shoot growth is basically the same as we saw in root growth: apical meristem cells divide, and those toward the tip remain meristematic, whereas those nearer the base of the plant elongate and finally mature as specialized cells.

AXE-ill-air-ee (Be careful! Axillary is often mispronounced so that it sounds like the word auxiliary.)

Unlike the growing root tip, the shoot tip does not have a cap, and, of course, another difference is that the shoot tip produces leaves. As the apical meristem lays down the cells to start a new leaf, it also leaves behind a lump of meristematic cells just in the angle where the leaf stalk joins the stem. This is called an **axillary meristem** (axilla = armpit), **axillary bud**, or **lateral bud**. Under certain conditions it can break its initial dormancy and form a new shoot tip, which grows out and forms a new branch with its own leaves (Fig. 19-4).

pry-MORE-dee-um

Each leaf begins as a **leaf primordium**, a slender pencil of meristematic

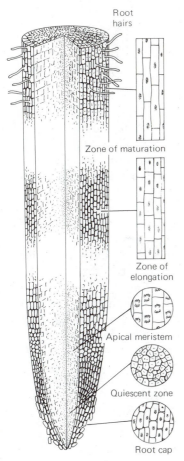

Root hairs

Zone of maturation

Zone of elongation

Apical meristem

Quiescent zone

Root cap

Fig. 19–2

A growing root tip. Above the labels for each area are shown the cells in that area as they would appear through the microscope.

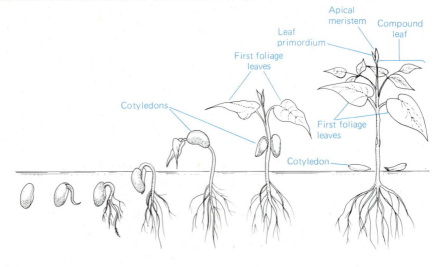

Apical meristem
Compound leaf
Leaf primordium
First foliage leaves
Cotyledons
First foliage leaves
Cotyledon

Fig. 19–3

Stages in the growth of a bean seed into a young bean plant. The lower part of the embryonic axis forms the beginnings of the root system before the upper part starts to develop.

Fig. 19–4

Growth in a shoot tip. (a) Section through the tip of a Coleus plant, showing the central mound of apical meristem, the leaf primordia it has recently laid down, and the axillary buds left in the angles between the leaf primordia and the stem. (Carolina Biological Supply Company) (b) Drawing to assist in identifying the structures shown in the photograph.

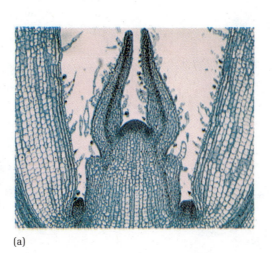

(a)

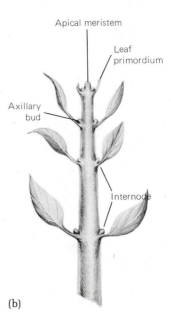

Apical meristem
Leaf primordium
Axillary bud
Internode

(b)

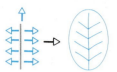 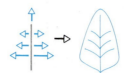 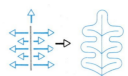 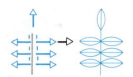

Fig. 19–5

Leaf shape is determined by the pattern of division and growth of cells along different parts of the original axis (gray) of the leaf primordium. In extreme cases (far right), a compound leaf forms, with the blade separated into discrete parts.

cells, which divide, enlarge, and specialize, forming the mature leaf. The pattern of these activities determines the final shape of the leaf (Fig. 19-5). In the bean plant, the first foliage leaves are **simple leaves**, with just one flat blade, whereas leaves produced later are **compound**, with the blade divided into several parts (Fig. 19-3).

Reproduction. For a time, the shoot meristem produces only stem growth and new leaves. Eventually, though, the balance of hormones in the plant changes, and the shoot meristems then produce flowers. Actually, a flower is an abbreviated shoot with many highly modified leaves—the flower parts—clustered together.

Beginning at the outside of the flower, the first flower parts are the green SEE-puls **sepals**, which protect the developing flower while it is still a bud (Fig. 19-6). Just inside the sepals are the **petals**, which in the bean flower have very specialized shapes and enclose the inner flower parts. If we pull the petals aside, we can see

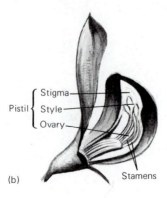

(a)

Petals

Sepals

Fig. 19–6

(a, b) The flower of a bean is very similar to the sweet pea flower shown here, but the bean flower's petals would not be so large and showy. The petals completely enclose the male parts (stamens) and the single female part, the pistil. (c) Arrangement of flower parts in a more generalized flower.

(b)

Pistil { Stigma / Style / Ovary }

Stamens

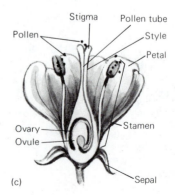

(c)

Stigma
Pollen tube
Pollen
Style
Petal
Ovary
Ovule
Stamen
Sepal

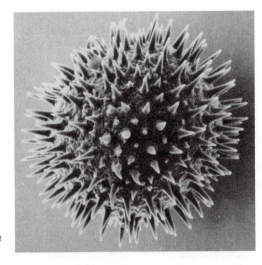

Fig. 19–7

Scanning electron micrograph of a pollen grain. (Biophoto Associates)

the **stamens**, which are the male, or pollen-producing, parts, and the single   STAY-menz
female **pistil**. At the tip of the pistil is the sticky **stigma**, which traps pollen   PISS-tul
grains from the stamens. Next is the long **style**, and at the base of the pistil is the
ovary. The ovary contains **ovules**, each enclosing a haploid **egg nucleus** and   OH-vuels
seven other haploid nuclei (Fig. 19-8).

A pollen grain landing on the stigma germinates, grows a pollen tube down
the style to an ovule in the ovary, and releases two sperm nuclei. These two
nuclei take part in **double fertilization**, a reproductive act unique to flowering
plants. One sperm nucleus fertilizes the egg nucleus, forming a zygote that
divides and becomes an embryo. The other sperm nucleus joins with two other
nuclei (the central nuclei) in the ovule; because each of these three nuclei is
haploid (see Section 16–F), the nucleus formed by this second fertilization has
three sets of chromosomes, making it **triploid**. This nucleus divides and gives   TRIP-loid
rise to the **endosperm**, a tissue that absorbs and stores food; in the bean the

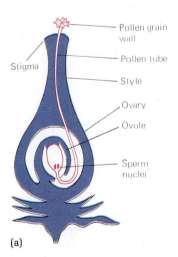

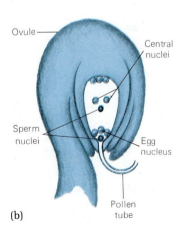

Fig. 19–8

Events from pollination to fertilization in a flower. (a) At pollination, the pollen grain is deposited on the stigma. Stimulated by the moist, nutrient-rich environment there, the pollen grows a pollen tube down the style to the ovule. Two sperm nuclei are released. (b) One sperm nucleus fertilizes the egg nucleus in the ovule; the other fertilizes the two central nuclei, forming an endosperm nucleus.

endosperm's food is eventually absorbed into the cotyledons and the endosperm does not appear in the mature seed. (Because nonflowering seed plants, such as gymnosperms, provide food for their embryos without using an endosperm, the adaptive advantage of double fertilization is not clear.)

While the embryo develops, the wall of the ovule around it also grows and becomes the seed coat, and the ovary enlarges around all the ovules as the **fruit**. (A string bean, or snap bean, is actually not a bean but a bean pod, which is a fruit; its seeds are the beans.) Seed coats and fruits are the parent plant's contribution to the protection of its offspring, shielding them from predators and loss of moisture. The parent plant also contributes the food absorbed by the endosperm.

Senescence and death.   When their seeds have ripened, the bean pods dry and split open, releasing the seeds. Now the bean plant enters the last phase of its life: **senescence,** or aging. New changes in its hormonal state trigger a general shutdown of its metabolism, and the plant undergoes "programmed death." The withered and dried bean plant returns to dust among its dormant seeds, which will someday germinate and continue the now-familiar life history of the bean.

senn-ESS-ence

19–B Woody Plants

Our bean plant lived for just a single season. Many plants, however, have life-spans measured in years, often attaining enormous size as they add new growth each year.

In woody trees and shrubs, the roots and stems grow thicker with the passing years. Each year, a layer of meristematic cells near the junction between the bark and the wood divides and adds a new ring of wood around that already present; by counting these **annual rings** we can tell the age of a woody stem (Fig. 19-9). In fact, the history of the tree is written in its wood, and by looking at the sizes of the cells and the thickness of the rings we can tell something about the weather during the year in which each ring grew.

ZYE-lum

Wood is **xylem**, or sap-conducting tissue (see Section 18-C). The outer annual ring, or several outer rings (depending on the species), actually transport water from roots to leaves and so are called **sapwood**. The inner **heartwood**, which no longer conducts sap, becomes a waste depository. Sapwood and heartwood often differ in color (Fig. 19-10).

At the tips of woody stems, the apical meristems, or **terminal buds**, grow and produce new portions of stem, with leaves. Each summer, the year's new stems lay down their first annual rings of wood, and cover themselves with bark. As winter approaches, new leaves produced by the terminal bud develop, not as the familiar green photosynthetic structures, but as a protective phalanx of small, tough **bud scales** around the terminal bud (Fig. 19-11). At the base of each photosynthetic leaf, cells become corky and seal the leaf off from the rest of the plant. When a leaf falls, a **leaf scar** shows where the leaf stalk was once attached to the twig. As in the senescence and death of the bean plant, death and detachment of the leaf are a programmed part of life.

Bud scale scars, left as bud scales fall off in the spring, form a ring around

(a)

(b)

Fig. 19–9

(a) Annual rings are very noticeable in a tree trunk cut straight across. (b) We can see that each annual ring section of a 3-year-old basswood twig has large cells at the inner rim, where the spring growth occurs, and that cells laid down as the season progresses are increasingly smaller. The thick layer of dark cells outside the third ring is the bark, containing phloem near the inner side and nonliving waterproof cells on the outside. (b, Carolina Biological Supply Company.)

Fig. 19–10

The sapwood and heartwood in an elm tree differ greatly in color.

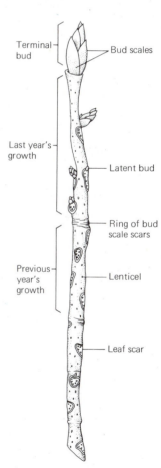

Terminal bud

Bud scales

Last year's growth

Latent bud

Ring of bud scale scars

Previous year's growth

Lenticel

Leaf scar

Fig. 19–11

A hickory twig.

the twig at the point where the terminal bud lay dormant during the winter. In the spring, the terminal bud resumes activity, lengthening the twig and laying down a new crop of leaves before overwintering once more in a protective nest of bud scales. We can tell the age of a twig without cutting it off the tree by counting the rings of bud scale scars in its bark (Fig. 19-11).

19–C Monocotyledons

Beans, many other **herbaceous** (soft-stemmed) plants, and most woody plants are dicotyledons, plants whose embryos have two cotyledons, or seed leaves. The other important group of flowering plants is the **monocotyledons**,

TABLE 19-1 COMPARISON OF MONOCOTYLEDONS AND DICOTYLEDONS

Characteristic	Monocots	Dicots
Seed leaf (cotyledon)	One	Two
Flower parts	In 3's or multiples of 3, or irregular	4, 5, their multiples, or irregular
Leaf venation	Parallel	Netted or fanlike
Vascular tissue	Bundles scattered through stem	Single ring of bundles or continuous ring
Form	Mostly herbs; few trees, e.g., palms	Herbs, shrubs, and trees
Examples	Lilies, grasses, grains, orchids, iris, onion, palms, crocus, daffodils, etc.	Oaks, maples, legumes, roses, mints, squashes, daisies, walnuts, cacti, violets, buttercups, poppies, etc.

whose embryos have just one cotyledon (Fig. 19–12). The lily·family (including onions), the iris and crocus family, the daffodils, and the orchids are members of this group. By far its most important members, though, from the human point of view, are the grasses, including the cereals (wheat, rice, corn, oats, barley, rye) that sustain human life in most parts of the world, and the palms, which produce dates, coconuts, betel nuts, oils, raffia, and rattan.

Monocotyledons and dicotyledons (or monocots and dicots, as they are usually abbreviated) can often be told apart by their appearance. Monocots tend to have long, thin leaves and parallel leaf veins, whereas dicot leaves have veins in a netlike or fanlike arrangement. A cross section of a monocot stem shows

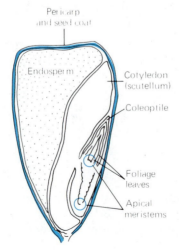

Pericarp and seed coat

Endosperm

Cotyledon (scutellum)

Coleoptile

Foliage leaves

Apical meristems

Fig. 19–12

The seed of corn, a monocotyledon. Compare with Figure 19-1, the bean (dicotyledon) seed. In corn, the embryo has only one cotyledon. The coleoptile is a sheath covering the foliage leaves until it has broken through the surface of the soil. In the corn seed, the seed coat is fused to the pericarp, the fruit, formed from the ovary wall. The nutritive endosperm tissue is separate from the embryo, whereas in the bean the cotyledons had absorbed all the food from the endosperm (this feature of beans is not, however, typical of all dicots).

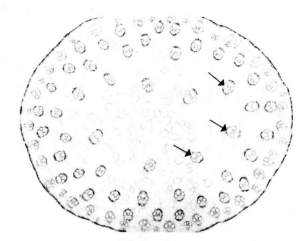

Fig. 19–13

Cross section of a corn stem, showing the vascular bundles scattered throughout (arrows) rather than arranged in a single ring as in herbaceous dicots (compare Figure 18-5). (Carolina Biological Supply Company)

scattered vascular bundles rather than the single ring of bundles in a dicot stem (Fig. 19-13). In addition, most monocots have flower parts in threes or multiples of three, whereas dicots have flower parts in fours, fives, or multiples of these, or in irregular numbers.

Monocots also have an interesting method of growth not seen in dicots. Besides the apical meristems that produce new growth at the tips of roots and shoots, monocots have **intercalary meristems**, which produce new cells at the bases of leaves. Hence monocot leaves can grow after their tops have been damaged, a phenomenon all too familiar to the homeowner whose weekends seem to be filled with the chore of mowing the grass!

in-TER-cul-air-ee

19–D Plant Hormones

Plant hormones are chemical messengers produced in certain parts of the plant and moving throughout the body, where they influence the growth or maturation of various cells and, in fact, regulate virtually everything that happens in a plant.

Tropisms.   The first root of a seedling grows downward into the soil, no matter what position the seed lands in as it is planted; a houseplant bends toward a window and intercepts the light entering from one side. These are examples of **tropisms**, growth patterns that orient the plant's parts favorably with respect to important environmental factors.

TROPE-ism

Tropisms are named after the responsible stimulus; thus a root growing toward the center of the earth, in response to the pull of gravity, shows **positive geotropism** (geo = earth), while a shoot shows **negative geotropism** by growing away from the pull of gravity (Fig. 19-14). The houseplant growing toward light shows **positive phototropism** (photo = light), and the weeping willow roots that grow all the way across the front lawn and clog the sewer drain pipes are demonstrating **positive hydrotropism** (hydro = water) by growing toward water.

Fig. 19–14

Tropisms in a corn seedling. The root (right) grows downward (positive geo-tropism, whereas the coleoptile sheath covering the shoot grows up (negative geotropism). Note the root hairs on the root. (Biophoto Associates)

Fig. 19–15

Phototropism in the coleoptile of an oat seedling. (a) The coleoptile grows straight up when uniformly lit, and toward the light when lit from one side. (b) An experiment showing that a chemical produced in the coleoptile tip is responsible for the bending. A block of gelatin-like agar is used to collect the chemical. (c) Since a plain agar block causes no bending, the phototropism is attributed to a chemical, named auxin, produced in the coleoptile tip and transferred to the severed coleoptile via the agar.

These tropic responses are mediated by plant hormones. The classic studies of phototropism were done with the oat **coleoptile**, the sheath over a seedling's leaves (see Figs. 19-12 and 19-14). Light falling on one side of the coleoptile causes **auxin**, a hormone responsible for cell elongation, to reach higher concentrations in the cells on the darker side. (Some auxin seems to be transported to the dark side from the light side, and some is destroyed by light.) Cells on the dark side then elongate more than cells on the illuminated side, and the growing coleoptile tip bends toward the light (Fig. 19-15).

Apical dominance. Many kinds of plants exhibit **apical dominance,** in which the growing apical meristem represses the growth of axillary buds on the same stem, causing them to remain dormant. This is due to auxin produced in the apical meristem and transported toward the base of the plant; if the apical meristem is removed, the axillary buds grow out, each becoming a new apical meristem laying down its own stem tissue (Fig. 19-16). A plant with strong apical dominance has a distinct main axis, with few side branches, whereas a plant with weak apical dominance has many well developed branches and a

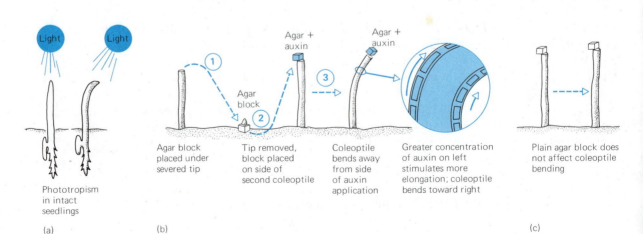

Phototropism in intact seedlings

(a)

Agar block placed under severed tip

Tip removed, block placed on side of second coleoptile

Coleoptile bends away from side of auxin application

Greater concentration of auxin on left stimulates more elongation; coleoptile bends toward right

(b)

Plain agar block does not affect coleoptile bending

(c)

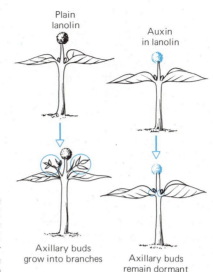

Fig. 19–16

Experiment showing that apical dominance is mediated by auxin. Removal of the apical meristem permits growth of axillary buds (left), but if the apical meristem is replaced with an auxin solution (right), the axillary buds are still inhibited from growing. (Why is plain lanolin applied to the plant on the left?)

bushy appearance (Fig. 19-17). Gardeners who favor the latter growth form often pinch out the apical meristems of their plants to encourage the growth of more branches. When such pinching is done, the stem should be removed to just above the axillary meristem that is desired to produce a new branch; if part of the main stem is left above the axillary meristem, it will become an unattractive stub.

Herbicides.   Some synthetic plant hormones, such as 2,4-D and TIBA, are used as herbicides, weed killers, in lawns. Monocots, which include most desirable lawn grasses, are much less sensitive to these hormones than are dicots, which include the broadleaved weeds. Applying the correct dosages to a lawn causes the dicots literally to "grow themselves to death"; they pass through uncontrolled growth resulting in gross deformities before they die. To be most effective, weed killers must be applied when the plants are actively growing and therefore sensitive to extra doses of hormones.

(a)

(b)

(c)

Fig. 19–17

Auxins play a part in determining plant shape by regulating the relative amounts of growth in the main stem and branches. All of these trees grew in the open on a golf course, but (a) shagbark hickory has little branch growth compared to trunk growth, whereas (b) sycamore and (c) American elm show progressively more branch growth.

Synthetic plant hormones were used as herbicides in Vietnam. Spraying from airplanes defoliated (removed the leaves from) vast areas of the countryside. However, a contaminant in the spray was found to produce malformations of developing animal fetuses, and it was discovered that this chemical had worked its way into the drinking water and into fish, the chief source of protein in the Vietnamese diet. Public protest finally caused the defoliation program to be abandoned.

Flowering.   The change from the vegetative phase of growth (production of stems and leaves) to the reproductive phase (production of flowers) is associated with a shift in the balance of plant hormones.

Some familiar plants, such as carrots and parsley, flower the second year after the seed began to grow—following a period of cold in the winter. Such plants can be forced to flower by placing them in the cold for several weeks, or by certain hormone treatments. Evidently a cold period somehow triggers a hormonal change in these plants, switching them from the vegetative to the reproductive phase.

Many plants flower only when exposed to a certain regimen of light and dark periods. Surprisingly, this was not clearly recognized until 1920, when researchers in the U.S. Department of Agriculture studied two flowering problems. The first involved "Maryland mammoth" tobacco, which reached prodigious heights in the field but did not flower, whereas cuttings rooted and grown in greenhouses during the winter flowered even at much smaller sizes. The second

Fig. 19–18

A Maryland tobacco plant in flower.
(Biophoto Associates)

problem concerned soybeans that flowered on the same date even when farmers staggered their planting schedules in order to try to stagger the harvest. Since all of the tobacco plants and all of the soybean plants flowered at the same time, regardless of size or age, the workers decided that flowering must be determined by some cue in the environment.

Experiments with different temperatures and light intensities failed to relate these factors to flowering, so the researchers turned to another hypothesis—that plants could respond to varying lengths of daylight—which seemed silly because it meant that plants would have to be able to measure time. But, sure enough, when the workers artificially changed the length of the plants' daily exposure to light, or **photoperiods**, they found that the tobacco and soybean plants would flower only if exposed to light for less than a certain amount of time each day! Accordingly, these were dubbed **short-day plants**.

We now know that they should actually be called "long-night" plants, because what plants need to flower is a certain minimum period of uninterrupted *darkness* (which in a normal 24-hour day naturally occurs if the days are sufficiently short). Interrupting the dark period with a flash of light destroys the plants' change to the flowering mode. That is why, if you own a Christmas cactus or poinsettia that you wish to flower during the holiday season, you must put the plant in a room or closet kept dark from sunset to sunrise. It usually takes several days of such treatment to make the plant flower.

Further experiments have shown that there are also **long-day plants**, requiring a certain minimum length of light period for flowering; examples are spinach, radishes, and barley. Short-day and long-day plants are defined not by the actual length of light administered to induce flowering, but by whether the "critical length" of the photoperiod is a maximum (for short-day plants) or a minimum (for long-day plants).

There are also many day-neutral plants, such as tomatoes and cucumbers, which will begin to flower at a certain age regardless of day length.

Ripening of fruit.   The ripening of fruit occurs under the influence of a hormone with a unique property: it is a gas, and thus it can travel through the air from one plant or fruit to another. Pineapple and mango growers have long known that setting fires in the fruit groves would synchronize ripening of their crops; we now know that this is so because the gaseous hormone, **ethylene**, is a   ETH-ill-een
common product of burning organic materials.

In ripening fruits, ethylene stimulates production of enzymes that reduce the fruit's acidity and other enzymes that break down the stiff cellulose cell walls in the fruit, thereby causing the fruit to soften.

The production of ethylene stimulates ripening, and a ripening fruit in turn produces still more ethylene. Hence the entire fruit ripens at once, and one ripening fruit will stimulate ripening of its neighbors. Over-ripening is also "contagious," and it is literally true that "one bad apple spoils all the good ones." In fact, apples are now often stored in airtight rooms filled with carbon dioxide, a gas that blocks the action of any ethylene produced by the fruits and prevents further production of ethylene.

Fig. 19–19

Sugar bush watermelon—the result of 20 years of breeding for compact vines, good flavor, and color. (W. Atlee Burpee Co.)

On the other hand, fruits such as bananas, pineapples, and citrus fruits are harvested green in tropical plantations and shipped to market. Here these fruits are ripened artificially by applications of ethylene. This avoids the risk of financial loss due to overripening of fruits during their travels; however, artificially ripened fruits have less flavor than the ones Mother Nature makes.

19–E Plant Breeding Programs

Since the beginning of agriculture, humans have been breeding plants selectively in order to increase their usefulness. Selection by early peoples who knew nothing of genetics or the fine points of modern breeding programs produced cultivated plants that were strikingly different from their wild ancestors. With an understanding of genetics (Chapters 16 and 17) came the cross-pollination of plants with various desirable features, in an attempt to obtain some offspring combining all the desirable genes in the same individual.

The genetics of some crop plants are fairly well understood. Corn, one of the most important crops in the United States, has been intensively studied. Corn is easy to breed. Male and female flowers form large, separate clusters, making artificial pollination easy. Corn is also an **annual** plant, completing a generation each year; this makes it possible to develop new strains of corn rapidly.

Other species pose greater problems. For example, a breeder trying to produce new strains of apples must wait 4 to 10 years until planted seeds grow into fruit-bearing trees. Furthermore, only about 1% of such trees will bear fruit even equal in quality to presently available varieties; most new genetic combinations will produce inferior fruits. Any promising new tree must be screened for another 10 years before it is ready for marketing . . . or for the woodpile. Even in annual plants, it may take many years to develop new varieties if the genetics of the species have not been extensively studied (Fig. 19-19).

19–F Vegetative Reproduction

Sexual reproduction occurs in most flowering plants; it increases the genetic variation in a population, and so provides many "trial" assortments that can be acted upon by natural selection. Asexual reproduction is also common in plants; indeed, many plants always reproduce asexually by **vegetative reproduction**—making new individuals from roots, stems, or leaves of the parent (Fig. 19-20). Asexual reproduction allows these plants to perpetuate combinations of genes that are well-adapted to the environment.

Vegetative reproduction is often desirable from the human point of view. Home gardeners root cuttings of *Coleus*, geranium, or ivy stems by placing them in water, or place leaves of African violets or jade plants on moist soil until they grow roots. We can help spread plants such as daffodils, tulips, and onions by digging up the bulbs when the tops die back and separating those that have multiplied, so that each has more room to grow.

Potatoes are an important crop produced vegetatively. A potato is an underground stem, or **tuber**; by digging up "seed potatoes" (tubers of good quality), cutting them into pieces—each with an "eye" (bud)—and replanting them, farmers can obtain many offspring from each potato plant. Potatoes grown from the "true seed" produced by potato flowers are almost always sorry affairs because sexual reproduction in potato plants generally produces genetic combinations that make inferior food plants.

Fig. 19–20

Vegetative reproduction in plants. (a) Strawberry plants put out runners which root where they touch the ground and form new plants. (b) Kalanchoe leaves produce tiny plants which eventually fall off and grow on their own. (c) Bits of Streptocarpus *leaf which have rooted and will form new plants where they touch the ground. (Biophoto Associates)*

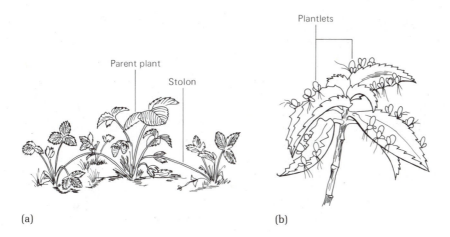

(a)   (b)   (c)

Fig. 19–21

Grafting. (a) A successful graft in an apple tree. Scion (above) and stock (below) formed a bulging callus as they grew together several years before this picture was taken. (b) A pink-flowered scion has been grafted onto a white-flowered stock apple to yield two different varieties on the same tree. (Biophoto Associates)

(a)

(b)

SIGH-on

Grafting is another means of artificial vegetative propagation. A **scion**, a twig or bud of a desirable plant, is attached to a **stock**—the root system or stem of another plant, from which a twig or bud similar to the scion has just been removed. Scion and stock are then wrapped closely together, and their cut areas soon produce new cells that merge the two parts into a functioning unit. Each part of the graft retains its own genetic identity, however.

Grafts work only between plants of the same or closely related species. Grafting is commonly used to produce rose bushes, grape vines, or fruit trees, coupling the desirable flowers or fruits of the scion with sturdy, disease-resistant rootstocks. For example, all the red delicious, golden delicious, and McIntosh apple trees now in existence are derived, by grafting, from single fencerow "volunteer" trees with fortuitous sets of desirable traits. To produce dwarf fruit trees, scions are grafted onto rootstocks of related species; thus, dwarf pear trees have pear scions grafted onto quince rootstocks. We can now buy dwarf apple or pear trees with five or six grafted branches, each bearing fruits of a different variety (Fig. 19-21).

SUMMARY

Early in a plant's life, all its cells can divide. Later, most cells mature, specialize, and lose their ability to divide. Cells that retain the capacity to divide form the meristems at various locations in the plant body. Roots and shoots grow in length through the division of cells in apical meristems, with subsequent elongation and maturation of the newly produced cells. Axillary meristems give rise to new stem branches, whereas meristems just under the bark of woody stems and roots add new cells that increase the diameter of these woody parts.

Plant hormones control growth and development, including tropisms, apical dominance, production of flowers, and production and ripening of fruits.

Sexual reproduction results in new genetic combinations in the offspring. These combinations are often less desirable than the parental genetic makeup, from either the human or the plant point of view. Many plants have means of vegetative reproduction, which perpetuates a particularly favorable combination of genes unchanged; vegetative reproduction may augment or replace sexual reproduction. Humans propagate many plants vegetatively by artificial means such as rooting and grafting. We also manipulate the sexual reproduction of many kinds of plants to suit our own tastes.

SELF-QUIZ ——————————————————————————————

1. Arrange the following events during the growth of a particular section of a shoot in proper order:
   a. cell division
   b. growth of a side branch from the main shoot
   c. cell maturation
   d. cell elongation

2. Which part of a bean has the largest total weight?
   a. embryonic axis
   b. cotyledon
   c. seed coat
   d. first foliage leaves

3. One year's growth in length of a young woody shoot is the distance between successive:
   a. rings of bud scale scars
   b. leaf scars
   c. axillary buds
   d. branches
   e. any of the above

4. For each characteristic listed below, tell whether it is characteristic of monocotyledons, dicotyledons, or both:
   _____ a. roots with root caps
   _____ b. flowers
   _____ c. netlike venation in leaves
   _____ d. two seed leaves
   _____ e. intercalary meristems

5. A flower part whose primary role is protection is the:
   a. stamen
   b. ovary
   c. sepal
   d. seed coat
   e. style

6. Fill in the blanks, using words from the list below the passage.

   In sexual reproduction in a flowering plant, pollen produced in the _____ lands on the _____ and grows a _____ to the egg, located in the _____ inside the _____. Union of the egg and pollen nuclei is called _____. The young plant of the next generation, or _____, develops, surrounded by a _____, which develops from the _____; this in turn is surrounded by a _____, which develops from the _____.

cotyledon	ovary	pollen tube	stamen
embryo	ovule	pollination	stigma
fertilization	pistil	seed coat	style
fruit	pollen	sepal	

7. The flowering of certain plants only under "short-day" conditions is an example of:
   a. apical dominance
   b. positive phototropism
   c. negative phototropism
   d. photoperiodism

8. Grafting is used to propagate plants because:
   a. it is faster than growing seeds
   b. it maintains a desired set of genetic characteristics
   c. it produces fruits that combine the genetic characteristics of two desirable strains of plants
   d. healthy plants will graft by themselves, thus reproducing profusely
   e. a plant can produce many more scions than seeds

QUESTIONS FOR DISCUSSION

1. Explain the statement, "The more a plant grows, the more it *can* grow." What are the limitations of this assertion?

2. Why is it important that root hairs form in the zone of maturation, rather than in the zone of elongation, apical meristem, or root cap?

3. On their honeymoon, Phil and Rhoda Dendron visited a forest. Phil carved their initials into the bark of a tree 10 cm in diameter, 1.5 m above the ground. If the tree grows 0.5 m taller each year, how far up will their initials be on their 10th wedding anniversary?

4. In a flower, the sepals usually become well developed first, followed by the petals, then the stamens, and last by the pistil(s). What are some advantages of this sequential development, rather than a program in which all the flower parts matured at once?

5. Explain how archaeologists could trace the growth of an ancient village as they excavated remains of timbers that were part of various buildings in the town.

6. Why are apples, oranges, and grapefruits sold in plastic bags with holes in them rather than in unperforated bags? Why is it that produce packaged in market trays with clear wrap often is too soft on the underside which you couldn't see in the store?

7. In some species of trees, individuals growing near streetlights become dormant later in the fall than other individuals. How could you account for this?

8. What is the adaptive value to a woody plant of the senescence and death of its leaves in the autumn? What is the value of the senescence and death of the entire bean plant?

9. Plants given large amounts of fertilizer, especially fertilizer with much nitrogen, often flower poorly or not at all, and do not accumulate food reserves; instead they engage in vigorous vegetative growth. Is there an adaptive advantage to this?

10. Some plants, such as dandelions and hawkweeds, have lost the ability to reproduce sexually but still produce flowers and set seed by development of the ovule without meiosis or fertilization. What is the advantage of this system over a more orthodox means of vegetative reproduction?

11. Seed for hybrid sweet corn must be obtained from breeders each year; planting seeds saved from hybrid ears will result in a motley assortment of traits rather than the desired uniformity. Using your knowledge of Mendel's laws (Chapter 16), explain why this is so.

SUGGESTED READINGS

General:

Weier, T. E., C. R. Stocking, and M. G. Barbour. *Botany: an Introduction to Plant Biology*, 5th ed. New York: John Wiley and Sons, 1974. An especially well-illustrated botany text.

Wilson, C. L., W. Loomis, and T. Steeves. *Botany*, 5th ed. New York: Holt, Rinehart, and Winston, 1971. A very clearly written introduction to botany.

Reproduction and Growth:

Echlin, P. "Pollen." *Scientific American*, April, 1968. Many interesting facts and illustrations, plus a discussion of how pollen develops.

Koller, D. "Germination." *Scientific American*, April 1959. Describes germination requirements of various species, showing how widely these needs may vary.

van Overbeek, J. "The control of plant growth." *Scientific American*, July 1968. A brief history of the discovery of some plant hormones.

Transport:

Biddulph, S., and O. Biddulph. "The circulatory system of plants." *Scientific American*, February 1959. An interesting account of experiments showing how various substances move in plants.

Cohen, I. B. "Stephen Hales." *Scientific American,* May 1976. A biographical sketch of the life and experiments of an energetic pioneer in plant physiology.

Zimmerman, M. H. "How sap moves in trees." *Scientific American,* March 1963. Outlines methods used to explore how transport occurs in xylem and phloem.

Special Interests:

Elias, T. S., and H. S. Irwin. "Urban trees." *Scientific American,* November 1976. Discusses effects of pollution and other aspects of city life on trees.

Heslop-Harrison, Y. "Carnivorous plants." *Scientific American,* February 1978. Briefly describes various types of carnivorous plants and presents recent experiments and photographs on capture and digestion of prey.

Part Five

ANIMALS

HUMAN REPRODUCTION AND DEVELOPMENT

20 20

OBJECTIVES

When you have studied this chapter, you should be able to:

1. List the main organs in the human female and male reproductive systems and give their functions.

2. Describe how gametes are formed, and explain how pregnancy begins.

3. Explain how "the pill," IUD, diaphragm and jelly, condom, rhythm, induced abortion, and sterilization of males or females operate as birth con-

trol methods; point out which part of the reproductive process is blocked by each method.

4. Name the three stages of early embryonic development, and state the important events of each stage.

5. Explain what is meant by the term cellular differentiation, and describe how it is believed to occur.

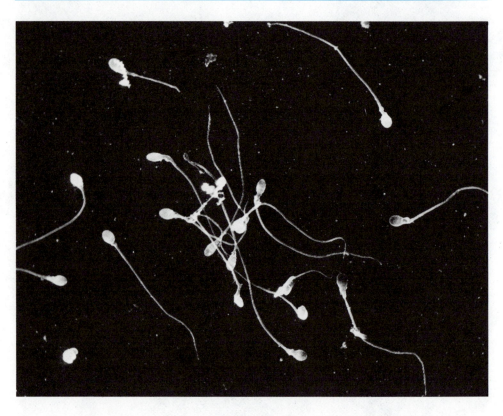

Biophoto Associates, S.P.L.

In the middle ages scientists thought that a man's sperm contained all the information needed for development, in the form of a miniature person, or "homunculus." The mother's role was merely to provide an environment in which the sperm could develop. We now know that the mother contributes an egg, which is fertilized by the sperm, and that both mother and father provide genetic information to the fertilized egg; just how that information is expressed during development, however, is still an intriguing problem.

Embryonic development consists of two main processes: cell division and differentiation. Cell division produces the trillions of adult cells from one original fertilized egg. **Differentiation** is the process by which these cells come to differ from one another so that the adult contains skin, liver, bone, and all the other kinds of cells. Since experiments using human embryos are seldom possible, much of what we know of our own development comes from studies on other animals and even on plants; differentiation has many common features in all animals and plants.

In this chapter we shall consider human reproduction and the development of the embryo.

20–A Reproductive Organs of the Human Female

The external sex organs of a woman are shown in Figure 20-2. Note that the openings of the urinary and reproductive tracts are separate; among the vertebrates, only females of higher mammals show this characteristic. In virgin women, the vagina contains the **hymen**, a membrane with a slit that allows passage of the menstrual flow.

The internal female reproductive organs consist of the ovaries and fallopian

vertebrate: animal with a backbone.

higher mammals: mammals except monotremes and marsupials (see Table 2-3).

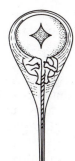

Fig. 20–1

A homunculus. Seventeenth century microscopists imagined that they could see little men within sperm cells and drew pictures of sperm like this one. They thought that babies developed by the growth of preformed human beings. This view was later displaced by that of embryologists who maintained that babies develop from sperm and egg by differentiation as well as growth.

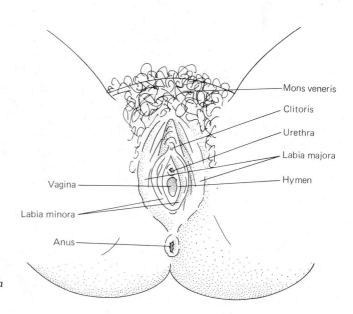

Fig. 20–2

The external genitalia of a woman.

Mons veneris

Clitoris

Urethra

Labia majora

Hymen

Vagina

Labia minora

Anus

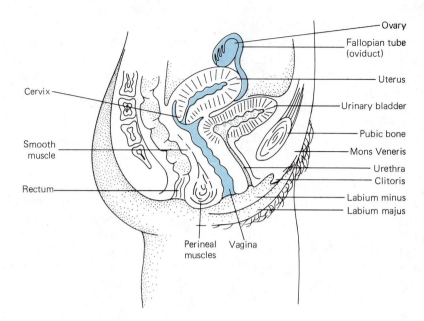

Fig. 20–3

The internal reproductive organs of a woman.

tubes (or oviducts), uterus, and vagina (Fig. 20-3). The two **ovaries** produce eggs. Girls are born with all the eggs they will ever have (about 400,000), and so damage to an ovary is potentially more harmful to fertility than is damage to a testis, which continues to produce sperm throughout a man's adult life. When a mature egg is released by the ovary, the beating of cilia lining the nearby **fallopian tubes** draws the egg into one of the tubes and on toward the uterus. Fertilization usually occurs in the fallopian tubes. An unfertilized egg disintegrates as it passes through the uterus.

fal-OPE-ee-un

The **uterus** is a very elastic organ with muscular walls; its main function is to carry a developing embryo and expel it during childbirth. The uterus is about the same size and shape as a pear. The lower opening of the uterus is the **cervix**, which protrudes into the vagina. The cervix contains the most powerful sphincter muscle in the body. (**Sphincters** are ring-shaped muscles that can contract to diminish the opening of a tube; other sphincters occur around the openings of the anus and the urinary bladder.) The strength of the cervical muscle is necessary to hold about 15 pounds of fetus and fluid in the uterus against the pull of gravity during pregnancy.

SERVE-ix

SFINK-ter

The muscular, elastic **vagina** is the receptacle for the penis during copulation, and the pathway to the exterior for the baby during childbirth.

20–B Reproductive Organs of the Human Male

The **testes** (sing., **testis**) of a human male produce **spermatozoa**, or **sperm**, from puberty (sexual maturity) until death. For unknown reasons, sperm production in mammals requires a lower temperature than that of the body. Mammalian testes usually lie in the cooler **scrotum**, a sac outside the body cavity. The

SPERM-at-oh-ZOH-uh

SCROH-tum

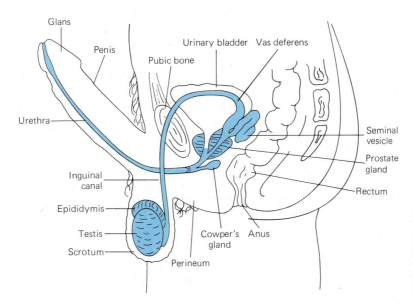

Fig. 20–4

Cross section through the pelvic region of a man during erection. The reproductive organs and path of the sperm are in color.

testes form in the abdominal cavity and descend into the scrotum through a canal in the abdominal wall before birth. (The presence of this canal between the abdominal cavity and scrotum is one reason more men than women experience inguinal hernias, which are splits in the sheet of muscle surrounding the abdomen.)

Muscular action carries mature sperm from the testes through the **vasa deferentia** (sing., **vas deferens**) (Fig. 20-4). Secretions from the **prostate** and other glands join the sperm to form **semen**, which leaves the body via the **urethra** (an outlet shared by the urinary and reproductive tracts in men). Most of the length of the urethra lies inside the **penis**, the organ that releases semen into the female's body.

VAH-sah   DEFF-er-EN-she-uh (sing., vahz DEFF-er-enz)

PROSS-tate (note that there is no ''r'' in the second syllable)

SEA-men

your-EETH-ruh

20–C   Physiology of Sexual Intercourse

During **copulation** (often called sexual intercourse in humans), the male's penis introduces sperm into the body of the female. Before the penis can enter the vagina, it must become at least partially erect under the influence of sexual stimulation. Sexual stimulation can be brought about by any of the senses, usually most effectively by touch. The external genitals, and especially the glans of the penis, are the most sexually sensitive areas (Fig. 20-4). Stimulation causes the spongy tissue of the penis to become engorged with blood and expand to its erect state.

Sexual arousal of a human female can also result from many different stimuli. The **clitoris** is the organ most sensitive to touch, and it is the focus of sexual sensation. In a sexually aroused woman, the vulva becomes swollen because of an increased blood supply, and the walls of the vagina secrete fluid which acts as a lubricant for entrance of the penis.

Fig. 20–5

Cross section of a human penis.

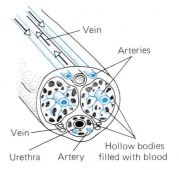

Orgasm is the physiological culmination of tactile stimulation, particularly of the glans and of the clitoris. Orgasm is characterized by an increase in heart rate, breathing rate, and blood pressure. It culminates in a burst of involuntary muscular contractions. In men, the muscles of the penis contract forcefully to eject semen, a process called **ejaculation**. Neither sex need experience orgasm for fertilization to occur; often, a small amount of semen is released before ejaculation, and this may contain enough sperm to fertilize an egg.

20–D Hormones and Reproduction

Hormones are of great importance in controlling the reproductive physiology of nearly all animals. For instance, a variety of hormones maintains the reproductive state of the adult human male. The most important is **testosterone**, which is produced by the testes and is needed for sperm production. Its presence also maintains male secondary sexual characteristics, such as beard growth and deep voice. Testosterone is also responsible for the greater muscular strength of men. In medieval times, when music written for boys' soprano voices was popular, some nobles maintained choirs of castrati (men castrated by removing their testes when they were boys). Because they produce no testosterone, castrati keep the high voices of childhood into adult life. Eunuchs, once used as male attendants in harems, were also castrated men. Men isolated from most sexual stimulation, for instance on expeditions to Antarctica, produce less testosterone and have slower beard growth than at other times in their lives.

Female hormones control women's menstrual cycles. The interaction of hormones produced by the pituitary gland, just beneath the brain, and by the ovaries produce this series of cycles starting at puberty. The cycles continue until menopause some 35 years later. The hormonal progress of the menstrual cycle determines a woman's fertility (Fig. 20-6). **Ovulation**, the release of a mature egg by the ovary, occurs about day 14 of the cycle. After ovulation, the ovary secretes increasing quantities of the hormones **progesterone** and **estrogen**, whose presence in the blood promotes thickening of the **endometrium**, the lining of the uterus. About 12 days later, the secretion of progesterone slows down until its concentration in the blood is so low that the endometrium sloughs off in **menstruation**.

Pregnancy is controlled by hormones produced by the **placenta**, an organ formed partly by the uterine wall and partly by the embryo. The placenta secretes a hormone which prevents the next menstrual cycle after fertilization and ensures that no new embryo can form while the first one develops. If the developing embryo is abnormal or dies, the placenta usually stops secreting its hormones, permitting the body to restart the menstrual cycle so that the pregnancy aborts (miscarries). It is estimated that as many as three out of five human embryos abort in this manner, often so early that the mother never knows she was pregnant.

By the same token, it is not surprising that hormones produced by the developing baby determine when birth shall occur. It is important that the baby be born when it is mature, rather than at some time determined by the mother's glands. When the fetus is mature it secretes hormones from its now-developed adrenal glands, a signal that triggers childbirth.

tess-TOSS-ter-own

OH-view-LAY-shun

proe-JEST-er-own ESS-troe-jin
en-doe-MEET-ree-um

pluh-SENT-uh

adrenal glands (uh-DREEN-al): glands that sit atop the kidneys (see Fig. 23-25).

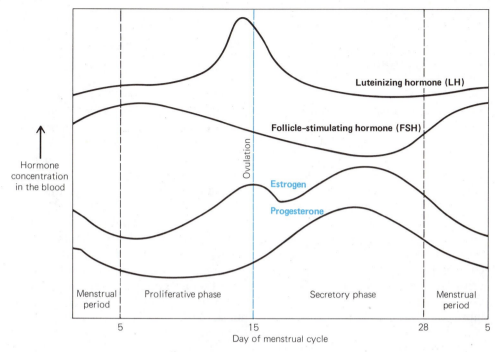

Hormone concentration in the blood

Luteinizing hormone (LH)

Follicle–stimulating hormone (FSH)

Ovulation

Estrogen

Progesterone

| Menstrual period | Proliferative phase | Secretory phase | Menstrual period |

5   15   28   5

Day of menstrual cycle

Fig. 20–6

Hormonal changes during the menstrual cycle of a woman. Hormones produced by the pituitary gland in the brain are labeled in black, those produced in the ovaries in color.

The placenta is expelled within an hour after the baby is born. Many mammalian mothers eat their placentas. One important result of this is that hormones eaten in the placenta trigger maternal behavior. Whether or not women would make better mothers if they ate their placentas has not, as far as we know, been put to the test!

Expulsion of the placenta during childbirth triggers **lactation** (milk production) by the mammary glands. Once this has occurred, milk production is thereafter on a supply-and-demand basis; the more a baby suckles, the more milk is produced about 24 hours later—an inconvenience for nursing mothers who leave home on business trips. Even a postmenopausal woman who has suckled a baby years earlier will produce an adequate milk supply if she suckles a baby, and some human populations use this ability to permit grandmothers to care for infants while their mothers work.

Fig. 20–7

A lactating sow and her litter. (Biophoto Associates)

20–E Birth Control

Since the dawn of history, human beings and other (particularly social) animals have regulated their reproduction so that the number of offspring bears some relation to the parent's ability to raise them. For instance, carnivorous birds with uncertain food supplies usually feed the weakest nestling last. If food is abundant, this nestling receives enough to live; if not, its nestmates will receive all the food and the weakling will be the first to die. Thus, valuable energy has not been spent on offspring that may not survive.

Abortion, prolonged breast-feeding, late marriage, and infanticide have been the most common methods of human population control over the years. Sponges placed in the vagina and condoms made of leather and pigs' bladders are also recorded in ancient Roman writings.

Strictly speaking, **contraception** refers to birth control methods which prevent fertilization. One of the most effective methods of contraception (although none except abstention or sterilization is 100% reliable) is "the pill." Birth control pills contain synthetic estrogen and progesterone, hormones whose presence at the given dosage prevents ovulation. Pregnancy is avoided because there is no egg to fertilize. The decrease in hormone levels when a woman stops taking the pills for part of the month causes menstruation.

The medical risks of using the pill are largely the same as those of pregnancy. The elevated level of progesterone in the blood produces a state similar to pregnancy and an increased risk of such things as high blood pressure and excessive blood-clotting. It appears from recent research that the extra estrogen in the body from the pill probably does not increase the risk of cancer as was once thought. Interestingly, there is no form of birth control or legal abortion that carries nearly as high a risk of maternal death nowadays as does pregnancy itself, except that cigarette-smokers over 35 are at greater risk from taking the pill than they would be from pregnancy (for unknown reasons).

SPERM-ih-SIDE-ul

Another method of contraception is the use of a rubber **diaphragm**, which is smeared with a **spermicidal** (sperm-killing) jelly or cream each time it is used. The diaphragm blocks the cervix so that sperm cannot reach the uterus.

The **condom** is the contraceptive device most commonly used by men, although pills, nasal-sprays and long-lasting hormone injections loom on the horizon. A condom is rolled onto the erect penis shortly before intercourse. It catches the semen so that sperm do not enter the female reproductive tract (in case of breakage, many women use spermicidal foam while their partners use condoms, and this combination rivals "the pill" in reliability). The condom is also the only form of birth control that can prevent the spread of venereal disease from one partner to the other.

The much less reliable **rhythm method** consists of avoiding intercourse during the "fertile period" of the menstrual cycle; that is, when there is an egg present to be fertilized. Since eggs and sperm remain alive for up to 2 days in a woman, this means avoiding intercourse for 2 or 3 days before and after ovulation. The difficulty is in deciding when ovulation will occur. It usually occurs 14 days before the start of the next menstrual period, but knowing that is no help in *predicting* the time of ovulation. Furthermore, ovulation can occur at any time during the menstrual cycle.

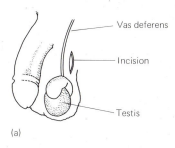

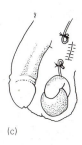

Vas deferens

Incision

Testis

(a)   (b)   (c)

Fig. 20–8

Vasectomy, the surgical procedure in which the vas deferens is severed so that sperm can no longer enter the semen.

Intrauterine devices (IUDs) are small plastic or metal objects inserted into the uterus by a doctor. IUDs are not true contraceptives; instead they somehow act on the uterine lining so that the already developing embryo cannot implant; just how this happens is still controversial. IUDs are as effective as contraceptive pills in preventing pregnancy. On the other hand, a disquieting number of IUDs become deeply imbedded in the wall of the uterus, causing dangerous abdominal infections and requiring surgery to remove the device.

20–F Sterilization and Abortion

Sterilization is any more or less permanent change that prevents an animal from reproducing sexually. Sterilization of either sex is the fastest-growing method of contraception in the world. The most common procedure for men is **vasectomy,** severing and tying off the vasa deferentia (Fig. 20–8). This is a simple operation, usually performed under local anesthesia. Afterward, sperm are still produced but are resorbed into the body. The fluid ejaculated contains the secretions of various glands but no sperm. Vasectomy has gained popularity as sperm banks have become more available. A man may store a frozen sample of his sperm in such a bank before undergoing vasectomy (which is not easily reversed). Sperm stored in sperm banks have already been used to produce hundreds of normal children by **artificial insemination**—the introduction of sperm into the female reproductive tract by means other than intercourse.

Sterilization of a woman usually involves **tubal ligation**, cutting and tying the fallopian tubes. This operation also can now be performed under local anesthesia. After tubal ligation, the ovaries continue to function as they did before, but sperm cannot reach the egg, and so fertilization cannot occur.

Induced abortion is probably the oldest birth control method known. Today, religious and ethical considerations dominate peoples' opposition to abortion. This, however, is a recent development. Abortion was accepted, and fairly common, in the United States and Europe until the early nineteenth century, when doctors began to oppose the practice on the ground that it led to unnecessary maternal deaths at a time when the risk of dying in childbirth was decreasing. Religious opposition to abortion developed later. The situation has changed quite rapidly in the last few decades and abortion is now, once again, legal in many countries.

There are three main methods of abortion. In **vacuum curettage** and **dilation and curettage** ("D and C") the fetus is sucked or scraped out of the uterus, usually

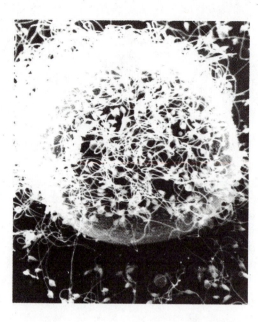

Fig. 20–9

The egg of a sea urchin surrounded by a tangle of sperm with spherical heads and threadlike tails (Biophoto Associates, S.P.L.)

under local anesthesia. ("D and C" is used for diagnosis and treatment of a number of uterine disorders as well as for aborting early pregnancies.) These methods are used early in pregnancy. After the fifteenth week of pregnancy, **saline injection** can be used. An injection of salt solution kills the fetus, and the uterus subsequently expels the fetus and the placenta.

Today, a properly performed, legal abortion is less likely to cause a woman's death than is a completed pregnancy. However, the dangers of illegal "backstreet" abortions and attempts to abort oneself cannot be overemphasized.

20–G Eggs and Sperm

GAM-eets

UH-oh-sites; sper-MAT-oh-sites

Eggs and sperm are **gametes**, or reproductive cells, produced in the ovaries and testes from precursor cells called **oocytes** and **spermatocytes**. A gamete contains only half as many chromosomes as does an ordinary body cell; when an egg is fertilized by a sperm, the fertilized egg ends up with the normal body-cell number of chromosomes again. The nuclear divisions which halve the chromo-

my-OH-sis

some number are collectively known as **meiosis**, and meiosis must occur some-time in the life history of every sexually reproducing organism if the number of chromosomes is not to double in each generation (Section 16–F). The major steps in formation of a sperm are shown in Figure 20-10.

OH-vum (pl., OH-vuh)

The formation of an egg, or **ovum** (pl., **ova**), is somewhat different from sperm formation. Here, meiotic nuclear division is accompanied by unequal division of the cytoplasm so that one **oocyte,** or egg mother cell, produces one large ovum and a number of smaller cells called **polar bodies** (Fig. 20-11). During evolution, the egg became the main source of stored food, ribosomes, messenger RNA, and other cytoplasmic components which support the early development

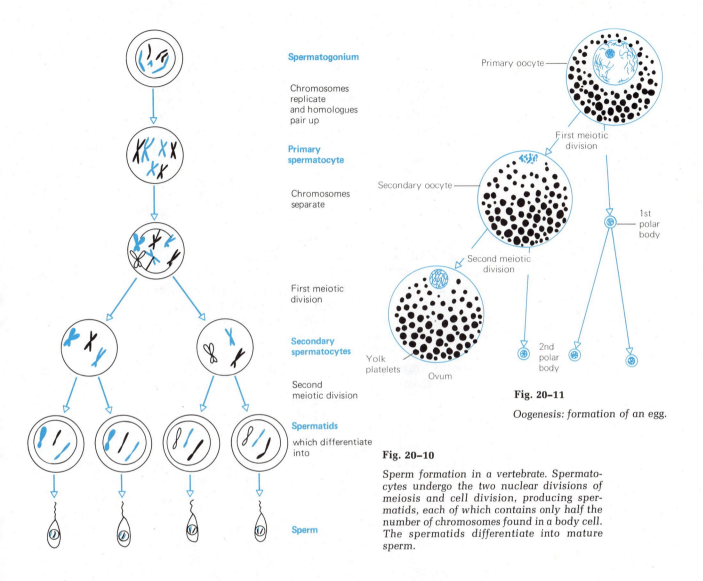

Spermatogonium

Chromosomes
replicate
and homologues
pair up

Primary
spermatocyte

Chromosomes
separate

First meiotic
division

Secondary
spermatocytes

Second
meiotic division

Spermatids

which differentiate
into

Sperm

Primary oocyte

First meiotic
division

Secondary oocyte

1st
polar
body

Second meiotic
division

Yolk
platelets

Ovum

2nd
polar
body

Fig. 20–11

Oogenesis: formation of an egg.

Fig. 20–10

*Sperm formation in a vertebrate. Spermato-
cytes undergo the two nuclear divisions of
meiosis and cell division, producing spermatids, each of which contains only half the
number of chromosomes found in a body cell.
The spermatids differentiate into mature
sperm.*

of the embryo. So egg formation consists of producing a large cell with a lot of
cytoplasm and only half the chromosomes found in body cells. Nearly all the
cytoplasm goes into the future egg at each cell division of egg formation, and
other vital substances move into the egg-to-be from surrounding cells in the
ovary. The polar bodies are really just a means of shedding unneeded chromosomes from the developing egg cell, and they soon die.

Although sperm formation always proceeds to completion whether or not
there is any chance of the resultant sperms' fertilizing an egg, egg formation
usually stops somewhere. In humans an oocyte, rather than an ovum, is actually
released from the ovary. Only if the oocyte is penetrated by a sperm will the
second meiotic division occur, producing a polar body and an ovum whose
chromosomes will then combine with those from the sperm.

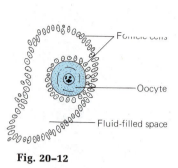

Follicle cells

Oocyte

Fluid-filled space

Fig. 20–12

*A mammalian follicle. A follicle
is part of an ovary containing
a developing egg surrounded
by cells that supply it with food
and other substances.*

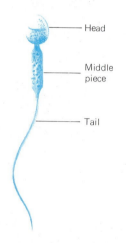

Head

Middle
piece

Tail

Fig. 20–13

The structure of a sperm.

AM-nee-on

AM-nee-oh-sen-TEE-sis

AL-an-toys

CORE-ee-on

20–H Fertilization and Implantation

The human egg is fertilized by one sperm, which penetrates the egg membrane and finally disintegrates inside the egg, leaving only its chromosomes to combine with the chromosomes of the egg. When a sperm penetrates an egg it causes a reaction in the egg membrane that prevents the entrance of further sperm, so that fertilization by more than one sperm is very rare.

In humans, fertilization usually occurs high up in a fallopian tube and the fertilized egg begins to develop as it descends the tube to the uterus, a process that takes several days. "Test tube" babies, first produced in the 1970s, do not really develop in test tubes; only fertilization occurs outside the mother's body. Some women have blocked fallopian tubes or spermicidal secretions in the vagina, and fertilization cannot occur naturally in their bodies. In a few cases, doctors have removed eggs from such women, added the husband's sperm in a test tube, and then reimplanted the fertilized egg, as it began to develop, into the woman's uterus. The success of this technique caused excitement because it proved that scientists had finally found a chemical solution in which eggs and sperm would survive and begin to develop. Before this time, mammalian eggs, in particular, had been damaged by all attempts to manipulate them outside the body. (Sperm survive better because they are protected by the semen—their own private reservoir of fluid.)

During the early stages of development, the embryos of all vertebrates look very much alike, as a result of their common evolutionary histories. Not until it acquires uniquely human characteristics, about 9 weeks after fertilization, is the human embryo called a **fetus** (not until birth is it called a baby).

When the embryo reaches the wall of the uterus, it consists of a little ball of cells, which will develop into the embryo proper; these are surrounded by a sphere of cells (Fig. 20-14). Some of these outer cells invade the wall of the uterus, anchoring the embryo and forming the first part of the placenta. Other cells in this outer sphere give rise to the membranes which surround the embryo as it develops, some of which also form part of the placenta.

The three membranes that are important in human development are the amnion, chorion, and allantois (Fig. 20-15). The **amnion** is a fluid-filled sac around the embryo, cushioning it against bumps; it is the sac that bursts when the "waters break" during childbirth. Since the amniotic fluid contains some cells sloughed off by the embryo, drawing fluid from the sac into a hypodermic needle (**amniocentesis**) permits a doctor to check for chromosomal defects in the embryo.

The **allantois** is a sac connected to the embryo's gut. In reptiles and birds it stores the embryo's waste products until the egg hatches. In humans it becomes riddled with blood vessels and makes up most of the embryonic side of the placenta.

The third membrane, the **chorion**, which surrounds the fetus outside the amnion, also makes up a part of the placenta. The chorion secretes the progesterone-like hormone which maintains pregnancy and can be detected in pregnancy tests.

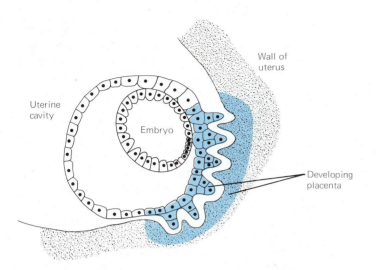

Fig. 20–14

A developing mammalian embryo implanting in the wall of the uterus.

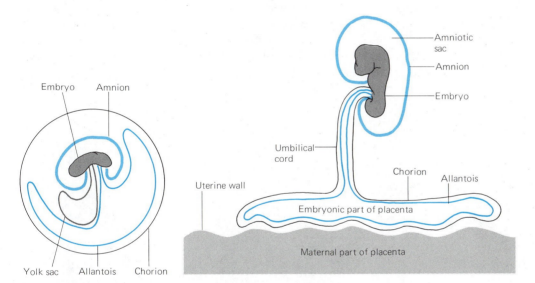

(a) The four extra-embryonic membranes found in all amniotes—reptiles, birds and mammals. (The yolk sac is small and of little importance in most mammals).

(b) Arrangement of the extra-embryonic membranes to form umbilical cord and placenta in mammals

Fig. 20–15

A mammalian embryo with the membranes that surround it during development.

20–I   Stages of Early Development

Because the human embryo undergoes its early development while it is traveling down the fallopian tubes, most of what we know about the first stages of embryonic development comes from work on other animals. The development of sea urchins has been intensively studied. Because sea urchin eggs are fertilized externally (outside the body) and are almost transparent, they are easy to observe. (Another reason for their popularity is that summers on the coasts where sea urchins abound are irresistible to many embryologists who live inland for most of the year.)

After fertilization, the single fertilized egg cell divides into 2, 4, 8, 16 cells, and so on. This period of cell division is known as **cleavage** and the end product of cleavage is a **blastula**, a hollow ball of cells with a cavity in the center (Fig. 20-16).

BLAST-you-luh

Identical twins are produced when the first cell division of cleavage produces two separate cells, each of which develops into a separate embryo instead of into parts of a single embryo. Since these cells contain identical genes the resulting embryos are genetically identical.

GAS-true-LAY-shun;
GAS-true-luh

During the next developmental stage, **gastrulation**, the cells of the blastula continue to divide, and they also rearrange themselves into three main groups of cells from which all later tissues arise (Fig. 20-17). The end product of gastrulation is the **gastrula**. The cells of the gastrula begin to develop in such a way that different tissues become visibly distinct. In the sea urchin, for instance, some cells start to secrete calcium carbonate particles, which will become part of the skeleton.

noor-you-LAY-shun

Gastrulation is followed by **neurulation**, the initiation of the nervous system. The **neural tube** forms as a pair of ridges rise up and then join together along the back of the embryo (Fig. 20-18).

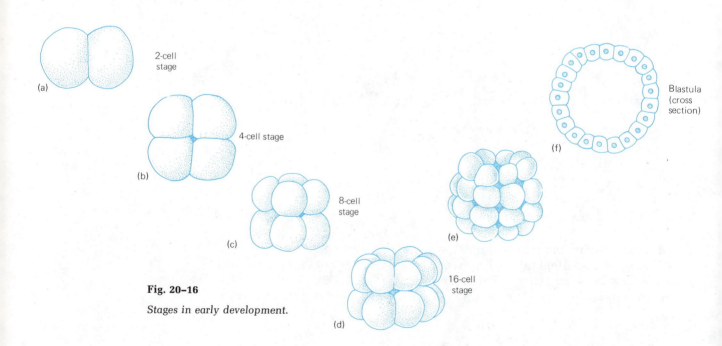

2-cell
stage

(a)

4-cell stage

(b)

8-cell
stage

(c)

16-cell
stage

(d)

(e)

Blastula
(cross
section)

(f)

Fig. 20–16

Stages in early development.

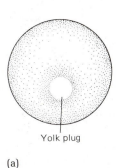

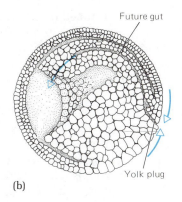

Future gut

Yolk plug

(a)

(b)

Fig. 20–17

An embryo during gastrulation. (a) External view. The yolk plug remains where cells have moved into the embryo. (b) Cross section to show how the cells move and the formation of several layers of cells.

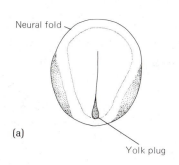

Neural fold

(a)

Yolk plug

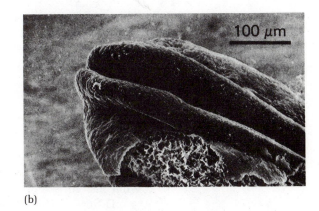

100 μm

(b)

Fig. 20–18

An embryo undergoing neurulation. (a) The neural folds will rise up and join to form the tube that turns into the nervous system. (b) A scanning electron micrograph of a later stage with the neural folds joining to form the neural tube. (Kathryn Tosney)

20–J Cell Differentiation

The differentiation of one cell type from another is one of the most obvious processes that occurs during development. How the single egg cell gives rise to adult cells as different as, say, a nerve and a liver cell, is one of the most intriguing questions in biology today. Cells differ because they contain different proteins. A cell's proteins determine what biochemical reactions can occur in the cell, what substances will cross its cell membrane, whether or not it can move, and everything else about it. The problem of differentiation is this: how do cells which are all descended from one fertilized egg acquire different proteins during development?

Since each protein is produced indirectly from a gene, one possible answer would be that during development different cells end up with different genes and so with different proteins. But biologists have become increasingly convinced that this is not usually the case. In most species, the evidence strongly suggests that the cells of an adult contain the same genes that were present in the fertilized egg.

One example of the evidence for this conclusion is that whole plants can be grown from cuttings. If a leafy stem of a *Coleus,* geranium, or grape vine is planted, it will form roots and become a complete plant. This means that all the genetic information needed to make root tissues exists somewhere in the stem cutting and that root cells with their genes are not necessary to form new root tissue. Even more dramatically, a few cells taken from a carrot or chrysanthemum and grown in nutrient fluid in a test tube, will develop into a complete plant (Fig. 20-19). In fact, commercial growers now produce choice varieties of chrysanthemums on a large scale by this method. Most of the new plants are genetically identical to the plant from which the original cells were taken. The original cells and all the root, stem, flower, and leaf cells produced from them apparently contain the same genes.

Experiments in which cell nuclei are transplanted into other cells suggest that the composition of the cytoplasm determines which proteins a cell produces. Consider this experiment: if a frog red blood cell nucleus is transplanted into frog egg cytoplasm, the nucleus immediately stops producing the RNA it needs to make red blood cell proteins, such as hemoglobin, and starts producing the RNA necessary for early embryonic development. This switch must be triggered by something in the egg cytoplasm.

Differentiation can be considered as a sort of chain reaction. Fertilization activates messengers in the cytoplasm which cause cell division, and it switches on certain genes. Some of the proteins (or products made by proteins) produced by these genes migrate to neighboring cells and activate different genes in those cells, and so on. The result is a complex series of reactions, each dependent on what has gone before.

Differentiation is a poorly understood process and a very active area of research. Cancer workers are interested in differentiation because something goes wrong in the process of differentiation when cancer occurs (see Chapter 14). An expectant mother's use of cigarettes, alcohol, and drugs may damage her unborn baby because these substances interfere with differentiation. All in all, this is a fascinating and mysterious area of biology, in which we can expect many new discoveries within the next decades.

20–K Later Human Development

The human embryo undergoes neurulation during the third week after fertilization. At this stage, the embryo is only about 2 millimetres long and is almost indistinguishable from the embryo of a frog or a sea urchin, but already it is surrounded by its membranes and attached to the uterus at the placenta. The placenta is now tiny, but by the time of birth it will develop into an organ which

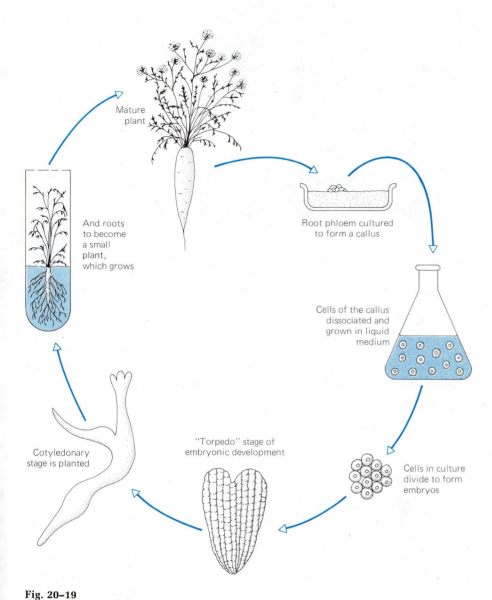

Mature plant

Root phloem cultured to form a callus

Cells of the callus dissociated and grown in liquid medium

And roots to become a small plant, which grows

Cotyledonary stage is planted

"Torpedo" stage of embryonic development

Cells in culture divide to form embryos

Fig. 20–19

Stages in the development of a mature carrot plant from cells isolated from a mature root.

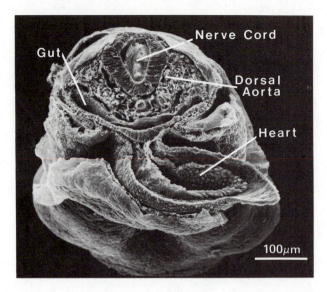

Fig. 20–20

Scanning electron micrograph of organs beginning to form in a chick embryo. (Kathryn Tosney)

looks like a raw hamburger, about the size of a stack of four dinner plates. The membranes joining the embryo to the placenta develop into a cord, the **umbilicus**, which grows thicker and longer as development proceeds.

By the end of the third week the embryo has entered the fourth stage of development, **organogenesis**; in this stage the major organ systems begin to form: the nervous system, gut, and blood vessels. The heart, shaped like a lumpy tube, starts to pulsate. Drugs and diseases are most apt to damage the embryo at these early stages when the organ systems are forming. After about 3 months of development, the embryo, although it is still small (about 30 millimetres long), is more or less fully formed and much less susceptible to malformations. The drug thalidomide, prescribed as a tranquilizer for pregnant women during the 1960s, stunted development of the limbs from the tiny limb buds during the fourth and fifth weeks of development. Similarly, if a pregnant woman has German measles during the fourth through twelfth weeks of pregnancy, the disease may damage the embryo's heart, eyes, ears, or brain, which are developing at that time.

During the third month, the fetus begins to move, and the mother may feel its movements. From this point onward, the most obvious progress is growth in size. There are still changes taking place, however. The nervous system is still immature, and so are the circulatory and respiratory systems; the fetus cannot survive outside the mother. The youngest fetus known to have survived a premature birth was about 23 weeks old, and it required continuous assistance to breathe, feed, and maintain its body temperature for many weeks thereafter.

20–L Birth

In humans, the date of birth averages about 270 days after conception, but there is much variation in the time a baby takes to develop. The uterine contractions which expel the baby and the placenta are called **labor**; the whole process can be divided into three main stages. The first stage is **dilation**, which lasts from

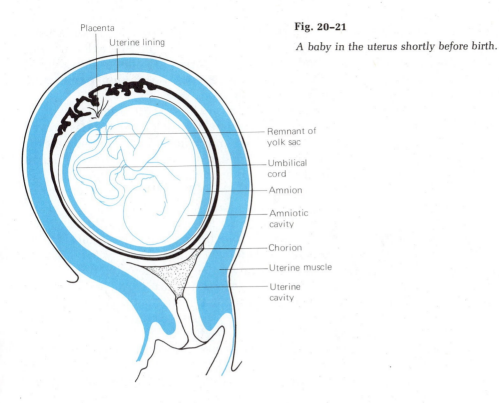

Fig. 20–21

A baby in the uterus shortly before birth.

Placenta
Uterine lining
Remnant of yolk sac
Umbilical cord
Amnion
Amniotic cavity
Chorion
Uterine muscle
Uterine cavity

about 2 to 20 hours and ends when the cervix of the uterus is fully open or dilated. The second stage, **expulsion**, which lasts from about 2 to 100 minutes, begins with **full crowning,** the appearance of the baby's head in the cervix, and continues while the baby is pushed, head first, down through the vagina into the outside world where it draws its first breath.

The third, or **placental**, stage begins when the baby is born. The uterus continues to contract while the umbilical cord is clamped, and some minutes after the baby is born the uterus expels the placenta. The umbilical cord can now be severed, and the baby's independent existence begins.

20–M Aging

Aging may be defined as changes that accumulate with time and make an organism more likely to die. Aging is inseparable from other developmental processes, and begins even before an individual officially "completes" development at sexual maturity. Slower healing, for instance, is one sign of aging, and even a human teenager's broken bones heal less rapidly than those of a young child.

Aging and death are genetically programmed. Species have characteristic life spans that can be altered by selection. An elephant dies of old age when it is about 200 years old, a human at about 70 to 100 years, and many insects at a few days or weeks. Aging assures death at this characteristic age if disease or predators do not kill the animal (or plant) first. The selective advantage of death caused by aging is much debated. Probably aging, by eliminating older individuals, leaves food, shelter, and other limited resources to younger individuals, who will usually bear many of the same genes, and whose reproduction is more important in evolutionary terms (Section 6–A).

We do not know how aging occurs; there are at least 20 different theories. Although we can describe many processes characteristic of aging, it has proved impossible to decide which of these is cause and which effect. For instance, the immune system, which defends the body against disease, becomes less efficient with increasing age. Is this why disease becomes more common with age, or does the immune system become less efficient because it is debilitated by disease as age increases?

As people grow older, their bodies cope less effectively with stress and disease. The ability to survive changing conditions depends largely on the functioning of the immune, nervous, and endocrine (hormonal) systems. Therefore, researchers have concentrated on the hypothesis that degeneration in these systems results in the loss of adaptability seen in the rest of the body, such as slower healing, hardening connective tissue, and brittle bones.

The suggestion is often made that aging starts after a particular cell line has divided a certain number of times. However, when this hypothesis has been tested in detail for any particular cell type, it has usually been disproven. For example, when cells removed from the bodies of very old animals are kept in appropriate environments, they may live and divide without any signs of aging for generations beyond the animal's own time of death. This suggests that the environment of a cell determines whether or not it ages. In this case, is there a particular set of cells in the body that actually *does* age spontaneously and so creates an environment in which all the rest of the body ages? We do not know.

One popular theory holds that aging is a cellular process resulting from the accumulation of mutations in all the body's cells. This does not explain aging in all animals, however. Mammals accumulate proportionately more mutations in the DNA of their cells during their lifetimes than do insects during their much shorter lives. Similarly, cells die throughout our lives but this alone cannot explain aging. Human beings who lose cells through disease or accident have survived with many more cells missing from vital organs (such as the liver, kidney, brain, and lungs) than are normally destroyed during aging.

Searches for a single cause of aging have all ended in failure. It seems likely that the "aging genes" do not control one single system but instead control a multiplicity of minor degenerations. Because the body's systems all interact with one another, minor deficiencies anywhere in the body can accumulate to produce aging of the body as a whole and of its individual systems.

SUMMARY

Human reproduction is sexual. The male and female parents produce gametes (eggs and sperm) by meiosis in the testes and ovaries. Sexual maturation during puberty, maturation of sperm and egg, and the female menstrual cycle are controlled by hormones. Hormones secreted by the placenta control pregnancy, and hormones also control birth and lactation.

Humans commonly control their reproduction by techniques that either prevent fertilization or abort a developing embryo from the uterus.

In the male, the testes, which produce the sperm, lie in the scrotum. Semen forms as glandular secretions join the sperm in the tubes leaving the testes. The penis is composed of spongy tissue which becomes engorged with blood during sexual stimulation; it can then be introduced into the vagina. Sexual stimulation may eventually result in orgasm by both sexes and in ejaculation of semen by the male.

When a mature egg is released from the ovary, it travels through a fallopian tube, where it may be fertilized by a sperm. The fertilized, developing egg then descends into the uterus where it implants in the uterine wall. Part of the embryo develops into the embryo proper and part forms the embryonic part of the placenta and the membranes surrounding the embryo.

The early stages of development are cleavage, gastrulation, and neurulation. Differentiation of one cell type from another first becomes apparent during gastrulation. Differentiation occurs because different genes become active, and so direct production of different proteins, in different cells during development. The control of differentiation is not understood, but the egg cytoplasm is clearly important.

Organogenesis follows neurulation, and after about 3 months of development the major organ systems have formed. The fetus cannot survive outside the uterus until about 6 months after fertilization. Birth occurs at about 9 months and involves powerful uterine contractions that expel the fetus and placenta from the uterus through the vagina.

Aging is an important part of development after birth. We understand very little about aging. It is possible that aging has many causes whose interactions are complex and difficult to disentangle or, less likely, that there is some major key to the problem of aging that has yet to be discovered.

SELF-QUIZ ——

Match the reproductive organs on the right with the descriptions on the left:

____1. Tubes that conduct sperm in male
____2. Receptacle for penis
____3. Production of seminal secretions
____4. Conducts egg
____5. Holds baby in uterus
____6. Produces sperm
____7. Prepares nutritive lining for embryo

a. cervix
b. fallopian tube
c. ovary
d. prostate gland
e. testis
f. urethra
g. uterus
h. vagina
i. vasa deferentia

8. From the list of reproductive organs and passages above, construct the correct route for the passage of sperm from the site of production to the site of fertilization.

For each of the following birth-control methods, choose the correct means of interference with reproduction:

____9.   Diaphragm and jelly
____10. Vasectomy
____11. The pill
____12. IUD
____13. Tubal ligation
____14. Induced abortion
____15. Condom

a. prevents fertilization of egg
b. prevents implantation of embryo
c. prevents completion of embryonic development of implanted embryo
d. prevents ovulation
e. prevents formation of sperm
f. prevents release of sperm into seminal fluid

16. Both egg and sperm formation involve equal _____ division; however, _____ division in production of _____ is unequal, whereas in production of _____ it is equal.

Match the correct stage of development listed on the right with each of the following characteristics:

___17. Cell division forming a single-layered, hollow ball

    C = Cleavage
    G = Gastrulation
    N = Neurulation

___18. Formation of tube that becomes nervous system

___19. Cell division and grouping into three types of cells

20. During differentiation, cells with the same DNA:

    a. develop similarly
    b. divide at equal rates
    c. contain different genes
    d. activate different genes

QUESTIONS FOR DISCUSSION

1. Does vasectomy affect male potency?

2. Does abortion affect a woman's subsequent fertility? Why?

3. Can a woman become pregnant the first time she has sexual intercourse? Why?

4. The venereal disease gonorrhea is now epidemic in the United States and Western Europe. Why do you think this is so?

Essay:
HUMAN "CLONES"

There has been some discussion recently of the moral problems raised by the possibility of controlling the genetic makeup of future populations by producing human beings with chosen and identical genes. Such a group of people is commonly referred to as a clone. A **clone** is a group of genetically identical individuals produced asexually from each other or from a common parent. The simplest example is a group of bacteria descended, by repeated cell division, from one original bacterium. The bacteria in the clone all inherit the genes of the parental cell.

Discussion of human clones has been stimulated by experiments similar to those described in Section 20–J. If a nucleus from a tadpole intestine cell is introduced into a frog egg with its own nucleus removed, the egg will develop into a normal, fertile, frog (Fig. 20-22). If each of many different nuclei from the same individual were injected into its own egg and the resulting individuals raised to maturity, one would have a clone of frogs. The members of the clone would be genetically identical because each would contain only the genes contained in the body cells of one individual. Research on test-tube babies (section 20-H) shows that human eggs can now be induced to develop normally outside the mother's body, at least for a few days. This suggests that the kind of nuclear transplantation successfully performed on frogs might also be possible with human beings.

In theory, at least, nuclear transplantation in humans may one day be possible. All the individuals produced by nuclear transplantation would have the genetic characteristics of whoever donated the nuclei. (Remember, however, that so many non-genetic influences affect human development, that the members of a clone would be no more alike than are identical twins; studies of identical twins separated from birth show that they are often quite dissimilar.)

The technical barriers to nuclear transplantation are still so immense, however, that we shall probably not be confronted by this reality in our lifetimes. For one thing, the mammalian egg is surrounded by a number of fragile membranes whose integrity is vital to the egg. These membranes have to be removed or penetrated to inject a new nucleus into the egg. The original egg nucleus also has to be removed. In frogs, this is done by shining ultraviolet light on one end of the egg where the egg chromosomes lie just under the surface; the ultraviolet rays destroy the chromosomes. A mammalian egg is much smaller than a frog's egg and the chromosomes occupy much of the cytoplasm. Destroying the egg chromosomes without damaging much of the cytoplasm needed for development looks, at the moment, like a hopeless task.

However distant human cloning may be as a technique, many people feel that we should be wise to confront the attendant moral problems before we are confronted with the technical reality. For this reason, research on human embryos has been restricted and even completely banned in the United States during much of the 1970s. The ban has now been lifted somewhat to permit experiments with fertilization outside the body; these have permitted some previously infertile women to become pregnant. Those troubled by the moral implications of cloning, however, urge legislative caution. What do you think? As a voter for the next 50 years, you will probably have to decide.

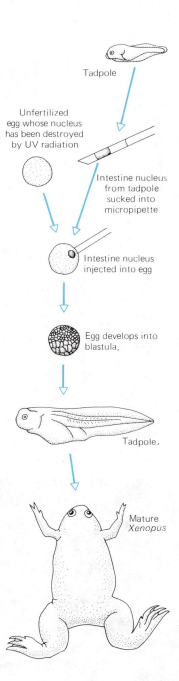

Fig. 20–22

A nuclear transplantation experiment in which a nucleus from a tadpole intestine cell is injected into an egg and supports the development of an adult frog.

NUTRITION AND DIGESTION

21

OBJECTIVES

OBJECTIVES When you have studied this chapter you should be able to:

1. Name the major classes of macronutrients and micronutrients and list the general functions of each class in the body.

2. List the parts of the human digestive tract in order and state what happens to the food in each part of the tract.

3. List the organs that secrete digestive enzymes in humans, and state the type of molecules broken down by enzymes from each of these organs.

4. List the parts of the digestive tract that carry out absorption, and the kinds of substances absorbed in each part.

5. Explain how symbiotic microorganisms contribute to the nutrition of their hosts.

6. List the functions of the mammalian liver, and explain the importance of this organ.

7. Describe what is known of how an animal regulates what it eats and the amount it eats.

Biophoto Associates

A nimals cannot make their own food from inorganic substances; rather, they must obtain organic molecules from their surroundings and process them into forms which their bodies can use. Feeding is a necessary evil; the more time and energy an animal spends feeding, the less it has left for activities of greater selective advantage, such as reproduction. As a result, there is strong selective pressure for an animal to feed as rapidly as possible. It has long been assumed that an animal's feeding habits were determined largely by its energy needs, but recent work makes it clear that many animals, including human beings, have more difficulty in obtaining enough of specific nutrients than they do in acquiring enough calories to supply their energy needs.

In this chapter we shall consider what nutrients an animal needs and how it obtains them and then turn to how food is processed by the digestive system so that it can be used by the body.

21–A Nutrients

The nutrients that any animal must ingest as food may be divided, for convenience, into those that are needed in large quantities—**macronutrients**—and those required in lesser amounts—**micronutrients.**

Macronutrients.   The macronutrients are fats, carbohydrates, and proteins. All three can serve as energy sources because they can be broken down and used to produce ATP (see Chapter 11). The amount of energy available from a given amount of a macronutrient is commonly measured as the number of **Calories** of heat which the nutrient will produce when fully oxidized (Table 21-1). All three classes of macronutrients can also supply carbon atoms used to build the various bits and pieces of a cell; proteins also supply amino acids for building the body's own, unique proteins.

The body stores macronutrient molecules that are not needed immediately for energy, growth, or repair. Carbohydrates are stored as glycogen in muscles and the liver, and fats are stored as fat. There is no protein storage, but excess protein is relieved of its nitrogen atoms and so converted into fats or carbohydrates, which can be stored.

The most common dietary problem of people in industrialized countries (especially the United States) is obesity: if more calories are ingested than are

ingested: taken into the body by way of the mouth.

TABLE 21-1 **THE CALORIC VALUES OF MACRONUTRIENTS**

Macronutrient	Calories* (Kcal) per gm
Protein	4.4
Fat	9.3
Carbohydrates	4.1

*A calorie is the amount of heat needed to raise the temperature of one gram of water by 1°C. The "Calories" in food are actually kilocalories (Kcal) which may be indicated by a capital C.
1 kilocalorie (Kcal) = 1 Calorie = 1000 calories

(a)

(b)

(c)

(d)

Fig. 21–1

Animals feed in a variety of different ways (a) A kingfisher; (b) a blowfly; (c) lions; (d) a mare with her foal. (Biophoto Associates)

used up, the excess is stored as fat. Overeating and lack of exercise are virtually the only causes of obesity, although it is clear, from the billion-dollar trade in reducing drugs and gadgets, that millions of people would rather not believe this.

In the rest of the world, a much more common problem than obesity, for animals and most people, is protein deficiency. This usually occurs, not because the diet contains too little total protein, but because it does not contain enough of the essential amino acids. The **essential amino acids** are those that must be supplied in the diet because an animal cannot synthesize them from other amino acids, although all animals can convert some amino acids into others.

KWASH-ee-OR-core

TRIP-toe-fan

ed-EE-muh

One of the best-known protein deficiency diseases is **kwashiorkor**, found mainly in African populations where the diet consists primarily of cornmeal. Such a diet contains very little of the essential amino acid tryptophan. Victims of kwashiorkor, particularly growing children who need much protein, are lethargic, have edema (swelling due to excess body fluids), and fail to grow normally. Another protein deficiency disease has been produced by the makers of baby formula who enlarged their market through advertisements persuading women in developing countries that bottle-feeding instead of breast-feeding was the chic, Western thing to do. Millions of women in these countries began buying

TABLE 21-2   **SOURCES AND DEFICIENCY SYMPTOMS OF WATER-SOLUBLE VITAMINS**

Vitamin	Major sources	Deficiency symptoms
Thiamine (B_1)	Liver, kidney, yeast, whole grains	Beriberi, loss of appetite, indigestion, fatigue, weakening of the heart and blood vessel walls
Riboflavin (B_2)	Milk, eggs, liver, spinach	Inflammation and breakdown of skin, swollen tongue, eye irritation
Niacin	Whole grains, chicken, yeast	Pellagra, fatigue, skin eruptions
Pyridoxine (B_6)	Whole grains, liver, yeast	Anemia, irritability, convulsions
Pantothenic acid	Eggs, liver, yeast	Similar to other B vitamins
Biotin	Egg white, intestinal bacteria	Rare; minute amounts required
Folic acid	Meat, vegetables	Anemia
Cobalamin (B_{12})	Kidneys, liver, intestinal bacteria	Inadequate red blood cell production
Ascorbic acid (C)	Fruits, potatoes, butter	Scurvy, anemia, slow wound healing

baby formula, despite the fact that breast milk is cheaper and better for the baby. Many of these women, who don't have enough money even to feed the adults in the family, who can't read instructions, and who live in unhygienic conditions, end up feeding the baby diluted, contaminated baby formula. Dilute milk does not contain enough essential amino acids to nourish a baby properly, and nutritionists ascribe millions of infant deaths to this advertising campaign.

On the other side of the coin, the Coca Cola Company has developed a version of its famous drink enriched, for a trivial cost, with essential amino acids. There are political and economic barriers to its distribution, and the new drink has not yet been marketed. Marketing this fortified drink might reduce starvation dramatically, because protein deficiency is a much more common cause of human death than is actual calorie deficiency.

Fat deficiencies can also cause nutritional disease, although these are almost unknown in the Western world. Certain fatty acids are vital constituents of cell membranes and of some hormones and, like essential amino acids, are necessary parts of the diet. In the Canadian north and in some parts of Asia, people living on diets of fish, rice, or fruit, which are all very low in fat, develop a craving for fats and treat them as a delicacy.

Micronutrients.   Micronutrients are the substances which an animal needs in small quantities and which must be supplied by its diet, either because the animal cannot synthesize them or because it cannot make them as fast as they are used up. Micronutrients can be divided into **vitamins**, which are organic compounds, and **minerals**, which are inorganic micronutrients.

Vitamins needed in the human diet usually are divided into two groups: water-soluble and fat-soluble. Water-soluble vitamins generally participate in metabolic reactions, such as those of cellular respiration; since they are easily excreted by the kidney, overdoses of these vitamins are usually harmless.

There are fewer fat-soluble vitamins and their functions are less well understood. Because these vitamins are not soluble in water, the body's enzymes must process them before they can be excreted by the kidneys. As a result, some of them can accumulate to toxic levels if eaten in larger amounts than the body can use. Some modern "diets" recommend dangerous doses of certain fat-soluble vitamins. (In 1978, for instance, two people on high-vitamin diets died of vitamin A poisoning.) The key to avoiding such dangers is common sense and moderation. Almost anything can be poisonous if consumed in sufficient quantity. In 1977 a woman actually drank so much water she killed herself, something most biologists would have sworn was impossible, granted the capacity of the kidneys to dispose of water. But in this case the woman drank water faster than the kidneys could process it and diluted the contents of her cells, with fatal results.

We need some minerals in relatively large amounts. Sodium and potassium, for instance, are vital to the working of every nerve and muscle, and large amounts of them are lost every day in urine, sweat, and feces. Modern reducing diets frequently advocate a reduction in salt (sodium chloride) intake since excess salt holds water in the body and ridding the body of water is a fast way to

TABLE 21-3 FUNCTIONS, SOURCES AND DEFICIENCY SYMPTOMS OF FAT-SOLUBLE VITAMINS*

Vitamin	Physiological function	Major sources	Deficiency symptoms
A (retinol)	Part of visual pigments in eye	Egg yolk, vegetables, liver, butter	Night blindness; drying and damage of skin and mucous membranes
D (calciferol)	Increases absorption of calcium and phosphorus and their deposition in bones	Fish oils, liver, milk, sunlight	Rickets (in children)
E (tocopherol)	Protects red blood cells, vitamin A, and unsaturated fatty acids from attack by oxygen; important in muscle maintenance	Green leafy vegetables, wheat germ oil	Breakdown of red blood cells; sterility (in rats)
K (menadione)	Needed for proper blood clotting	Intestinal bacteria, liver, leafy vegetables	Hemorrhage in the newborn, who lack the intestinal bacteria which synthesize vitamin K

*Vitamins A, D, and K are toxic in large amounts.

TABLE 21-4 **PHYSIOLOGICAL ROLES AND SOURCES OF THE IMPORTANT MINERALS**

Mineral	Physiological roles	Food sources
Sodium (Na)	Water balance Acid–base balance Absorption of glucose into cells Transmission of impulse in muscles and nerves	Table salt (NaCl), milk, meat, eggs, baking soda, baking powder, carrots, beets, spinach, celery
Potassium (K)	Blood salt and acid balance Nerve and muscle function Glycogen formation Protein synthesis	Whole grains, meat, legumes, fruits, vegetables.
Calcium (Ca)	Component of bones and teeth Blood clotting Muscle contraction Nerve impulse transmission Cell membrane permeability Enzyme activation	Milk, cheese, green leafy vegetables, whole grains, egg yolk, legumes, nuts
Phosphorus (P)	Bone formation Absorption and transport of glucose, glycerol, fatty acids Energy metabolism Acid–base balance	Milk, cheese, meat, egg yolk, whole grains, legumes, nuts
Magnesium (Mg)	Constituent of bones and teeth Carbohydrate and protein metabolism	Whole grains, nuts, meat, milk, legumes
Chlorine (Cl)	Water and acid–base balance Hydrochloric acid in stomach	Table salt
Sulfur (S)	Part of proteins Activates enzymes Energy metabolism Detoxification reactions	Meat, eggs, cheese, milk, nuts, legumes

lose a few pounds without shedding even an ounce of fat. This strategy works, however, only if the body's salt metabolism is somewhat unbalanced in the first place. In most normal people, the body's salt content alters very little with fluctuations in the amount of salt eaten. If you eat more than the body needs, the excess is excreted by the kidneys and if you eat less than you need the kidneys excrete less and conserve the body's supply. A recent diet book recommends eating less salt and says you don't need salt in the summer or when you exercise because sweat contains no salt—which is ridiculous; you have only to taste sweat to know that it contains a lot of salt.

Calcium is another mineral needed in relatively large quantities. It is necessary for muscle action and, with phosphorus, in bone and tooth formation. **Trace minerals**, on the other hand, are needed only in tiny amounts, and the functions of some of them are unknown.

21–B Digestion

Digestion is the breakdown of food macromolecules into smaller molecules, particularly amino acids, monosaccharides (simple sugars), and fatty acids.

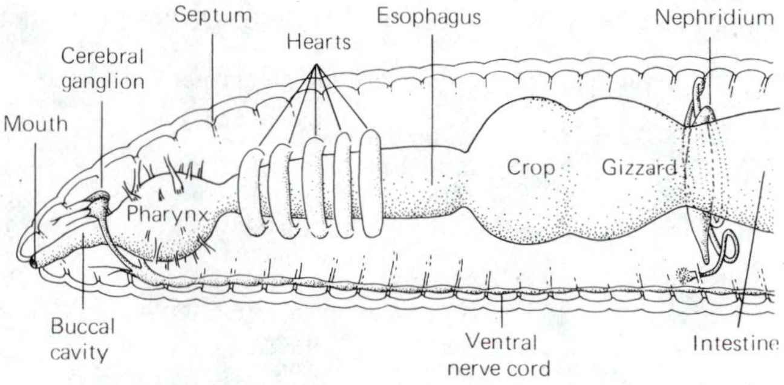

Fig. 21–2

Cutaway section of part of an earthworm, to show the structure of a fairly simple digestive tract.

Digestive enzymes carry out digestion either within the body's cells or outside cells in the cavity of the digestive tract.

All higher animals digest most of their food in the tube from mouth to anus which is variously known as an **alimentary canal**, **digestive tract**, or **gut**. A tubular digestive system permits food to move in only one direction; additional food can be taken in while previously eaten food is being digested, and different parts of the tube are specialized for different functions.

From a food's-eye-view, a digestive tract usually starts with a mouth, followed by a muscular, sucking pharynx (Fig. 21-2). Somewhere near the front end most animals have a means to break the food into smaller parts—teeth in mammals, fish, and sharks; a bill in birds; a muscular gizzard containing stones or grinding edges in earthworms, insects, most birds, and crocodiles.

Since humans are **omnivorous** (omni = all; vorous = eating), feeding on both plants and animals, our digestive tract lacks the extreme specializations found in many other vertebrates. Our teeth, for example, are all about the same size. The chisel-shaped **incisors**, in the front of the mouth, cut off bites of food from larger items; the pointed **canines** aid in holding; and the flat-topped, grinding **molars** in the rear mash food into a pulp. Meanwhile, the salivary glands pour saliva into the mouth through ducts under the tongue. Saliva contains a starch-digesting enzyme which may accomplish some digestion if the food is held long enough in the mouth before swallowing; probably more importantly, it also lubricates the food so that it does not scratch the delicate lining of the tract. Chewed, lubricated food is swallowed and passes through the **pharynx** (throat), gliding over the **epiglottis**, a sort of trapdoor over the entrance into the **trachea** (windpipe), and enters the **esophagus**.

The next port of call, the stomach, serves several functions. It is the widest part of the gut, in keeping with its function as a holding chamber that collects food and releases it into the **intestine** in small doses (Fig. 21-3). Having a stomach enables us to eat infrequent meals and spend the intervening hours in other ways. The muscular walls of the stomach churn the contents until they have the consistency of thick soup. Glands in the stomach secrete **hydrochloric acid** (HCl), which makes the stomach's contents strongly acidic, and the protein-digesting enzyme **pepsin**. The extreme acidity of the stomach is necessary for pepsin to work. Pepsin digests collagen, the fibrous tissue in meat. Only vertebrates can make pepsin, and without it they would not be able to digest much of the meat they eat.

FA-rinks (pronouce the *a* as in *cat*)

epp-ee-GLOT-tiss

TRAKE-ee-uh

ee-SOFF-uh-gus

COLL-uh-jen

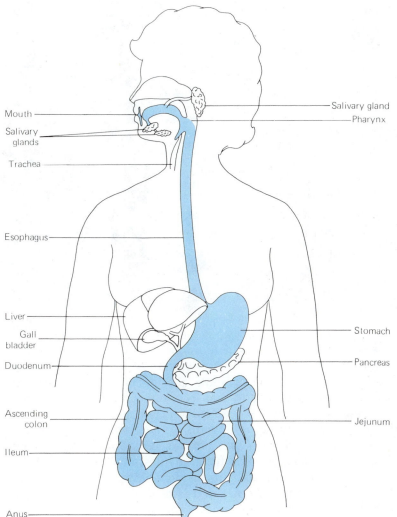

Mouth

Salivary glands

Trachea

Esophagus

Liver

Gall bladder

Duodenum

Ascending colon

Ileum

Anus

Salivary gland

Pharynx

Stomach

Pancreas

Jejunum

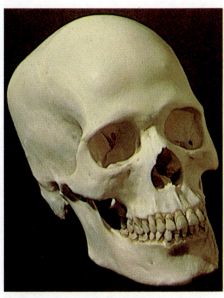

Fig. 21–4

A human skull showing teeth of different shapes and sizes. (Biophoto Associates)

Fig. 21–3

The human digestive tract.

It is remarkable that the stomach does not digest itself, because the hydrochloric acid and pepsin it secretes can digest most flesh. The stomach protects itself by secreting an acid- and enzyme-proof mucus which coats the stomach wall. Even so, life for a stomach cell is short; the stomach surface loses about half a million cells a minute, and all its cells are replaced every 3 days. Ulcers occur when the stomach does not secrete enough mucus to protect itself. Similarly, the esophagus and the duodenum on either side of the stomach (Fig. 21-3) can become ulcerated if they are overexposed to stomach acid. People are more prone to ulcers if they live for some time under stressful conditions which they cannot control, but why this psychological condition should cause the stomach to malfunction is unknown. Many expensive remedies and operations for ulcers are available, but the usual cure is merely to relieve the stress and then rest the stomach by eating small meals of foods that do not aggravate the ulcer.

TABLE 21-5 SOME MAMMALIAN DIGESTIVE ENZYMES

Origin	Enzyme	Action
Salivary glands	Amylase	Breaks down starch and glycogen
Stomach	Pepsin	Breaks down proteins
Pancreas	Lipase	Breaks down triglycerides
	Amylase	Breaks down starch and glycogen
	Trypsin Chymotrypsin Carboxypeptidases }	Break down proteins
	Ribonuclease	Breaks down RNA
	Deoxyribonuclease	Breaks down DNA
Small intestine	Aminopeptidases	Break down proteins
	Lipases	Break down triglycerides
	Glucoamylase Lactase Sucrase }	Break down disaccharides

dew-uh-DEE-num or
dew-ODD-en-um

PAN-cree-us

The stomach's contents are released in spurts into the **duodenum**, the first part of the small intestine. In the duodenum, secretion of sodium bicarbonate neutralizes the acidity of the stomach's contents, and it is here that digestion of most food takes place, carried out by enzymes from the pancreas and from the cells lining the duodenum. Much of the digested food and most of the water are absorbed into the bloodstream through the walls of the small intestine.

The remaining unabsorbed food passes into the **large intestine**, or **colon**, where millions of bacteria live and work. The large intestine absorbs water and minerals, as well as vitamin K produced by its resident bacteria, and pushes the remaining fecal matter into the **rectum**, where it is held till voided. **Defecation** (expulsion of the feces from the body) depends on contraction of the walls of the rectum and relaxation of the **anal sphincter**, a circular muscle surrounding the very end of the gut.

SFINK-ter

21–C Absorption

About 10 litres of fluid are absorbed from the intestines into the blood each day. Of this, about 1.5 litres is fluid we have drunk and about 8.5 litres is fluid full of mucus and digestive enzymes secreted by the body itself. Most of the absorption occurs in the small intestine. Only about 100 ml (less than half a cup) of water and small amounts of inorganic ions are lost in the feces every day. The bulk of the feces consists of bacteria, from the population of bacteria living in the large intestine, and water. The digestive system is so efficient that only small amounts of undigested food remain to be voided.

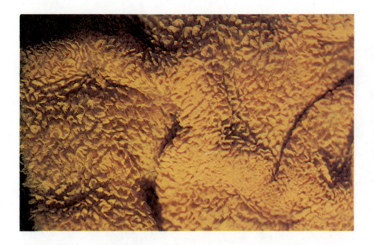

Fig. 21-5

The lining of part of the small intestine of a mouse. (Biophoto Associates)

Fig. 21-6

A sabellid, a marine segmented worm, feeds by filtering food out of the water with its tentacles. (Biophoto Associates)

21-D Feeding and Digestion in Herbivores

Most of the aquatic plant life on earth consists of tiny floating plants. It is, therefore, not surprising that most aquatic herbivores are **filter-feeders**, straining these minute plants out of large volumes of water (Fig. 21-6).

Animals that eat land plants have special problems. Unlike aquatic plants, which are buoyed up by the water and thus have little supporting tissue, land plants must support themselves, and they have evolved tough cell walls containing a high proportion of hard-to-digest cellulose. Breaking cell walls and digesting cellulose are major problems for most terrestrial herbivores.

Most herbivorous insects cannot digest cellulose. Their mouthparts are adapted to breaking and piercing cell walls so that they can feed on the cell contents. One reason why herbivorous insects such as locusts and grasshoppers are so destructive is that they use only a small fraction of the food they eat, so they have to eat a lot to obtain the nourishment they need. About half the plant

Fig. 21–7

Sheep, the animals used in most experiments on ruminant digestion. (Biophoto Associates)

material they eat—particularly cell walls and starch grains—appear in their feces unchanged. (There are some exceptions, such as termites, which eat wood. Their guts contain unicellular organisms which secrete wood-digesting enzymes.)

In contrast to herbivorous insects, most herbivorous mammals have a collection of symbiotic microorganisms housed in a specialized portion of the gut. These microorganisms include bacteria which can digest cellulose into simple sugars. The bacteria break down these sugars in anaerobic respiration, releasing smaller organic molecules that nourish the herbivore they live in. Since these bacteria live under anaerobic conditions (in the absence of oxygen), the part of the gut they live in is often called a fermentation chamber. In some herbivores this is a **caecum** (= blind), a sac set off to one side of the gut at the junction of the small and large intestines. The human appendix is probably a vestigial caecum which became reduced in size as we evolved toward a less herbivorous diet.

Digestion aided by symbiotic microorganisms has reached its greatest complexity in the ruminants, such as cattle, deer, and sheep. **Ruminants** are mammals in which the stomach contents are alkaline rather than acidic and the stomach is divided into several fermentation chambers (Fig. 21-8). Food passes through the gut slowly since the bacteria break down cellulose most completely if they are given a long time to work on it. Food descends first to fermentation chambers called the rumen and reticulum, where it is mixed with the alkaline saliva and with microorganisms (Fig. 21-8). The contents of the rumen are then regurgitated into the mouth and the animal "chews the cud." On its second descent, the food goes straight to the omasum and finally enters the abomasum, where the ruminant recaptures many of the nutrients which its microorganisms had used up by digesting the microorganisms themselves.

SEEK-um

ROOM-in-ants

alkaline: lying on the basic (rather than acidic) side of pH scale (see Section 9–F).

oh-MASE-um;
AB-oh-MASE-um

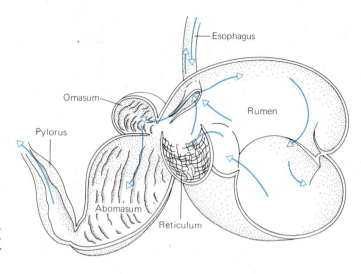

Fig. 21–8

A ruminant's stomach. The arrows show the complicated path followed by food.

21–E Symbiotic Bacteria

Symbiosis, which means "living together," describes the situation in which two organisms live in close association, each benefiting from the presence of the other. For example, the oxpecker bird spends much of its life searching for parasites cn African herbivorous mammals. The bird gets food from the association and the mammal is relieved of parasites that might damage its health.

Symbiotic bacteria, or **symbionts**, live both in the digestive tract and on the outer surfaces of the bodies of nearly all vertebrates; these bacteria are vital to a healthy life. A few human babies have been raised in sterile conditions because they lack the immune system which would normally protect them from harmful microorganisms, and other animals are sometimes raised in sterile conditions for research. Studies of such animals have shown that animals without the normal complement of symbiotic bacteria are at a major disadvantage compared with their normal relatives; many sterile animals die of bacterial infections that would not endanger a normal animal. What are the advantages of living so intimately with all these bacteria?

Symbiotic bacteria in our digestive systems ("gut symbionts") provide us with vitamins, calories, and amino acids. Some bacteria can synthesize amino acids using urea and ammonia, which animal enzymes cannot do. This is particularly valuable when the diet is low in protein. Ruminants use gut symbionts to best effect, and because ruminants contain so many bacteria which synthesize vitamins, they need many fewer vitamins in their diets than those of us who are less well inhabited. It has been calculated that a ruminant's symbionts use about 6% of the calories the ruminant ingests (as mentioned before, some of this is recaptured when the host digests the microorganisms.) In return, the ruminant receives vitamins, amino acids, fatty acids, and calories which it could not otherwise digest. (In addition, 18% of a ruminant's food calories end up in the form of the gas methane—which makes opening a cow's stomach a repulsive task.)

In addition to gut bacteria, animals carry bacteria that live on dead skin cells and parasitic bacteria that live on living tissues. The so-called "flora" of a human being includes *Staphylococcus epidermidis* on the skin, *Staphylococcus aureus* in the nostrils, and *Bacteroides fragilis* and *Escherichia coli* in the intestine. A major advantage of these normal inhabitants is that they compete with **pathogenic** (disease-causing) bacteria for space and food, reducing the number of harmful bacteria that can become established in the body. Some bacteria also produce antibiotics that inhibit the growth of other microorganisms. Foreign bacteria invade the skin, gut, or mucous membranes more easily if the resident bacteria have been removed by douching or in any other way. On the other hand, members of an animal's flora may cause disease if they become established in tissues where they are not normally found. *Escherichia coli* can cause cystitis (inflammation of the bladder) and *Staphylococcus aureus* can cause serious infection if it invades a wound; this is one reason why surgeons wear masks. But by and large, the bacteria that live with us are guests that we should welcome.

21–F Functions of the Mammalian Liver

Although the liver is not part of the digestive tube, it plays a vital role in nutrition. Most of the liver's work involves moving things into and out of the bloodstream. However, a small portion of the liver, the **gallbladder**, connects directly with the small intestine via the bile duct. Bile secreted by the gallbladder enters the duodenum through this duct. **Bile** is a mixture of salts, bilirubin (made when hemoglobin from the blood is broken down in the liver), cholesterol, and fatty acids. (Gallstones form when cholesterol settles out of the bile in the gallbladder.) Bile has two functions in the intestine: it acts as a detergent, breaking fat into smaller globules, and it aids in the absorption of fats from the digestive tract. This is why removal of the gallbladder sometimes causes difficulty with lipid absorption from the intestine.

bill-uh-RUBE-in

Digested food molecules absorbed into the bloodstream from the intestine travel directly to the liver by way of the hepatic portal vein. Before the food-rich blood circulates to the rest of the body, the liver may change the concentration, and even the chemical structure, of the food molecules in transit. The liver performs a vital role in detoxifying otherwise poisonous substances. In addition, it stores food molecules coming from the intestine, converts them biochemically, and releases them back into the blood at a controlled rate. For instance, the liver removes glucose from the blood under the influence of the hormone insulin and stores it as the polysaccharide glycogen. When the level of glucose in the blood falls, the liver breaks down glycogen and releases glucose into the blood.

The liver also synthesizes many of the blood proteins and releases them into the blood when they are needed, and it destroys old blood cells and recycles some of their constituents. In addition, the liver removes nitrogen from superfluous amino acids and converts it into the nitrogenous waste urea, which is excreted by the kidneys. The liver and the kidneys together exert most of the body's control over what stays in the bloodstream. For this reason severe liver damage or loss of the liver is rapidly fatal.

21–G Stored Food and Its Uses

Muscular activity uses up considerably more than half the calories we normally expend every day. Most of that muscular activity goes to keep the body's temperature at its normal level. Tiny muscles that we don't even notice wiggle and twitch all the time, performing biochemical reactions which release heat and so maintain the body temperature. This is why we expend more calories in winter, especially in an under-heated house. Most of our other daily calorie usage goes for functions like urine formation and nervous activity, whose caloric requirements vary little from day to day.

The body's carbohydrate stores (glycogen in the liver and muscles) would supply its energy needs for only about 12 hours if they alone were used. On the other hand, a human being of normal weight can usually survive without food for at least 6 weeks, using fat reserves for energy (Fig. 21-9). A hibernating animal, with a lowered temperature and metabolic rate, can survive for months on the fat reserves that it built up by eating ravenously during the autumn.

A given weight of fat provides about twice as many calories as the same weight of carbohydrate or protein, so energy is stored more compactly when it is in the form of fat. Fat is stored in the fat cells of **adipose tissue**, a storage tissue found under the skin, between muscle fibers, in the breasts and buttocks, between folds of the intestines in the abdomen, and elsewhere in the body. Fat is constantly exchanged between the bloodstream and adipose tissue: every molecule of fat in adipose tissue is replaced about every 3 weeks.

ADD-ih-poce

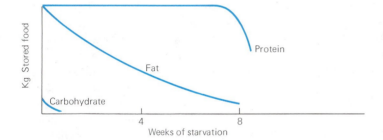

Fig. 21–9

The fate of stored food in a starving human being whose initial body weight was 15% fat.

Fig. 21–10

Feeding in animals is precisely regulated by hormones and other, unknown, mechanisms. Birds, like this blue tit, feed so that they almost double their weights before migration and before the breeding season. (Biophoto Associates)

21–H Regulation of Feeding

Feeding is governed by two kinds of controls: long-term and short-term. Long-term regulation ensures that enough stored food is maintained in the body. It can be altered by the action of hormones which ensure, for instance, that an animal builds up its fat reserves before its hungry young are born or hatched, and before hibernation. Short-term regulation ensures that an animal eats regularly on a day-to-day basis, so that food passes through the gut more or less continuously.

The control of feeding is poorly understood. The brain contains various centers which, when stimulated, start or stop feeding and control the selection of food. There are various hypotheses, but no real answers, as to what stimulates these brain centers in a normal animal.

Habit is one factor that has a major effect on short-term feeding. People accustomed to three meals a day become hungry and experience muscular contractions ("hunger pangs") of the stomach if they miss one meal. On the other hand, people living alone, or working intensely so that they are not reminded of mealtimes, frequently miss meals without feeling hunger. Distention of the stomach inhibits the feeding center in the brain and is the main reason we stop eating after a large meal. On the other hand, people and animals whose food never reaches the stomach (because a tube has been inserted into the esophagus to divert food to the outside of the body instead of the stomach) also stop eating after awhile. This suggests that the quantity of food which has passed the mouth is monitored in some unknown way.

The "ideal weights" on life insurance tables (Table 21-6) have been adjusted upward since the 1960s. This happened because studies showed that many people as much as 20% heavier than their "ideal weights" lived longer than those of ideal weight! There is in fact considerable variation in how much body fat an individual carries or should carry. If you have struggled to lose weight over and over and always seem to return to a weight somewhat over your ideal weight, it is very likely that your long-term regulator is "set" to the higher weight and that, as a result, you are no less healthy than your slimmer friends. Those of us in that situation should probably give up fighting the battle of the bulge, exercise to maintain our health, and hope that seventeenth-century fashions in bodies will return in our lifetimes!

Why an animal eats *what* it does is even more complicated and less well understood than why an animal eats as much as it does. For carnivores, things are reasonably simple since carnivores eat food that is highly nutritious and generally non-toxic. Carnivores rapidly learn to avoid food that tastes unpleasant, as you realize if you have ever seen a cat attack a toad.

Omnivores and herbivores face a much more difficult situation. Many plants contain chemicals that are toxic to animals and even more (including many of our common foods) contain chemicals that are toxic if consumed in large quantities. Very few herbivores can eat just one kind of plant because no single plant species contains the complete mix of macronutrients and micronutrients that an animal requires in its diet. (Animals like pandas and koalas that feed on just one species for long periods obtain other nutrients from other species of

TABLE 21-6 **DESIRABLE BODY WEIGHTS (IN LBS.)**

Height	Small frame WOMEN	Small frame MEN	Medium frame WOMEN	Medium frame MEN	Large frame WOMEN	Large frame MEN
4'9"	106		114		122	
4'10"	108		116		124	
4'11"	110		118		126	
5'0"	113	118	121	126	129	134
5'1"	116	121	124	129	132	137
5'2"	120	124	128	132	136	140
5'3"	123	127	132	135	140	143
5'4"	127	131	136	139	144	147
5'5"	130	134	139	142	148	150
5'6"	134	138	142	146	152	154
5'7"	138	142	146	150	156	158
5'8"	142	146	150	154	160	162
5'9"	146	150	154	158	163	166
5'10"	150	154	158	162	166	170
5'11"	154	158	162	166	170	176
6'0"	158	164	166	172	174	182
6'1"		170		178		188
6'2"		179		184		194
6'3"		184		190		200

plants at various stages in their lives.) Presented with a field of mixed plants, a sheep adds one new species of plant to its diet at a time and eats very little of it. Presumably, this gives the long-term regulation system time to "tell" the sheep whether the new species is nutritious or toxic. Using this system, a herbivore can work its way up to a nutritionally balanced diet containing a variety of different species. Omnivores such as pigs and humans, and even carnivores such as dogs, can do the same thing. In one experiment, human 2-year-olds were given a large variety of foods every day and permitted to eat whatever they wanted. On some

Fig. 21–11

Butterflies "puddling," sucking up salt from soil where minerals are concentrated. (Mark Rausher)

days their diets looked most peculiar—one would eat nothing but beans, another nothing but boiled eggs—but when the food they had eaten for a month was totalled up, the children were found to have selected nutritionally perfect diets!

One reason animals can select an adequate diet is that they develop specific appetites for things that their bodies need. An animal with a long-term deficit of fat in its diet will develop a craving that can be satisfied only by fat. Children with calcium deficiencies may eat plaster (mainly calcium carbonate) from walls. When people working in the tropics develop salt deficiencies because they sweat so much, they may drink salty water and find it delicious, although it has a repulsive taste to anyone who is not short of salt. Specific appetites are obviously very valuable since they lead animals to search out and eat nutrients that the body needs. Many human beings have an almost insatiable specific appetite for sugar. This is probably an evolutionary relic from the time when fruit was our main source of sugar. Until about the eighteenth century, an appetite for sugar was valuable because it induced people to eat fruit, which contains minerals and vitamins as well as carbohydrates. Unfortunately, pure sugar, with little nutritional value, is so readily available today that our appetite for sugar is a major contributor to obesity and tooth decay.

SUMMARY

Since animals, unlike plants, cannot manufacture their own food, their diets must contain all the organic and inorganic substances they need for metabolism, growth, and energy. Every animal requires fats, carbohydrates, and proteins (macronutrients), and vitamins and minerals (micronutrients) in its diet.

Digestion breaks food down into molecules that can be absorbed from the gut into the body. In vertebrates, digestive enzymes are secreted by the salivary glands, pancreas, and lining of the stomach and small intestine. Many animals, particularly vertebrate herbivores, harbor microorganisms which live as symbionts in the alimentary canal and also digest food. Digested food is absorbed into the bloodstream and body across the enormous surface area of the small intestine.

The liver plays a major role in controlling the fate of newly absorbed food molecules. It stores excess glucose as glycogen, synthesizes many blood proteins from amino acids in digested food, detoxifies poisonous substances, and converts nitrogenous and other wastes into forms that can be excreted by the kidneys. Excess carbohydrate and protein in the body are converted into fat and stored in the cells of adipose tissue.

Feeding is regulated by long-term and short-term control mechanisms that are not well understood. These controls ensure that the alimentary canal is efficiently occupied most of the time and that an animal maintains its body reserves of fat. Animals have complicated regulatory systems that control what, as well as how much, they eat.

SELF-QUIZ

Matching: For each numbered phrase below choose the letter of the correct class of nutrient on the right. More than one letter may be correct: choose all that apply.

——1. Inorganic nutrients
——2. Macronutrient that cannot be stored in the body
——3. May be the source of energy for the body's metabolism
——4. Stored in adipose tissue
——5. Organic substances needed in small amounts to aid in metabolic reactions
——6. Digested by enzyme in saliva
——7. Absorbed in large intestine

a. protein
b. carbohydrates
c. fat
d. water-soluble vitamins
e. fat-soluble vitamins
f. minerals

8. In humans, digestion of food is completed in the:
   a. mouth
   b. stomach
   c. small intestine
   d. large intestine
   e. rectum

9. In humans, protein digestion is carried out by the enzymes secreted by the:
   a. stomach, pancreas, and salivary glands
   b. liver, salivary glands, pancreas, and small intestine
   c. salivary glands, stomach, pancreas, and small intestine
   d. liver, stomach, pancreas, and small intestine
   e. stomach, small intestine, and pancreas

10. Which of the following is probably *not* an activity of symbiotic microorganisms of the gut?
   a. utilization of the host's food for its own nutrition
   b. reproduction
   c. manufacture of bile
   d. breakdown of substances that the host cannot digest
   e. manufacture of vitamins needed by the host animal

11. Which of the following is *not* a function of the mammalian liver?
   a. secretion of digestive enzymes for export to the gut
   b. regulation of blood glucose and amino acid content
   c. removal of nitrogen from excess amino acids and production of urea
   d. production of plasma proteins for the blood
   e. destruction of old red blood cells
   f. detoxification of poisonous substances.

QUESTIONS FOR DISCUSSION

1. Herbivores can seldom survive by eating only one species of plant (for example, corn lacks the amino acid tryptophan, which animals need in their diets, and many plants contain too little sodium). This is probably no evolutionary accident. What's "in it" for the plant?

2. Why does it take longer to become hungry after a protein-rich meal than after a meal that is mostly carbohydrate?

3. Trace the fate of a piece of pepper pizza through the human digestive tract. (Contents of pizza: crust—carbohydrate and various B vitamins; cheese—protein, fat, calcium, phosphorus; tomato—vitamin C, potassium; pepper—vitamin A, iron, and cellulose)

4. Various kinds of stress on an organism can upset the normal nutrient balance. For example, infection increases the rate at which vitamin C is used up. How might an organism respond to compensate for this disturbance? Will such compensation invariably involve changing the amount of the nutrient needed in the diet?

Essay:
DIET AND CARDIOVASCULAR DISEASE

Cardiovascular diseases, those that affect the heart and blood system, cause about half the deaths in the United States. They include **hypertension** (high blood pressure) and **atherosclerosis** (hardening and blockage of the arteries—blood vessels that carry blood from the heart to the tissues). Cardiovascular diseases cause death in many ways, such as strokes and heart attacks (including "coronaries," blockage of the arteries that supply blood to the heart muscles). Hypertension can be controlled by drugs; dietary measures such as losing weight and reducing the intake of sodium are also helpful. Reducing sodium intake may be difficult because meat contains quite a lot of sodium, and many commercially prepared foods contain sodium in added salt, preservatives, and monosodium glutamate (MSG). In one study, a hamburger, a cheese pizza, and a meat pie each contained more sodium than a portion of French fries liberally sprinkled with table salt!

In atherosclerosis, the artery walls develop deposits of lipids, including cholesterol, that reduce the internal diameter of the blood vessel and also make its walls less elastic. For many years, doctors have believed that the cholesterol level in the blood was correlated with atherosclerosis. Several studies have now shown that blood cholesterol levels have little or no correlation either with the amount of cholesterol in the diet or with death from cardiovascular disease.

New studies suggest that susceptibility to heart attacks is more closely correlated with levels of substances in the blood called HDLs and LDLs than with cholesterol. Cholesterol does not travel free in the blood but is combined with lipid-protein molecules. Most of the cholesterol is carried by low-density lipoproteins (LDLs); some is carried by high-density lipoproteins (HDLs). Studies of American whites and blacks, and of Israeli, Japanese, and Hawaiian men (that is, on a varied collection of people) have shown that the risk of heart attack is greater the lower the HDL concentration and the higher the LDL concentration in the blood. Researchers have also found that some people who have permanently high levels of HDL or low LDL because of their genetic makeup apparently never die of atherosclerosis.

The factors correlated with high HDL levels are those long known to be associated with a low risk of heart attack—being female (a woman's HDL level is about 10% higher than a man's), being slim, exercising, not smoking and consuming moderate amounts of alcohol (rather than consuming none or large amounts). However, where people have attempted to alter these factors, for instance by losing weight or by taking up running, there is no convincing evidence that they have reduced their risk of heart attack. You will notice that these studies talk about correlations, not about cause and effect. This is because cardiovascular disease takes a long time to develop and because few experiments are possible on human subjects, which makes scientific investigation—and determination of cause and effect—of such disease difficult. Even an obvious correlation can be misleading. For instance, runners have a low incidence of cardiovascular disease. Is this because people with high HDL levels are predisposed to run or does running raise HDL levels? Runners are also more likely to consume alcohol and less likely to smoke than are members of the general population. Is this cause or effect or neither? We do not know.

While research on this complex subject is still inconclusive, it is probably fair to say that the evidence to date shows that, unless you are obese, no major alteration in diet will have much effect on your chances of dying from a heart attack.

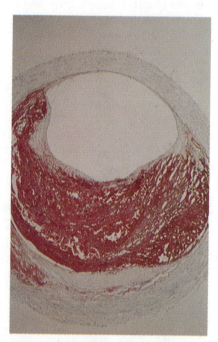

Atherosclerotic deposit (red) in an artery (grey). (Biophoto Associates)

REGULATION OF BODY FLUIDS

22222

Physiology is the study of the functioning of cells, tissues, and organs or, more generally, of the mechanisms that keep the body alive. In this chapter we shall consider the circulatory, immune, respiratory, and excretory systems, and the physiology of temperature regulation.

The conditions under which animal cells can live are very restricted. This is illustrated by the difficulty of culturing tissues outside the body and by the remarkable similarity of the body fluids of all animals. Controlling the environment in which its cells live is part of the body's general task of **homeostasis** (homeo = similar; stasis = staying), or "staying the same," which includes regulation of the temperature and of the food, salt, and gas content of the fluid bathing all cells. Physiology is largely concerned with homeostasis and the body's ability to adapt to changing conditions.

HOME-ee-oh-STAY-sis

389

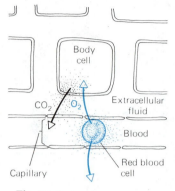

Fig. 22-1

A cell exchanging oxygen and carbon dioxide with the extracellular fluid from which it gets everything it needs and into which it expels its wastes.

22-A Body Fluids

There is much evidence that the sea was the home of the earliest organisms; from the sea these creatures obtained nutrients and oxygen, and into the ocean they discharged their wastes. The vastness of the sea provided a constant, unlimited supply of needed substances, and it diluted toxic wastes to negligible levels. The oceans also provided moisture and almost-constant temperature.

The blood and body fluids of a higher animal are often likened to a "captive sea," because they perform the same functions inside the animal's body that the sea performs for its one-celled inhabitants. However, the blood also plays some very sophisticated roles that cannot be duplicated by sea water.

Extracellular fluid (ECF) surrounds and bathes all the cells of multicellular animals. Cells take in the things they need—amino acids, sugars, oxygen, minerals—from the ECF, and into it they discharge their wastes—mainly carbon dioxide from respiration and nitrogenous wastes from the breakdown of proteins. The well-being of the cells depends not only on the presence of the right substances, but also on the correct proportions of these substances with respect to one another and to the water that dissolves them. For example, oxygen will diffuse into a cell only if oxygen is more concentrated outside the cell than inside.

We can summarize the function of most of the body's organs by saying that their role is to ensure homeostasis within the body, either directly or indirectly. An organism's homeostatic ability determines where it can live, and there are limits for all organisms. Humans, for instance, lose their ability to maintain the constancy of the extracellular fluid if exposed to extremely cold or dry conditions; unless protected from these extremes, we die. But, as warm-blooded vertebrates, we can control our body temperature and water balance better than many organisms can, and so we can survive in many more environments.

In this chapter we shall consider how substances enter and leave the extracellular fluid; the role of the circulatory, respiratory, and excretory systems in keeping things constant; and some of the regulatory mechanisms that control these systems.

Fig. 22-2

The three main types of blood vessels (not to scale).

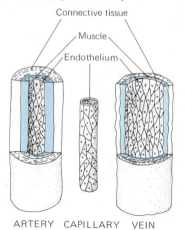

22-B The Circulatory System

The **circulatory system** is the body's multi-purpose fluid transport network, linking all parts of the body. It consists of the heart, blood vessels, and, of course, the blood. The heart is a muscular pump that contracts and pushes blood through a set of pipes, the blood vessels. **Arteries** are vessels that move blood away from the heart; **veins** are vessels that carry blood returning to the heart. Blood travels from arteries to veins through short, thin vessels called **capillaries**.

Extracellular fluid (ECF) and capillaries. Although the heart is the most obvious part of the circulatory system, the real action takes place, not in the heart, but in the capillaries. Capillaries are so narrow that blood cells must pass through them single-file, but they branch in dense networks throughout almost

Fig. 22–3

A capillary bed. Blood vessels are white and blood flows from the bottom to the top of the picture. Across the bottom is an artery with 3 main pairs of smaller arteries leading up from it. The arteries branch to form ever-smaller blood vessels, until the blood reaches the fine network of capillaries from which it flows into the veins at the top of the photo. (Biophoto Associates)

every organ in the body, including the walls of the heart and the large arteries and veins.

The walls of the capillaries are very thin, with tiny gaps between adjacent cells. The pumping of the heart exerts pressure on the blood, forcing water, dissolved salts, and food molecules out through these tiny gaps; since blood cells and proteins are too large to leave in this way, they remain in the blood. The fluid that leaves the blood becomes part of the extracellular fluid.

Obviously, the extracellular fluid must drain back into the blood at some point or we should swell up and burst. We *do* swell up at times. This water retention, experienced by some women who are pregnant or taking contraceptive pills (or by anyone who has been sitting still for a long time), is known as edema. Normally, however, the extracellular fluid drains steadily into tiny transparent vessels of the lymphatic system, which finally empty back into the blood system by way of a vein near the heart (see Section 22–C). ed-EE-muh

All the body's fluid compartments are in contact with one another. The extracellular fluid is in contact with both the blood and the cerebrospinal fluid (CSF), which bathes the brain and spinal cord. This means that regulating the composition of the body's fluid is a single job. If food and oxygen enter any part of the system, they will reach it all. This continuity of the body fluids makes it possible for single organs like the kidneys and lungs to control what enters and leaves the body's fluids. Although we commonly think of the body's transport system as blood traveling from the heart to the tissues and back again in blood vessels, parts of the system are actually much more circuitous than this. Fluids, as well as dissolved substances, move into and out of the blood, and even cells can move into and out of the blood in various places. SAIR-ih-broh-SPINE-al

Blood.   The familiar red fluid called blood is really a tissue whose volume is about half liquid (**plasma** with proteins dissolved in it) and half various types of cells (Table 22-1). Blood cells can be divided into three main groups: white

TABLE 22-1   **COMPOSITION OF HUMAN BLOOD**

Main components of the blood	Comments
Water	
Salts	
Sodium	
Potassium	
Calcium	
Magnesium	
Chloride	
Plasma Proteins	Blood minus its cells is called plasma
	Plasma minus its proteins is called serum
Blood Cells	
White cells (leukocytes)	
Red cells (erythrocytes)	
Platelets (thrombocytes)	
Substances Transported by the Blood	
Sugars	
Amino acids	
Fatty acids, glycerol	
Hormones	
Nitrogenous waste	From protein breakdown
Carbon dioxide	From respiration
Oxygen	

LUKE-oh-sites
air-ITH-roe-sites
PLATE-lets

blood cells (**leukocytes**), red blood cells (**erythrocytes**), and **platelets**. All are formed in the **bone marrow** (the soft tissue in the center of bone), although white blood cells may also reproduce after leaving the marrow. There are many different types of white blood cells; most of them are involved in protecting the body from disease.

Red blood cells are the most numerous blood cells; their main function is to transport oxygen. Conditions that increase the body's demand for oxygen, such as living at high altitudes or engaging in strenuous exercise, stimulate erythro-

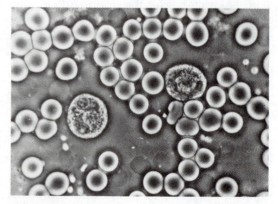

Fig. 22–4

A blood smear. The small, featureless cells are red blood cells, the big, granular ones are white blood cells and the faint, transparent circles are red blood cell ghosts— empty cell membranes. (Biophoto Associates)

cyte production. Erythrocytes live for about four months in the bloodstream and then disintegrate. **Anemia** is a condition in which the blood contains fewer erythrocytes than is normal, or the red cells contain less hemoglobin than is normal. Anemia is a symptom of many different diseases in which red blood cells are produced too slowly or destroyed too quickly.

The platelets are important in blood clotting. Damaged tissues release **histamine**, a substance that increases the blood supply to the area and permits fluid and clotting proteins to leak out of the bloodstream into the tissue. A blood clot soon forms, by a complex series of reactions, and slows down the bleeding. Clotting also blocks the entry of bacteria and other foreign substances into the wound. White blood cells attracted to the wound engulf bacteria and damaged cells and eventually die themselves. In a serious local inflammation, the dead white cells accumulate as **pus**.

HIST-uh-mean

Circulation.   Blood is pushed through the circulatory system by the heart. The human heart is really two pumps joined together. Each pump is made up of a thin-walled **atrium** or **auricle**, which receives blood from veins, and a thick-walled **ventricle**, which pumps blood into arteries. The right side of the heart receives blood from the body and pumps it to the lungs; the left side receives blood returning from the lungs and pumps it to the rest of the body (Fig. 22-5).

ATE-ree-um   AWR-ih-kul
VEN-trih-kul

What is the advantage of this double circulation? Blood loses pressure as it passes through the capillaries of the lungs; when it returns to the heart, it can be pumped back up to full pressure again before it sets out for the body. If this second pumping did not occur, the blood reaching the body would be at a much lower pressure and would travel too slowly to deliver needed substances at the required rate. Fairly high pressure is needed to push fluid (including dissolved

Fig. 22–5

The structure of a mammalian heart.

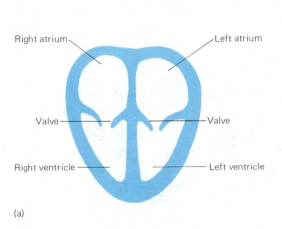

(a)

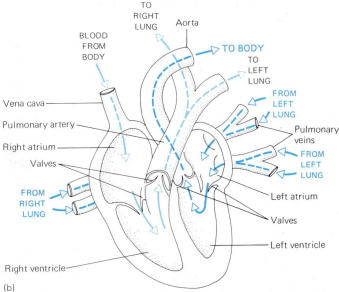

(b)

substances) out of the capillaries into the extracellular fluid and, as we shall see, to push blood plasma into the kidneys for cleaning. Second, the double circulation ensures that oxygenated blood coming from the lungs is not diluted in the heart by deoxygenated blood coming from the rest of the body.

The overall rate of circulation depends largely on how much and how often the heart pumps. The human heart pumps about 5 litres per minute during sleep and five times that amount during strenuous exercise. Most people's hearts pump about 75 times per minute (known as the pulse rate). Training, however, increases the volume of blood pumped by the heart at each stroke, and a marathon runner's resting pulse may be as low as 40 strokes per minute, pumping the usual 5 litres per minute of blood.

The two atria of the heart contract together, followed by contraction of the two ventricles. The pressure on the blood as the ventricles contract is called **systolic pressure**. The **diastolic pressure** is the lowest pressure of the blood, reached between heartbeats. Diastolic pressure indicates the degree of elasticity of the arteries. Blood pressure is usually measured in the artery of the upper arm. In a normal, resting person, the systolic pressure is about 120 millimetres of mercury and the diastolic pressure about 80 millimetres; this is conventionally expressed as 120/80. The blood pressure is highest in the arteries near the heart

sis-TALL-ic   DIE-us-TALL-ic

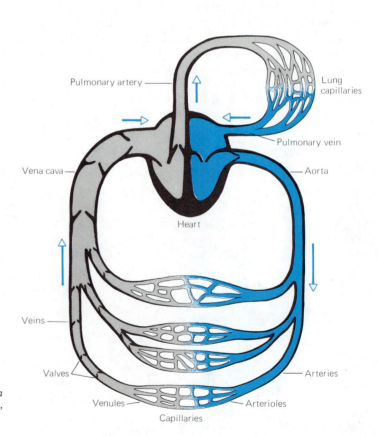

Fig. 22–6

Simplified diagram of the blood circulation in a mammal. Gray represents deoxygenated blood, color represents oxygenated blood.

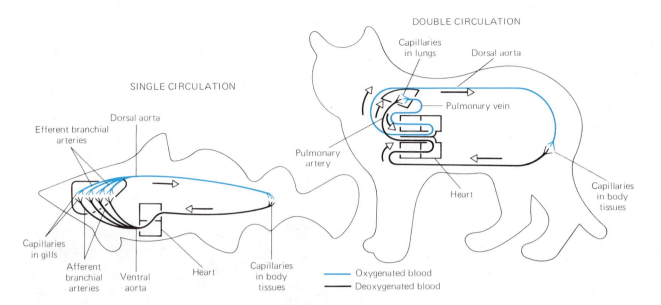

Fig. 22–7

Single and double circulation. In a fish, blood passes through the heart once during each circuit of the body. In birds and mammals (e.g., the cat), blood passes through the heart twice before it returns to any one point in the system.

and drops as the movement of the blood is slowed by the friction of the vessel walls on its circuit through the body.

Valves in the veins and the heart ensure that the blood travels in only one direction. When they close, the valves in the heart make a noise which can be heard with a stethoscope. Normal closing produces a sharp, snapping sound. **Heart murmurs** occur when a valve is narrowed, forcing blood to flow turbulently, or when there is a leak in a valve, allowing blood to regurgitate back through the valve.

Although the large arteries are quite muscular and elastic, the walls of veins are much thinner and stretch more easily. The blood pressure drops dramatically when blood passes through a capillary bed, and blood reaching the veins is under very little pressure. When you consider the vertical distance from your big toe to your heart (or the even greater distance from a giraffe's foot to its heart), the problem of returning blood to the heart against the force of gravity assumes formidable proportions. In fact, blood return is a remarkably passive process. The blood moves sluggishly upward in the legs, impelled somewhat by the push of blood from behind but moved mainly by pressure of the leg muscles, which push the blood inside veins when the muscles contract. When the muscles relax, valves in the veins prevent the blood from slipping back.

Because the body's muscles play a major role in returning blood to the heart, blood return is slower during periods of inactivity. Sitting or standing for a long time is more tiring than gentle walking because blood collects in the legs, shutting off some of the supply of oxygen that would normally be carried by the blood. Recent studies have shown that students who jiggle their feet—although they run the risk of irritating their neighbors—are more alert and perform better on long exams than their peers who sit still.

Varicose veins are veins whose valves have been damaged. Where a defective valve exists, the blood accumulates instead of proceeding on its journey to

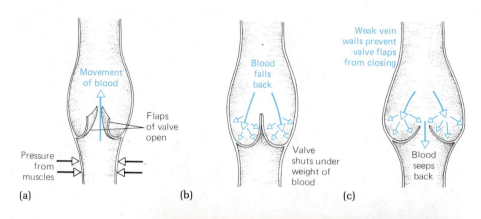

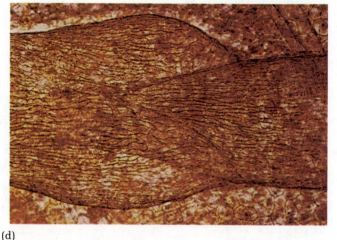

Fig. 22–8

*Valves in veins. (a) Valve open. (b) Valve closed.
(c) A varicose vein, a vein with weak walls; blood
causes the vessel walls to bulge until the valves
will not close. Blood will then flow backwards
through the valve instead of returning to the heart.
(d) Photo of a valve in a lymphatic vessel.
(Carolina Biological Supply Company)*

HEM-uh-roids

the heart. The pool of blood distends the thin walls of the vein so that it swells.
Hemorrhoids are varicose veins of the rectum, usually damaged by pressure
from constipation or pregnancy.

22–C The Lymphatic System

lim-FAT-iks

Fluid drains from the extracellular spaces back into the blood system via
lymphatics, vessels which have almost transparent walls and, like veins, contain
valves (Fig. 22-8d). Lymphatics also pick up fats from digested food absorbed by
cells lining the intestine and proteins from the liver destined to become blood
proteins (see Section 21-F). Large hormone molecules also enter the blood via
the lymphatic system. At various places in the body, the lymphatics form **lymph
nodes**, which act as filters to remove foreign particles or dead cell debris from the
body fluids (Fig. 22-9). White blood cells that defend the body against disease
congregate in the lymph nodes as part of the filter; swelling and pain in the
lymph nodes result from enlargement of these organs when the body is fighting
an infection.

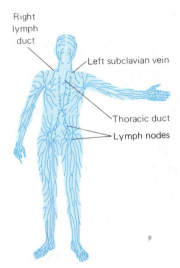

Right
lymph
duct

Left subclavian vein

Thoracic duct

Lymph nodes

Fig. 22–9

*The position of the major lymphatic vessels
and lymph nodes in the human body.*

22–D The Immune System

The circulatory and lymphatic systems of vertebrates house and transport
the molecules and cells that make up the **immune system**, whose role is to
recognize and destroy foreign organisms or chemicals that enter the body.

An invader is recognized by a reaction between an invading antigen and the
body's antibodies. An **antigen** is any substance, usually containing at least some   ANT-ih-jenn
protein, that can elicit an immune response. An **antibody** is a protein produced
by one of the immune system cells, with a shape and electric charge distribution
that allow it to bind with an antigen of complementary shape and electric charge.

An example of antigen-antibody reactions involves the human ABO blood
groups. People of different blood types have different antigens on their red blood
cell membranes, as outlined in Table 22-2. Each person has antibodies in the
blood plasma against any red blood cell antigens that are *not* present on his or her
own cells. If a person receives a blood transfusion containing red cells bearing
antigens not present on the recipient's cells, antibodies in the recipient's blood

TABLE 22-2	BLOOD ANTIGENS AND ANTIBODIES IN HUMAN ABO BLOOD GROUPS	
Blood type	**Antigen on red blood cell membranes**	**Antibody in blood plasma**
A	A	anti-B
B	B	anti-A
AB	A and B	—*
O	Neither	anti-A and anti-B

*Neither anti-A nor anti-B

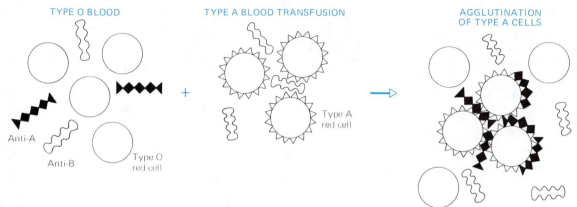

TYPE O BLOOD   TYPE A BLOOD TRANSFUSION   AGGLUTINATION OF TYPE A CELLS

Anti-A

Anti-B

Type O red cell

Type A red cell

Fig. 22–10

Agglutination of type A cells given in transfusion to a type O person. The recipient's anti-A antibodies attach to the antigen A on the surface of the type A red cells. Clumps of red cells bound together by antibody precipitate out of the blood-stream.

combine with the donated red cells and form clumps of cells and antibodies (Fig. 22-10). Such clumps will obstruct circulation in smaller blood vessels, which may be fatal. This is why blood from donor and recipient must be matched before a transfusion is given.

Various white blood cells make antibodies, and each cell makes only one kind of antibody, which can combine with only one or a few, similarly shaped antigens. The human body is believed to make antibodies to about a million different kinds of antigens. It produces small amounts of all these different antibodies even if it never encounters the corresponding antigens. An antibody-producing cell carries a sample of its antibody on its cell membrane, where it can combine with any foreign antigens it encounters.

In order to distinguish invaders as foreign, the immune system must recognize substances and cells which are legitimate members of the body. Most of the body's own cells produce antigen molecules attached to the cell membrane. Humans have a number of genes that code for these cell-surface antigens; the chance that two people will have the same set of cell-surface antigens is slim, although close relatives are likely to have similar sets, and identical twins have identical antigens. Early in development, the immune cells that make antibodies against the body's own antigens are inactivated or destroyed, so that the remaining antibodies are directed, not against "self," but against "not-self" antigens yet to be encountered.

Humoral and cellular immune responses.   There are two possible types of immune response to a foreign antigen, each mediated by a different type of cell. In a **humoral response**, the white blood cells involved release their antibodies into the blood. In a **cellular response**, there are poorly understood interactions among various immune system cells. Usually, both types of response occur to some degree but, for simplicity, let us consider each in a "pure" form.

As an example of a humoral immune response, let us consider an infant receiving its first diphtheria vaccination. The injection contains **toxoid**, a heat-weakened form of the bacterial toxin that causes diphtheria. White blood cells with surface antibodies that can react with the toxin combine with the toxoid.

This induces the cell to manufacture more antibody and release it into the blood, and also to grow and divide, so that the population of cells with anti-diphtheria antibodies increases. At this point, white blood cells called macrophages enter the picture. **Macrophages** (macro = large; phage = eater) constitute an army of nomadic garbage-collectors, ridding the body of dead cells and of antigen–antibody aggregations.

MAC-roe-fage

The cellular immune response predominates in the rejection of tissue transplants from another person. Since different people have different cell-surface antigens, the cells from another person are "foreign" to the recipient's body. Some of the recipient's cells react to antigens on the donated cells; this causes them to release a substance that attracts macrophages to the site, and also to grow and divide—again producing a large population of cells which will react to the particular invading antigen. The graft becomes loose and puffy, with areas of pus between the graft and the host tissue, and the grafted tissue dies. Drugs are used to prevent the rejection of grafted tissue by the immune system after organs are surgically transplanted. Kidney transplants have proved particularly successful, partly because the kidney's connections with the rest of the body consist of a few large tubes that are reasonably easy to connect. Only a shortage of donors prevents thousands more successful kidney transplants than are actually performed each year.

Immunological memory and vaccination.   The immune response produced when the body first encounters a particular antigen is called a **primary immune response**. After the body has successfully warded off a particular antigen, cells specific for that antigen gradually become less common in the circulation, but many remain in the lymph nodes, living for months or years in human beings. These cells serve as the body's **immunological memory**, primed to fight another round if the same invader should strike again. The second (or later) appearance of the same antigen evokes a **secondary immune response**, which is more rapid, more extensive, and longer-lasting than the primary immune response. This explains why a second tissue graft from the same donor is rejected much more rapidly and violently than the first, and why we seldom catch the same strain of cold or flu twice in the same winter. Furthermore, we can develop especially long immunological memories for many diseases, such as mumps, measles, or chicken pox, and so we tend to catch them only once in a lifetime.

Vaccination against specific diseases actually produces a primary immune response and thereby creates an immunological memory, although the practice of vaccination started long before scientists realized how it worked. Edward Jenner, an English physician, developed the first vaccine, against smallpox, after he noticed that dairy workers who had had the relatively mild disease cowpox seemed to be immune to smallpox. Jenner deliberately rubbed pus from cowpox sores into scratches made on the skin of people who had been exposed to smallpox, and he succeeded in preventing many expected cases of smallpox. In this case, the antigens of smallpox and cowpox are so similar that the same antibodies work against them. Since that time, medical researchers have developed a number of vaccines for bacterial diseases, and in recent decades have produced vaccines for viral diseases, such as polio and influenza, which was once believed impossible. "Booster shots" serve to jog the body's immunological memory into producing more antibodies and more cells, assuring that there are plenty of memory cells available if a diphtheria or whooping cough bacterium should invade.

Just the opposite approach is used when a dangerously large dose of antigen enters the body, as when tetanus organisms release toxin in a puncture wound, or when a poisonous snake injects venom. In such cases, the treatment is an injection of **antiserum** prepared from the blood of an animal, such as a horse, that has periodically received small doses of the toxin. The horse produces antibodies to the toxin, and these antibodies, when injected into a person, can inactivate the antigen.

coll-OSS-trum

Infants acquire this kind of passive immunity from their mothers. Some of the mother's circulating antibodies enter the fetal blood before birth, and these protect the newborn for a time, until its own immune system begins to function. The breast-fed newborn is also protected by **colostrum**, a thin fluid produced by the mammary glands after childbirth before the flow of milk begins. Colostrum contains antibodies believed to protect the human infant's digestive tract from infections. Once the normal bacterial inhabitants of the digestive tract become established, they themselves suppress the invasion of dangerous newcomers. Human babies do not absorb antibodies from colostrum into the blood, although the young of some other mammals do.

22–E The Lungs and Gas Exchange

One of the most important roles of the blood is to carry oxygen to the tissues and remove carbon dioxide. Some oxygen travels dissolved in the blood, but about 16 times as much binds loosely to the red pigment **hemoglobin** inside the

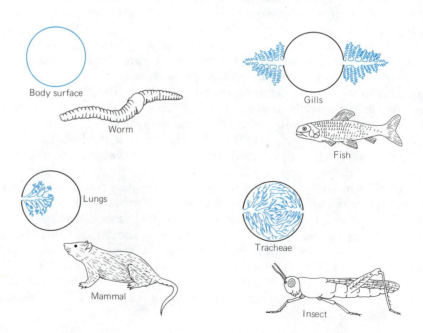

Fig. 22–11

The four main types of respiratory surfaces found in different animals.

Fig. 22–12

The human respiratory system. (a) General structure. (b) Closeup of the area where gas exchange takes place.

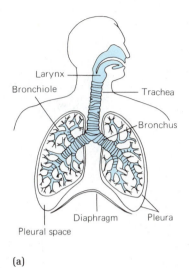

Larynx

Bronchiole

Trachea

Bronchus

Diaphragm

Pleura

Pleural space

(a)

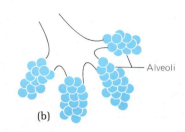

Alveoli

(b)

red blood cells. On the other hand, most carbon dioxide is dissolved in the blood fluid, with just a small fraction attached to hemoglobin and other blood proteins.

Animals have evolved several different types of respiratory surfaces by which they exchange gases with the environment (Fig. 22-11). The surface must always be kept moist because gases can enter and leave the body only when they are dissolved in water. In mammals, reptiles, and birds the respiratory surface of the lungs lies inside the body and exchanges gases with the inhaled air. The walls of delicate, frothy-looking air-filled sacs called **alveoli**, with an enormous total surface area, make up the respiratory surface.

al-VEE-uh-lee (sing.: alveolus)

In the lungs, millions of capillaries lie just under the moist respiratory surface. Blood flowing through these capillaries continually takes up oxygen and gives up some of its carbon dioxide to the air in the alveoli. Only about 10% of the carbon dioxide in the blood is released to the air in the lungs. The other 90% stays in the blood, where it performs a vital function as a **buffer**, a substance that prevents the acidity (pH) of the blood from fluctuating. In the tissues, where cells are continually using up oxygen in cellular respiration and releasing carbon dioxide as a waste product, the opposite exchange occurs: oxygen goes from the blood into the extracellular fluid and the cells, and carbon dioxide enters the blood.

The blood's ability to transport oxygen is reduced by two common pollutants that occupy hemoglobin's oxygen-binding sites. One is carbon monoxide, found in automobile exhausts and in cigarette smoke. Hemoglobin carrying carbon monoxide is a bright purple, and gives a characteristic violet color to the skin of a person with carbon monoxide poisoning. The other pollutant is nitrite; nitrite pollution has caused poisoning in agricultural areas, such as the San Joaquin Valley of California, where nitrate fertilizer leaches into the drinking water. Intestinal bacteria convert nitrates into poisonous nitrite.

If the concentrations of gases in the alveolar air and the blood were the same, no exchange would take place. It is essential that the air in the lungs be changed frequently, by breathing movements, or **ventilation**, if this gas exchange is to

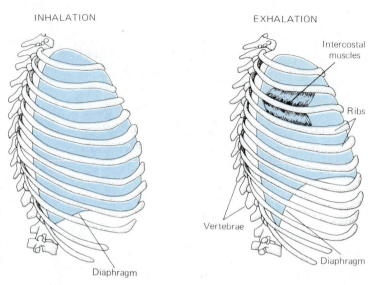

INHALATION   EXHALATION

Intercostal muscles

Ribs

Vertebrae

Diaphragm

Diaphragm

Fig. 22–13

Respiratory movements in a human being. As the ribs are lifted and the diaphragm is lowered, the size of the chest cavity increases. Air rushes in through the nose, mouth or both. Exhalation is a passive process and occurs when the breathing muscles relax.

DIE-uh-fram

continue. You inhale by contracting your diaphragm, which is a muscle sheet forming the floor of the chest cavity, and your rib muscles. These contractions lower the diaphragm and raise the ribs up and out, increasing the size of the chest cavity and so decreasing the pressure within it. As a result, air rushes into the lungs, equalizing the air pressure inside the lungs with the atmospheric air pressure. You can feel the suction if you close your mouth and hold your nose while you make the movements of inhaling. During exhalation, the diaphragm and rib muscles relax, allowing the chest cavity to return to normal size and pushing out some of the air in the lungs. There is always a fair amount of air in the lungs, keeping the alveoli walls from collapsing and sticking together.

The rate of ventilation determines the gas composition of the lungs: the faster you breathe the more oxygen and the less carbon dioxide in the alveolar air. Breathing must be regulated so that the amount of oxygen reaching the tissues from the blood remains constant. Sense organs in the blood vessels of the neck monitor the amount of carbon dioxide and, to a lesser extent, the amount of oxygen in the blood. If the amount of carbon dioxide rises, the sense organs signal the brain, which in turn sends more frequent nerve signals to the diaphragm and rib muscles to contract, thus increasing the breathing rate. This reduces the amount of carbon dioxide in the alveolar air and so in the blood.

If the carbon dioxide content of the blood drops below a certain critical level, breathing is inhibited. By **hyperventilating**, that is, taking several deep breaths in rapid succession, you can hold your breath for longer; swimmers often do this so that they can stay underwater for a longer time. At each breath, however, the carbon dioxide content of the blood falls, and if it falls too far, you lose consciousness and automatically start to breathe normally. This is an important body defense that prevents you from reducing the carbon dioxide level in your blood to a dangerous extent. Similarly, if you intentionally hold your breath, the carbon dioxide level of the blood will rise and, above a certain level, you will lose

TABLE 22-3 **IMPORTANT WASTE SUBSTANCES PRODUCED BY THE HUMAN BODY, AND SITES AT WHICH THEY ARE EXCRETED**

Substance excreted	Excreting organ(s)
Nitrogenous waste (from protein breakdown)	Kidneys Small amount in sweat
Water	Kidneys Skin (in sweat) Lungs (evaporation)
Salts	Kidneys Skin (in sweat)
Carbon dioxide	Lungs
Spices, drugs, etc.	Lungs Kidneys

consciousness and again start to breathe normally. It is impossible to kill yourself just by holding your breath, although some children whose parents do not know this learn that they may get their own way by holding their breath.

22–F The Kidneys

The kidneys control the composition of the body fluids; they are not the only organs that perform this function, but they are the most important. Blood is cleansed and altered as it flows through the kidneys. Because the blood is in contact with the extracellular fluid, the kidneys control the composition of all the body's fluids by controlling the composition of the blood. The importance of this function can hardly be overemphasized: the heart will stop if there is a small increase in the amount of potassium in the blood; too much water in the extracellular fluid or a slight increase in magnesium blocks nerve function. Consequently, major kidney damage is rapidly fatal.

The kidneys control the amounts of water and of salts (such as chloride, magnesium, and sodium) in the blood, and they also rid the body of various wastes. Some wastes (especially carbon dioxide) leave the body by way of the lungs, and others leave by way of the skin. Only the kidneys, however, can rid the body of appreciable amounts of toxic nitrogenous wastes, mainly urea, produced by breakdown of proteins. In addition, the kidneys detoxify various substances, converting them to forms that are not poisonous to the body and can be excreted. Hormones and various drugs are also altered and excreted by the kidneys before they accumulate to dangerous levels. Onions, garlic, and some other spices have volatile components that leave the body through the lungs—that is, "on the breath." Other parts of the same spices are excreted through the kidneys.

One of the most important functions of the excretory system in any organism is to control the water content of the body. We lose a great deal of water from our bodies every day. Some is needed to excrete nitrogenous waste in the urine, since urea is toxic in too concentrated a solution; some is inevitably lost by evapora-

Fig. 22–14

The human urinary system. The kidneys lie in the small of the back, against the spinal column.

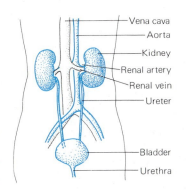

Vena cava
Aorta
Kidney
Renal artery
Renal vein
Ureter
Bladder
Urethra

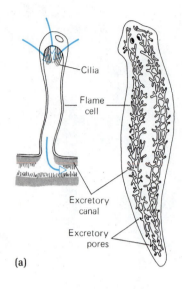

(a)

Cilia

Flame cell

Excretory canal

Excretory pores

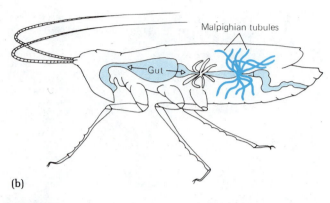

Malpighian tubules

Gut

(b)

Fig. 22–15

Excretory systems of two invertebrates. (a) In a planarian flatworm, flame cells collect body fluids and pass them down ducts to pores on the body surface. (b) In an insect, the Malpighian tubules discharge nitrogenous waste into the gut.

tion from the lungs as we breathe; a small amount leaves in the feces. In addition, a lot of water (and some salt) leaves the body every day as sweat—for example, a person working in the desert may lose 1 litre of sweat an hour. We must drink fluids to make up for all these losses, and the kidneys' job is to keep the water content of the blood and extracellular fluid the same in the face of these constant gains and losses.

The excretory organs of all animals work in essentially the same way: they take in body fluids, they alter their composition by reclaiming some substances and transporting others from the body into the collected fluid, and then they expel the fluid into the environment (Fig. 22-15). By controlling the composition of the urine that leaves the body, the kidneys also control what remains in the body's fluids.

NEFF-ron

The excretory organs of vertebrates—the kidneys—are made up of units called **nephrons**. Each nephron is a long, slender tubule, with one end bulging out to form a **renal capsule** around a knot of blood capillaries (Fig. 22-16). Because blood in the capillaries is under pressure, some of its fluid is pushed out through the spaces between the cells in the capillary walls (this is the main reason we need relatively high blood pressure). As we saw before, the spaces are too small to permit blood cells or proteins to leave the capillaries; the appearance of blood or proteins in the urine may indicate damage to these capillaries.

As the filtered fluid passes through the nephron tubule, cells that form the tubule walls **resorb** useful substances—glucose, water, certain salts—into the body, and **secrete** other substances—such as other salts, drugs, and hormones—from the body fluids into the forming urine. Blood capillaries winding around the nephrons exchange substances in either direction with the extracellular fluid, which in turn exchanges substances with the nephron cells. Urine leaving the nephrons flows through **collecting ducts** and leaves the kidneys via **ureters** to be stored in the **urinary bladder**. Urine eventually leaves the body via the **urethra**.

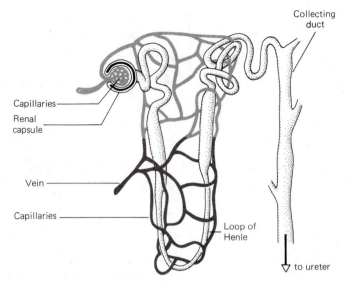

Fig. 22–16

A kidney nephron and its blood supply.

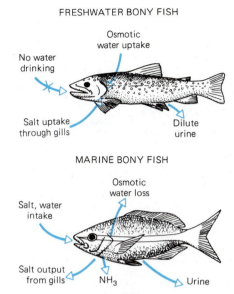

Fig.22-17

Salt and water balance in bony fish depends on the gills as well as the kidneys. The gills inevitably exchange salts and water, as well as gases, with the environment. The freshwater fish is in danger of losing body salts to the surrounding water, and it also gains water by osmosis through the gills. The gills actively transport salts into the body, and the kidneys resorb salts, expelling a dilute urine that carries excess water out of the body. The marine fish loses water to the sea, which is saltier than its body fluids. It drinks seawater and excretes excess salt through the gills. Most of its nitrogenous waste (NH_3) also leaves via the gills. Little water is lost in the urine.

Some figures will give a better idea of the magnitude of the kidneys' task. About 1700 litres of blood pass through the kidneys of an adult human every day; this is nearly a quarter of the heart's output. There are only 5.6 litres of blood in the human body; thus every drop of blood passes through the kidneys, where its contents are monitored and altered, almost 500 times a day. About 180 litres of fluid filters from the blood into the nephrons every day. Clearly we do not produce 180 litres of urine a day; most of the water filtered into the kidneys is absorbed back into the body through the kidney tubules.

The amount of water reclaimed by the kidneys can be altered, depending on the body's water content. If we drink more liquid than we need, we produce more urine. There are limits to the kidneys' ability to regulate water content, however, and some other animals' kidneys are better at it than ours are. Sea birds, marine mammals, laboratory rats, gerbils, and some desert animals can live indefinitely with only sea water to drink (Fig. 22-18).

Although salt water contains all the water the body needs, it also contains excess salts which must be excreted if they are not to accumulate and poison the body. Because salt excretion via the kidneys requires the use of precious water to dissolve the salts, this is a real problem. Some animals can excrete salts without using the kidneys. Sea birds excrete a concentrated salt solution through their nostrils (Fig. 22-19). Laboratory rats have no such device, but have kidneys efficient enough to excrete a very concentrated urine, disposing of a lot of salt with a minimum of water.

Human beings cannot survive by drinking sea water. A shipwrecked mariner is surrounded by "water, water everywhere, nor any drop to drink." By itself, the human kidney could just about break even in excreting the salt in sea

Fig. 22–18

Marine mammals drink sea water but excrete the salts in a very concentrated urine, conserving water.

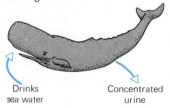

MARINE BIRDS

Salt water
excreted

Drinks
salt water

Concentrated
urine

Fig. 22–19

Osmotic adaptations of sea birds. To conserve water, a nasal gland excretes a concentrated solution of the salt taken in with sea water and the kidneys excrete a concentrated urine.

water. However, sea water contains enough magnesium to cause diarrhea, and so extra water is lost in the feces; this is why a human being loses more water by drinking sea water than by not drinking.

22–G Regulation and Homeostasis

Various regulatory systems are responsible for homeostasis; they constantly monitor the body's composition and keep it constant. They act essentially like thermostats, responding to a rise or drop in oxygen, blood pressure, water content, and so on, by adjusting the rate of the process that can correct the rise or drop. We shall consider three of these regulatory systems here, to give you an idea of how they work: the kidney's control of water content, the response of the circulatory system to exercise, and the changes in circulation and behavior that allow animals to regulate body temperature.

ant-ee-DIE-your-ETT-ic
VASS-oh-PRESS-in
pit-YOU-it-AIR-ee

Control of water loss. The main factor controlling the amount of water excreted by the kidneys is **antidiuretic hormone (ADH)**, also called **vasopressin**, which is released into the blood by the **pituitary gland**, just below the brain. ADH in the blood increases the kidneys' absorption of water back into the body from the kidney filtrate, reducing water loss in the urine. Secretion of antidiuretic hormone increases if the body loses water (Fig. 22-21). The body detects a decrease in its water content in one of two ways: either as a reduction in blood volume, such as might be caused by severe bleeding, or as an increase in concentration of substances in the extracellular fluid, which might be caused by ordinary water loss such as sweating.

Since alcohol inhibits the secretion of vasopressin, heavy drinking increases water loss in the urine. This leads to the general dehydration of the body which contributes largely to the thirst and discomfort of a hangover.

DIE-ub-BEE-tiss
in-SIP-uh-dus
mell-ITE-us

People who produce abnormally little vasopressin have the disease **diabetes insipidus**, characterized by thirst and the production of large quantities of dilute urine. (This type of diabetes is much less common than **diabetes mellitus**, characterized by sugar in the urine.)

Response to exercise.   When we start to exercise, the nervous system sends impulses to the **adrenal glands** which sit like caps atop the kidneys. These glands then release the hormone **adrenalin** into the bloodstream.

uh-DREN-uh-lin

Adrenalin causes the **spleen** (an organ near the stomach) to release some of its stored blood into the bloodstream, increasing the volume of circulating blood. Adrenalin also causes local dilation (widening) of the capillaries in the skin, muscles, and heart, increasing the blood supply to these organs. During exercise, the heart will have to work harder to pump more blood, and at a faster rate; the muscles will have to move the limbs; and the skin will have to sweat more than usual to give off the extra body heat produced by the activity. Adrenalin also causes constriction (narrowing) of blood vessels in the abdomen and kidneys, cutting down their blood supply. This trade-off helps to maintain the blood pressure. (There is not enough blood to fill the expanded circulatory system at normal pressure.)

Adrenalin also increases the breathing rate and the heartbeat rate. As a result, oxygen is taken in, and carbon dioxide is given off, faster than usual, and the heart sends blood through the body faster, speeding delivery of extra oxygen to the hard-working muscles and hastening removal of waste products.

During exercise, the muscles produce more carbon dioxide than usual, and this has its own regulatory effect. The carbon dioxide makes the blood more acidic, and an increase in acidity has three effects: it makes the blood give up more oxygen than usual in the muscles; it increases the widening of the blood vessels in the muscles; and it also stimulates the nervous system to increase the secretion of adrenalin, and the breathing and heartbeat rates, still further (Fig. 22-22).

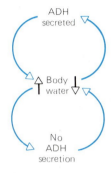

Fig. 22–21

Control of water retention by the kidneys. A decrease in body water causes increased secretion of antidiuretic hormone (ADH), which makes the kidney absorb more water. An increase in the volume of fluid in the body reduces ADH and permits loss of more water in the urine.

Although it would seem that all these adjustments to exercise are altering conditions in the body, what they actually do is to ensure that the composition of the extracellular fluid around the body cells, and particularly the brain, remains as it would be without exercise. Without all these adjustments, the extracellular fluid would heat up and would contain too little oxygen and too much acid. When the exercise is too violent for the regulatory system to cope with the changes, this is precisely what happens; acid accumulates around the muscles, causing "cramps." Cramps also serve a regulatory function since they prevent further exercise, forcing the body to slow down and permit things to return to normal.

These are only a few of the interactions involved in the body's reaction to exercise, but they illustrate the complexity and precision of the regulatory mechanisms that maintain the status quo within the body and thus maintain life itself.

Temperature regulation.   Homeostasis involves regulating not only the chemical composition of the body fluids, but also the body's temperature. Most enzymes function efficiently only when they are at or near their optimum temperature, and so the body must keep warm—but not too warm.

All organisms produce heat from their chemical reactions, collectively known as their **metabolism.** Mammals and birds have high metabolic rates, which permit them to maintain high body temperatures. A high metabolic rate requires a great deal of oxygen, used in the reactions of cellular respiration. To

muh-TAB-uh-lism
MET-uh-BOLL-ic

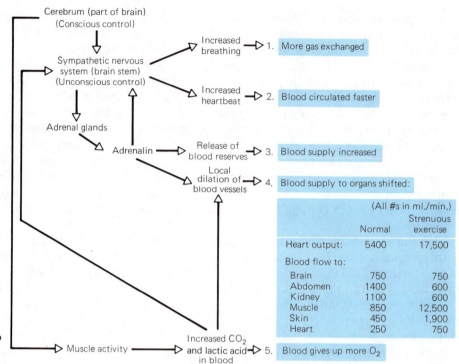

Fig. 22-22

Some of the body's responses to exercise.

deliver oxygen rapidly to all their cells, birds and mammals have high blood pressure, which forces oxygen-laden blood to the body's capillary beds; large numbers of red blood cells carrying oxygen; and large respiratory surfaces in the lungs, where oxygen is absorbed.

When the body's metabolic rate is not high enough to generate needed heat, the brain induces minute muscular movements that we do not notice, or even noticeable shivering. This muscular activity speeds up muscle metabolism and produces extra heat. When the brain detects a decrease in body temperature, the nervous system also constricts the blood vessels near the surface of the body, and less blood flows through the skin where it would lose heat to the environment (Fig. 22-24).

The body is further protected from heat loss by layers of fat just beneath the skin and by a covering of hair, less well-developed in humans than in most other mammals (Fig. 22-23 b). On a cold day, a tremendous amount of heat is lost from the head because the nerve tissue dies quickly if deprived of food and oxygen, and thus blood flow to the brain cannot be cut down as can the blood flow to the rest of the body surface. This may be why we have retained the hair on our heads, although we have lost most of our body hair during the course of evolution. We compensate for our relative hairlessness with clothes, which must be kept dry to keep us warm, because water absorbs a lot of body heat.

When the body temperature rises above normal, various physiological mechanisms dissipate heat in two main ways. First, blood flow to the skin is increased so that more heat is lost by radiation; second, and more important, most birds and mammals use evaporative cooling, based on the fact that water

(a)

Fig. 22–23

Regulation of body temperature. (a) A male emperor penguin in the frigid Antarctic keeps his chick warm by carrying it on his feet even when he walks. (U.S. Navy) (b) Scanning electron micrograph of human hairs. Muscles move mammalian hairs, altering the amount of heat that leaves the body. (Biophoto Associates). (c) Birds, like this tit with a peanut, have flat smooth feathers for flight and short fluffy feathers over the rest of the body to retain heat. (Biophoto Associates).

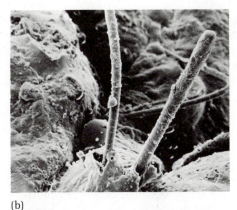

(b)

(c)

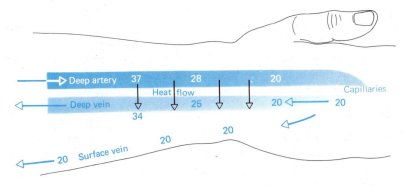

Colored arrows = blood flow
Black arrows = heat flow
Numbers = temperature in °C

Fig. 22–24

Countercurrent heat exchanger between an artery and a vein. Heat is lost in the hand. As the blood runs through the deep vein and artery it exchanges heat so that blood is cooled by the time it reaches the hand and warmed as it returns to the body. When the body needs to lose heat, blood returns from the hand to the body via a surface vein where it is not warmed.

takes up a lot of heat when it turns from a liquid into a gas. Human beings release moisture onto the skin in the form of sweat when the body overheats; the water evaporates, cooling the body. Other mammals and birds achieve the same effect by panting, which increases evaporation from the respiratory surface. They also rearrange their fur or feathers to increase air flow over the skin and thus increase heat loss to the air.

HOME-ee-oh-THERM-ic

POY-kill-oh-THERM-ic

The unscientific term "warm-blooded" corresponds approximately to the scientific term **homeothermic** (homeo = same; therm = heat), applied to animals like birds and mammals that maintain a constant temperature at all times. "Cold-blooded" corresponds roughly to **poikilothermic** (poikilo = variegated), applied to animals whose temperatures fluctuate with that of the environment. Confusion arises because most people think of all animals other than birds and mammals as poikilotherms, which is not necessarily correct. Aquatic invertebrates make up the biggest group of poikilotherms. Most, like a jellyfish in tropical seas at 20°C or a crayfish in a northern pond fluctuating from 4°C in winter to 20°C in summer, tolerate the temperature changes imposed by their environments. As the temperature changes from winter to summer, some of these animals synthesize new sets of enzymes that function better at the new temperatures. A few reptiles and amphibians, and most fish, are poikilotherms, but most reptiles, and some large fish, normally have body temperatures appreciably above those of their environment. Similarly, among invertebrates, bees die if they get too cold and moths cannot fly unless their wing muscles are warmer than the temperature of the human body. It is obvious that the distinction between poikilotherm and homeotherm is not clearcut.

Ectotherms (ecto = external) are those animals that regulate their temperatures largely by behavioral adaptations that control the amount of heat exchanged with the environment. For instance, reptiles and insects bask in the sun, and trout move into colder water; in these cases, heat exchange with the environment alters the animal's temperature. **Endotherms** (endo = internal) are those animals that depend largely on physiological mechanisms to control their temperatures. Most animals fall somewhere between the two, using both behavioral and physiological temperature regulation. Thus desert rodents, domestic dogs, and human beings (all endotherms) perform ectothermic behaviors such as seeking shade on a hot day while many ectothermic fish, amphibians, and insects show the endotherm's tendency to generate and retain enough metabolic heat to keep at least parts of their bodies at high temperatures nearly all the time.

Generating enough metabolic heat to keep the body at about 40°C uses a lot of energy (from food and oxygen). In evolutionary terms, there is no point in being warm enough to be active when neither predators, prey, nor mates are about, and many animals conserve energy by allowing their body temperatures to fall at these times. For instance, most land invertebrates, amphibians, reptiles, and small mammals have daily periods of **torpor** when they cease to move and permit their body temperatures to vary with the surrounding temperatures.

Because their bodies are usually warmer than the surrounding air, mammals and birds lose heat across their body surfaces. This loss is worse for a small endotherm because it has a large ratio of surface area, which loses heat, to body volume, which produces heat. To produce enough heat, a small endotherm such as a shrew or hummingbird must eat more than its own weight in food every day, and it is never more than a few hours away from starving to death. Allowing the body temperature to drop during a period of torpor relieves some of the drain on its energy resources. **Hibernation** (in winter) or **estivation** (in summer) are longer periods spent in a state of low body temperature.

ESS-tih-VAY-shun

SUMMARY

Cells can live only in a very restricted environment; the levels of food, gases, water, and salts—and, in some animals, temperature—must be maintained within very narrow limits. Because the cell is always taking in food and oxygen and releasing wastes, it constantly alters its environment of extracellular fluid. Continuous exchange of these substances between the ECF and blood in the capillary beds of the circulatory system maintains the ECF as a proper environment for cell life.

In turn, the blood is renewed as it passes through other capillary beds: in the capillaries at the respiratory surface in the lungs, the blood gives up carbon dioxide and takes in fresh oxygen; in the capillaries of the kidneys it gives up water and small molecules and, further on, exchanges various substances until its content of water, salts, and wastes has been readjusted; in the capillaries of metabolically active tissues the blood gains heat, and in the capillaries of the skin it can give off any excess heat to the air, either by direct radiation or by exuding water and salts as sweat.

Because the blood in capillary beds is under pressure from the pushing of the heart, the bloodstream loses fluid which, however, is collected and returned by the lymphatic system. The branching of the circulatory and lymphatic systems throughout the body makes them ideal highways for the cells and antibodies of the immune system to reach any part of the body invaded by foreign cells or particles.

Changing conditions in the body are met by changes in its physiology. The distribution of blood to various parts can meet the demands of exercise, overheating, or chilling. The formation of urine can adjust to temporary conditions of dehydration, blood loss, and changes in salt balance. In addition, the reproduction of cells of the immune system responds to particular foreign antigens invading the body. All of these mechanisms allow the body to keep the cellular environment constant and to reallocate its resources to meet changing needs.

SELF-QUIZ

1. Which system in the body is responsible for each of the following activities?

    _____a. ridding the blood of nitrogenous wastes

    _____b. collecting extracellular fluid and returning it to the blood

    _____c. exchanging gases with the environment

    _____d. carrying food throughout the body

2. Blood vessels that carry blood toward the heart are called:

    a. lymphatics
    b. veins
    c. arteries
    d. capillaries
    e. alveoli

3. The blood loses most oxygen when it is travelling through:

   a. the right atrium and the right ventricle
   b. the lungs
   c. a vein in the arm
   d. the capillaries in a muscle

4. The function of the blood platelets is:

   a. blood clotting
   b. fighting infections
   c. carrying oxygen
   d. exchanging gases with the environment
   e. preventing hardening of the arteries

5. The main function of hemoglobin is:

   a. transporting oxygen
   b. transporting carbon dioxide
   c. making the blood less acidic
   d. blood clotting

6. The introduction of an antigen into the body triggers a response against that antigen by:

   a. causing antibody molecules to assume a shape that permits them to bind the antigen
   b. causing mutations in cells so that they produce antibodies to the antigen
   c. killing cells with the proper antibody, so that they disintegrate and release the antibody
   d. stimulating reproduction of cells that make antibody to the antigen

7. Both hyperventilation and holding one's breath can result in loss of consciousness. Other conditions being normal, this happens because of:

   a. alteration of carbon dioxide levels in the blood
   b. loss of hemoglobin from the red blood cells
   c. distress of the lungs
   d. abnormally high oxygen loss from hemoglobin

8. Substances that the body must sometimes excrete are listed below. Indicate which organ listed on the right excretes each substance on the left. (There may be more than one organ that excretes some substances; give all correct answers.)

Substance	Organs
____Carbon dioxide	a. Lungs
____Water	b. Skin
____Salts	c. Kidneys
____Sugars	
____Hormones	
____Spices	

9. Heavy bleeding causes a decrease in the amount of urine produced. This results primarily because of:

   a. an increase in the amount of water filtered out of the blood into the nephrons
   b. a decrease in the amount of water filtered into the nephrons
   c. an increase in the amount of water absorbed back into the body from out of the nephrons
   d. a decrease in the amount of water absorbed back into the body from the nephrons

10. Increasing the adrenalin content of the blood would decrease the flow of blood to the:

    a. brain
    b. skin
    c. liver
    d. heart
    e. lungs

11. The heat of the human body comes mainly from:

    a. metabolism
    b. shivering
    c. warm clothing
    d. basking in the sun
    e. sweating

QUESTIONS FOR DISCUSSION

1. Ice fish are a family of bony fish that inhabit Antarctic waters. These fish, unlike all other adult vertebrates, have no hemoglobin in their blood. How do you think they survive without it? What other physiological characteristics would you expect to find in these fish?

2. Do you think that the respiratory surface of your lungs is exposed to the environment? Why or why not?

3. What would happen if a person with type B blood received a transfusion of type O blood?

4. What is the advantage of having a urinary bladder? Birds don't have urinary bladders; why not?

5. Medical science would find it very useful if cells or tissues of one individual could be transplanted to another without danger of rejection. What might be the advantages of the cell membrane antigens that distinguish the cells of different individuals?

6. In this chapter we saw that one cell can produce only one kind of antibody molecule. What might happen if a cell could produce two or more types of antibody at once?

7. Many mammals hibernate, but birds don't. Why not?

ANIMAL 23 COORDINATING MECHANISMS

OBJECTIVES   When you have studied this chapter, you should be able to:

1. Describe the roles of sense organs, the nervous system, and effectors in the activities of the body.

2. List the five kinds of stimuli that sense organs may detect, and some sense organs that detect each type of stimulus.

3. Draw a simple diagram of the eye, placing the following parts correctly: cornea, pupil, iris, lens, retina; give the function of each part; briefly describe the location and function of the rods and cones; describe how light striking a rod cell is detected by the brain.

4. Describe the basic structure and function of a neuron; explain how information about the intensity and type of a stimulus is transmitted in the nervous system.

5. Draw a model synapse with its principal components, explain the role of each component, and explain how information is transmitted across the synapse.

6. Describe the "fight or flight" response, listing at least four physiological changes that occur during the response.

7. Draw the unit between two successize Z lines in a skeletal muscle fiber as it would appear during contraction and relaxation of the fiber.

8. Name the two main proteins involved in contraction of muscle, indicate the position of each in your drawing from objective 6, and state the roles of these proteins and of calcium and ATP in muscle contraction.

9. Outline the sequence of events that takes place (on a chemical and cellular level) from the time a nerve impulse arrives at the neuromuscular junction in a skeletal muscle, through contraction of the muscle fiber, to subsequent relaxation of the fiber.

10. Explain how the strength and duration of contraction of a skeletal muscle are controlled.

11. Describe the relationship between the hypothalamus and the pituitary gland, and their role in the body's endocrine system.

An animal's digestion, respiration, reproduction, and behavior may be isolated to make them easier to study, but they are inseparable in a living animal. All must be coordinated in such a way that they work together. The nervous and hormonal systems are responsible for coordinating the body's activities. The sense organs (including, for example, the eyes and ears and the

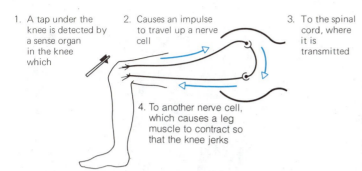

1. A tap under the knee is detected by a sense organ in the knee which

2. Causes an impulse to travel up a nerve cell

3. To the spinal cord, where it is transmitted

4. To another nerve cell, which causes a leg muscle to contract so that the knee jerks

Fig. 23–1

The knee jerk reflex. This reflex arc illustrates the simplest kind of control loop within the body. It includes (1) a sense organ that detects a stimulus; (2) nerves that transmit information from the sense organ to the central nervous system and back to the effector organ; (3) an effector organ (in this case, a muscle) which does something.

organs that detect changes in blood chemistry) receive stimuli, such as light and pressure, which give information about the body's external and internal environments. This information is relayed to the **central nervous system**, consisting of the brain and spinal cord, which directs the muscles or glands to make adjustments. These adjustments are of two main types: those that maintain homeostasis within the body, and those that produce responses to external events.

A reflex arc is a simple example of a control loop (Fig. 23-1). The arc consists of a sensory cell which detects the stimulus; a chain of nerve cells, including cells in the central nervous system, that pass along the signal that the stimulus has been received; and a muscle or gland, which performs an appropriate response to the stimulus. Stimulation of the sensory cell sends a message through the chain of cells. Because only a few nerve cells occur between the sense organ and the muscle or gland, the response to a stimulus detected by the sensory cell is rapid.

In this chapter, we shall consider the sense organs that receive stimuli, the nerves that convey information about stimuli, and the hormones and muscles that react to this information. Although we shall examine mainly human systems, similar systems are also present in the bodies of many other animals. The main difference between ourselves and other animals is the enormous development of our brains and the extraordinary variety of behavior this gives rise to.

23–A Sense Organs

We sometimes speak of human beings as having five senses—sight, hearing, touch, smell, and taste—but, in fact, we have at least a dozen types of sense organs. Each is specialized to detect a particular type of **stimulus**—some form of energy. The cells in a sense organ that react to stimuli are the **receptors**, and they are most usefully classified by the type of stimulus energy to which they respond: light, pressure, chemical changes, temperature, or electrical current. Humans have all these types of receptors except those that react to electrical currents. A shark is an example of an animal that has receptors reacting to electrical current; it can track its prey by detecting the weak electrical currents that all animals produce.

Fig. 23–2

Sense organs on an insect's head include the eyes (big honeycombed balls) and numerous hair-like pressure receptors which respond to air movement or touch. (Biophoto Associates)

Among our pressure receptors are our ears (which react both to pressure waves in the air and to the pull of gravity), touch and pain receptors in the skin and in some internal organs, and organs that detect stretch in our muscles. We have temperature receptors in our skin, on our tongues, and in various places inside the body. Similarly, there are chemical receptors inside the body as well as the more obvious ones of the nose and tongue.

The role of a sense organ is to detect a particular stimulus and relay information about it to the central nervous system. To do this, the receptor must first convert the energy of the stimulus into electrical energy—the only form of energy that can travel along a nerve. We will take a closer look at just one of our sense organs—the eye.

The eye. Almost all animals have photoreceptors (photo = light) that transform light energy into electrical energy using **pigments**, colored molecules that change temporarily when struck by light rays. In humans the photoreceptor cells lie in a delicate layer called the retina, lining the rear two thirds of the eyeball (Fig. 23-3). The rest of the eye is made up of structures that channel or filter light before it reaches the receptors.

Fig. 23–3

The human eye. The receptor cells are in the retina. The fovea is an area with especially good ability to resolve fine detail in an image.

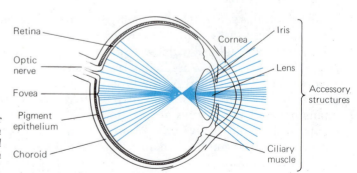

Fig. 23-4

Different eyes see different things. The human eye sees a marsh marigold with no markings (left). An insect sees big patches on the same flowers (right) because its eyes react to ultraviolet light energy, whereas ours do not. (Biophoto Associates)

The wall of the eyeball has three layers. The outer layer is the "white" of the eye. The front part of this layer is the bulging, transparent **cornea**. The cornea is   CORE-nee-uh
protected from dust and dryness by a constant bath of tears and by frequent blinking, which wipes the cornea. The very fast **blink reflex** causes the eyelids to close when anything touches the cornea or when the eye sees anything that might hit it. The cornea is extraordinarily resilient to damage. Slits in the cornea can actually be stitched up without loss of vision; also, if the cornea becomes opaque from disease or old age, corneal transplants can now be used to replace it. (Because the cornea lacks blood vessels, the body's antibodies do not come into contact with the foreign antigens in a donated cornea [see Section 22–D]).

The middle layer of the eyeball wall has an opening, the **pupil**, at the front. Pigments around the pupil in this middle layer give our eyes their color and also block out light; thus the only light that actually enters the eye goes through the pupil. Too much light could damage the retina; the ciliary muscles change the size of the pupil, dilating (expanding) it in dim light and constricting it in bright light. This prevents damage, while admitting enough light for vision. (Eye doctors use drops of muscle relaxant to block the pupil control reflex so that they can shine a beam of light into the eye to examine it.) Behind the pupil lies a rounded **lens**, attached to muscles that can change its shape and so focus light passing through it.

The inner layer of the eye is the retina. The retina contains the **rods** and **cones**, which are the receptor cells that respond to light, and it also contains the nerves that carry information from the eye to the brain (Fig. 23-5).

The cone cells are most abundant in the center of the retina. They are used for vision in bright light, and they give rise to detailed, well-defined images. Cones are also the most important receptors for color vision.

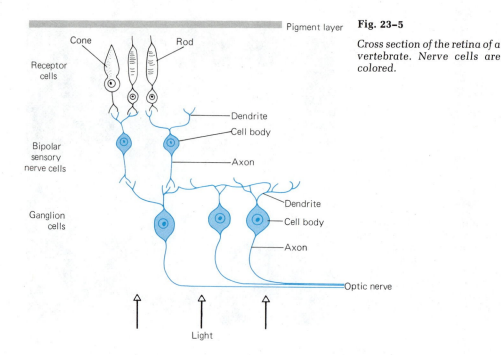

Fig. 23–5

Cross section of the retina of a vertebrate. Nerve cells are colored.

Rod cells are more abundant toward the edges of the retina. They are exceedingly sensitive and enable us to see in light too dim to permit vision by way of the cone cells; however, rods do not permit us to distinguish colors very well. Their location explains why we can see an object better at night if we look sideways, rather than directly at it.

Light striking a receptor cell in the retina is absorbed by pigment molecules, causing the molecules to change shape. In some way that is not understood, this change of shape causes an electrical discharge in the membrane of the receptor cell; this discharge is passed on to a nearby nerve cell, which transmits the impulse to the brain. Eating carrots helps us see in the dark because a pigment in carrots can be broken down to form vitamin A, which is further converted into part of the light-sensitive pigment in rod cells.

23–B The Nervous System

NOOR-onz

Neurons. **Neurons** are the cells of the nervous system that can transmit information in the form of electrical **nerve impulses**. These impulses are somewhat like those that travel in an electric wire, but there are important differences. First, electricity in a living organism moves in a fluid instead of in a solid medium. Furthermore, the impulse is carried, not by the metal in a wire, but by movements of charged ions across the neuron's cell membrane; as the impulse sweeps down the neuron, each point in the membrane in turn becomes temporarily "leaky" to ions. Another difference is that the neuron constantly expends its own energy to move ions back to where they started from; thus it renews its ability to carry electrical impulses.

Each neuron consists of a **cell body**, which contains the cell nucleus, and a number of filaments (Fig. 23-6). These filaments are classified by their part in information processing. Filaments that receive information from other cells are called **dendrites**, whereas filaments that transmit information away from the dendrites toward the next cell are called **axons**. Most neuron cell bodies lie within the central nervous system. The longest axons are found in neurons that pick up information from peripheral sense organs and convey it to the central nervous system. The axon of such a neuron may extend from, say, a toe to the cell body a metre or more away, just outside the spinal cord. Neurons that lie completely within the central nervous system have shorter dendrites and axons because the cells are closer together.

Before a nerve impulse can occur in a neuron, its dendrites must receive a certain minimum, or **threshold**, stimulation. Once this threshold stimulation is reached, the nerve impulse begins and travels down the entire length of the axon. Every nerve impulse is the same size and travels at the same speed as any other impulse in the same neuron. The nerve impulse is known as an "all-or-nothing" event because of these properties—its threshold and its unvarying nature.

A neuron is either "on" (transmitting an impulse) or "off" (not transmitting an impulse). Given such simple behavior, how can we account for the ability of the nervous system to recognize and process complex information? In other words, if a neuron in the retina is either sending the message "my photoreceptor sees light" or not sending this message, how do we detect color, shape, texture, pattern, brightness, and movement when we look at something?

Information about the magnitude of a stimulus (for example, brightness of light) may be coded in three basic ways:

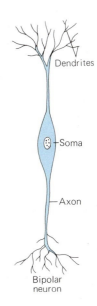

Fig. 23–6

Structure of a neuron.

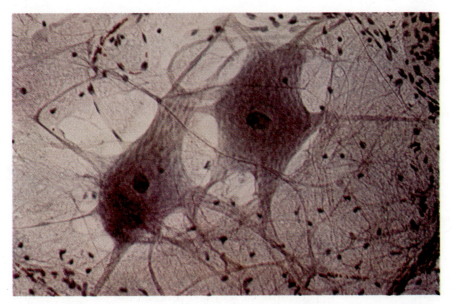

Fig. 23–7

Neurons as they appear under the light microscope. Two cell bodies lie in a tangled mass of dendrites and axons. (Carolina Biological Supply Company).

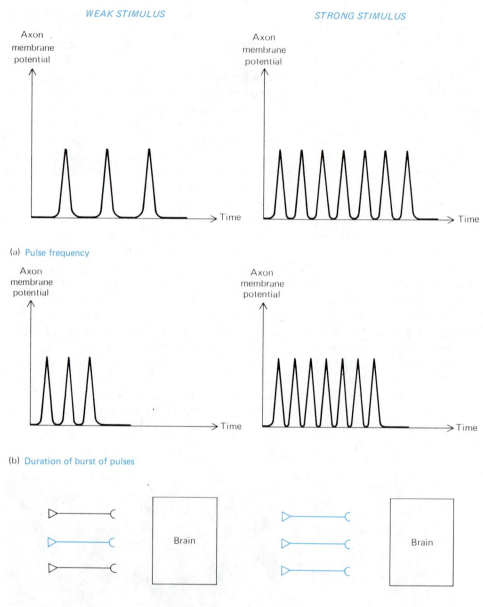

WEAK STIMULUS STRONG STIMULUS

Axon membrane potential

Time

(a) Pulse frequency

Axon membrane potential

Time

(b) Duration of burst of pulses

Brain

Brain

(c) Number of neurons firing

Fig. 23–8

Coding of stimulus intensity in neuron activity. The "spikes" in (a) and (b) are recordings (by an oscilloscope) of the electrical activity which reaches an electrode placed at one spot on a neuron as nerve impulses sweep along the neuron. A strong stimulus will cause the neuron (a) to fire more rapidly, (b) to fire a longer burst of impulses, or (c) cause more receptor cells to fire (color).

1.   *Frequency.* The stronger the stimulus, the more frequent the impulses will be—that is, the more quickly each impulse will follow the last. A weak stimulus initiates only a few impulses per second, whereas a strong stimulus initiates many (Fig. 23-8a).

2.   *Duration of a burst of impulses.* The stronger a stimulus, the longer the train of impulses that it initiates will last (Fig. 23-8b).

3.   *Number of neurons firing.* The stronger the stimulus, the more neurons will fire impulses. If several neurons have different threshold levels, a weak stimulus will cause only a few to fire, whereas a stronger stimulus will exceed the thresholds of more neurons and cause more cells to fire (Fig. 23-8c).

But the brain can tell more than just this information about the intensity of the stimulus. The "wiring" of neurons (their pattern of connections to each other) conveys additional information. For example, a particular part of the brain receives impulses only from neurons in a certain part of the retina; this "tells" the brain the angle at which light has entered the eye. Furthermore, the wiring is arranged such that particular neurons send out impulses only when the stimulus impinging on the retina is a certain size or shape, or is moving in a certain way.

All the impulses transmitted by neurons in the visual system are just like the nerve impulses from the neurons in the nose, skin, and ears. The brain interprets the input of retinal neurons as light because it knows that these neurons are connected to light receptors. If the neurons in the visual pathways send impulses for other reasons (for instance, when something hits you on the head or when you press your fingertips gently against the corners of your eyes) you "see stars" because the brain interprets these pressures as light. (The fact that you also feel the pressure shows that the eyes also have pressure receptors).

Synapses.   Because information about where a nerve impulse comes from is so important to the brain's functioning, the study of connections between neurons, and of how and why they are formed and destroyed, is a very active area of research. The junction where one neuron stops and another begins is called a **synapse**. A synapse is composed of a gap, or cleft, plus the two membranes on   SIN-aps either side of the gap (Fig. 23-9). Information travels across a synapse in one of two ways: the first, and most recently discovered, is by a simple electrical jump; the membranes of the two neurons are so close together that the electrical impulse can hop from one to the other. Electrical synapses are not at all common; at most synapses, the information travels across the synaptic cleft in the form of a

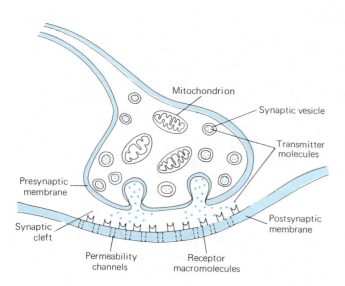

Fig. 23–9

A synapse.

uh-SEAT-ul-COE-lean

VEE-sick-uls

chemical messenger called a **transmitter**. The best-known transmitter is acetyl-choline, which occurs at many synapses in the brain, in other parts of the nervous system, and at the **neuromuscular junctions**, which are synapses between neurons and muscles.

In chemical synapses, a nerve impulse arriving at a synapse causes little sacs called vesicles in the **presynaptic** (= before the synapse) end of the axon to release the transmitter they contain into the synaptic gap. The transmitter crosses the gap and binds to special receptor molecules in the **postsynaptic** (= after the synapse) membrane of the receiving neuron. This binding alters the postsynaptic neuron's membrane, making it more permeable to certain ions—possibly to the extent that this neuron will fire a nerve impulse of its own.

The transmitter is rapidly destroyed by an enzyme or absorbed into one of the neurons. If it were not, the receiving neuron would go on firing indefinitely and eventually die of exhaustion. This is precisely how some nerve gases and insecticides work. DDT and the common garden spray Sevin inhibit the enzyme that normally breaks down acetylcholine. Because the acetylcholine is not destroyed, it goes on stimulating muscle contraction at the neuromuscular junctions, and the insect goes into uncontrolled muscular convulsions until it dies. (Such substances must be used with care; they are also toxic to other organisms, such as people and pets, that use acetylcholine as a transmitter.)

The structure of a chemical synapse explains why nerve impulses normally travel in only one direction along a neuron. The synapse works in only one direction because, of the two neuron membranes at the synapse, only one encloses sacs of transmitter and only the other has receptors for the transmitter. As a result, the postsynaptic membrane can conduct information only *away* from the synapse—which makes it part of a dendrite—and the presynaptic membrane can conduct information only *toward* the synapse—which makes it part of an axon.

So far, we have described an **excitatory synapse**, in which binding of the transmitter stimulates the dendrite beyond the synapse. In **inhibitory synapses**, binding of the transmitter to the postsynaptic membrane causes changes such that it takes a bigger excitatory stimulus than usual to trigger a nerve impulse. In other words, when an inhibitory synapse is activated, a stimulus that is ordinarily just at the threshold to make the neuron fire will evoke no such response.

A neuron may make more synapses with one neuron than with another, and may have excitatory synapses with some neurons and inhibitory synapses with others. The sum of excitatory and inhibitory effects on its dendrites determines whether or not the neuron will fire a nerve impulse. These arrangements allow extremely complex control of information processing, and the nervous system can vary its response to the same stimulus at different times. We are all familiar with such changes: the sounds the house makes seem louder and more threatening when we are alone late at night than when someone else is there; our response to a ringing telephone depends on the nature of the call we are expecting.

Chemical transmitters. In addition to acetylcholine, which we have already encountered, various neurons in the brain produce other chemical transmitters. One of these is noradrenalin, closely related to adrenalin, which is also found in the peripheral nervous system. Of similar chemical structure are dopamine and serotonin, found only in the brain. Dopamine is the transmitter for a small group of neurons concerned only with muscular activity. Parkinson's disease, characterized by bursts of uncontrollable muscular movement, may sometimes be caused by lack of this transmitter and is now often treated with dopamine. Serotonin is a transmitter produced by a group of cells whose cell bodies lie in the medulla, the "stem" of the brain just above the spinal cord. Serotonin is thought to be concerned with such basic functions as sleep, consciousness, and emotional states.

NOR-uh-DREN-uh-lin

DOPE-uh-mean
SAIR-uh-TONE-in

Recently, substances called endorphins have been isolated from various parts of the brain. Their natural function is not fully understood, but it is interesting that, chemically, they are similar to morphine. The existence in the brain of receptors for endorphin may explain why the brain is so sensitive to morphine, opium, and related drugs.

en-DOOR-fins

Drugs and neurons.   It is well known that many drugs affect the nervous system, but we understand the actions of only a few.

Opium, from the seedpod of a poppy, has been used as a drug since classical Greek times, not only because it is the most effective pain-killer ever discovered, but also because of the euphoric state it induces. Opiates were used as pain-killers in the Civil War in the United States, and addiction to opiates has been a social problem in the United States ever since. The search for a non-addictive opiate has been intense, but all the opium derivatives ever produced—including morphine, Demerol, methadone, codeine, and heroin—eventually produce debilitating addiction in most people who take them. Opiates bind to postsynaptic receptors in the brain and block the binding of any neurotransmitters that are released. This prevents the transmission of nervous impulses along a tract of nerves by which the body normally "tells" the brain that it is in pain. (Pain is a useful biological reaction, for when the brain is informed that some part of the

Fig. 23–10

An opium poppy. (Biophoto Associates).

Central nervous system

Brain
Spinal cord

Peripheral nervous system

Nerves
Ganglia

Divisions of peripheral nervous system:

Somatic (voluntary)
Autonomic (involuntary homeostatic control)

Fig. 23–11

The major divisions of the vertebrate nervous system.

body is in pain, perhaps from a cut or burn, it causes the body to move away from the source of the damage.)

Drugs such as LSD, psilocybin, mescaline, yohimbine, and barbiturates all act on synapses in the brain. Some of them produce hallucinations in ways that are not yet understood. The action of amphetamines (Benzadrine, Dexedrine, and "speed") is more easily understood. Like Sevin and other insecticides, they block enzymes that would normally break down transmitter molecules at synapses, so that the synapses continue to fire long after they would otherwise have stopped. As with the insecticides, overdoses are fatal.

Alcohol has no effect on synapses and the direct effect of alcohol on the nervous system is not understood. Alcohol does, however, kill neurons faster than they would otherwise die—at the rate of about 10,000 per ounce of alcohol consumed. This probably accounts for the acute mental deterioration of some alcoholics.

The autonomic nervous system. The autonomic nervous system, found in all vertebrates, is a system of nerves throughout the body which controls most of the body's homeostasis (Fig. 23-11). The autonomic system regulates the heartbeat and controls contraction of the muscles in the walls of the blood vessels and the digestive, urinary, and reproductive tracts. Autonomic nerves also stimulate glands to secrete mucus, tears, and digestive enzymes.

There are two main parts of the autonomic nervous system; the sympathetic and the parasympathetic systems. Each has different functions. The **sympathetic system** dominates in time of stress: it initiates the **"fight or flight" reaction**—increases in blood pressure, heartbeat rate, breathing, and blood flow to the muscles and skin and decreases in the flow of blood to the digestive organs and kidneys. These changes ensure an adequate oxygen supply for muscular exertion, and increased blood flow to the skin allows the body to lose muscle-generated heat. In contrast, the **parasympathetic** system acts as a counterbalance by stimulating the opposite reactions in these organs. When the parasympathetic system is in command, digestion and elimination are promoted.

Although the autonomic nervous system can carry out its function automatically, it is by no means completely independent of the part of the nervous system that is under an animal's voluntary control. For example, it is possible to decide to stop breathing, at least for a short time. Recent studies have also shown that humans and animals can be trained to change their heart rates, blood pressure, and digestive reflexes voluntarily. Any voluntary control that endangers life, however, will quickly disturb the homeostasis of the brain tissue, resulting in unconsciousness. When this happens, the autonomic system will take over again and restore normal functions.

The brain. The human brain and spinal cord make up the central nervous system, where information coming from the sense organs is processed; where decisions are made; and where commands to muscles and glands originate. Although most scientists today believe that the workings of the brain can be explained in terms of the interactions of its cells, we still know very little about the neuronal basis for the dreams, thoughts, ideas, and emotions generated by

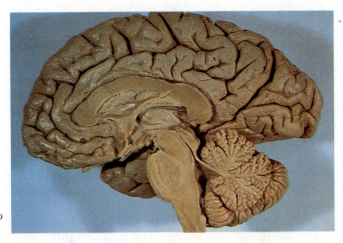

Fig. 23–12

Human brain. (Biophoto Associates).

this wonderful organ. The brain remains one of the last bastions of "vitalists," people who believe that human life cannot be explained entirely in terms of atoms and biochemical reactions.

The three most obvious parts of the brain are the **cerebrum** (including the cerebral cortex), the **cerebellum**, largely responsible for the subconscious coordination of muscular activity and balance, and the **medulla**, responsible for such homeostatic functions as heartbeat and respiration.

suh-REE-brum
SAIR-uh-BELL-um
med-ULL-uh

One of the reasons why it is so difficult to study the brain is that all of its parts

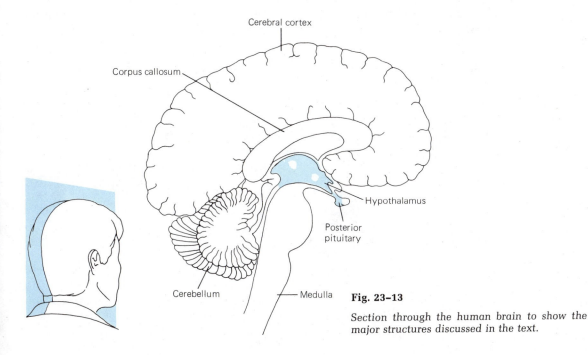

Fig. 23–13

Section through the human brain to show the major structures discussed in the text.

cortex: the outermost layer of an organ.

interact. Indeed, it is becoming clear that the connections between different parts of the brain are largely responsible for the complexity of brain function.

The **cerebral cortex** (Fig. 23-13) is, in evolutionary terms, the "newest" part of the brain. It is absent in fish and amphibians and is most complex in primates, especially humans. Because of this, it is the most-studied area of the human brain. By implanting electrodes in the cortex, experimenters have shown that particular areas of the cortex have particular functions (Fig. 23-14). For instance, stimulation of the ears and eyes causes impulse activity in the auditory and visual parts of the cortex, respectively. Then there are areas that control motor activity (voluntary muscular activity); damage to the motor speech area affects control of speech. Most mysterious are the "silent" areas of the cortex, with no obvious functions. It is possible, although far from certain, that these areas control memory and learning. The ratio of silent to active areas in the cortex is particularly large in human beings, so it is assumed that these areas have something to do with making the human brain so special.

ruh-TICK-you-lar

The **reticular formation** of the brain runs through the medulla (Fig. 23-15). It filters information arriving from a variety of sense organs and determines the brain's general level of arousal. Filtering of sensory information is part of our everyday experience. We may be awakened by an alarm clock but sleep through the equally loud noise of a radio or subway train; we may be quite unconscious of a nearby conversation until we overhear our own names, after which we are "aroused" and listen to the conversation.

Another important part of the brain is the hypothalamus, responsible for integrating nervous and hormonal control of the body; it is discussed in Section 23–D.

Memory. We still know very little about learning and memory, two of the more complex of the brain's functions. What is memory, the brain's information store? For many years scientists destroyed various parts of the brains of trained

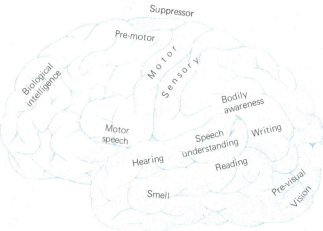

Fig. 23–14

The human cerebral cortex, showing functions assigned to different areas.

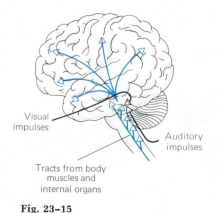

Fig. 23–15

Pathways through the reticular formation of the human brain (colored arrows).

animals to find out where memory is located. They concluded that memory is nowhere and everywhere.

There are two kinds of memory: short-term and long-term. When we take lecture notes, we use short-term memory, remembering what the lecturer has said, with any luck, just enough to write it down. If information is to be stored for any length of time, it must be transferred to the long-term memory where it may remain, much of it in subconscious form, for life. For many years, some researchers thought that long-term memory was laid down as RNA or protein molecules, because RNA and protein are synthesized when information is stored in the long-term memory, and because drugs which inhibit RNA synthesis destroy the memory. However, these experiments are inconclusive because RNA and protein synthesis are needed indirectly for many kinds of cell activity. You may have heard of experiments in which trained worms were fed to untrained worms which, supposedly, took over the memory of the worms they had eaten and so became instantly trained on the strength of their meal. Most people now mistrust these feeding experiments since behavioral experimentation has grown more sophisticated. The net result is that we know almost as little about memory as we did a hundred years ago. The best guess is probably that memory is stored in the form of electrical impulses constantly whizzing around the myriad circuits in our brains.

New synapses form and old ones deteriorate in the vertebrate brain throughout life. This is why human thought and behavior are constantly adaptable, although young brains are more adaptable than old. The neurons in our brains die at the rate of about 10,000 a day; since no new neurons are formed after early childhood, these can never be replaced. This does not produce total mental atrophy after the age of 30, however; if you live to be 100, less than 0.5% of your brain will die before you do—barring strokes, which can kill areas of the brain by cutting off the blood with its supply of needed food and oxygen.

Sleep.   Sleep is another mysterious nervous function. Most mammals and birds sleep, although many of them relax less than we do and remain upright while they sleep. Reptiles, amphibians, and fishes also have periods when they are very unresponsive to stimuli. Sleep is undoubtedly related to the equally mysterious daily periods of activity known as **circadian rhythms**, which all   sir-KAY-dee-an
animals and plants display (described in Fig. 23-16).

With all the progress in biology in the last 50 years, we still have little idea why we must sleep each night, or why sleeping has such a profound effect on our temper, alertness, and emotional stability. To make things more complicated, the period of sleep required varies considerably from one individual to another. Both guinea pigs and human beings with brain damage have lived for years without sleep, so sleep is obviously not necessary to survival. A sleeping animal is at the mercy of predators, and so it seems that sleep should have been at a selective disadvantage. What counteracting selective advantage has maintained it during the course of evolution? We don't yet know.

For now, all we can do is describe the physiological changes that occur during sleep. Heart rate and blood pressure drop, breathing becomes more shallow, and body temperature drops slightly. However, the temperature of the

Fig. 23–16

One example of (daily) circadian rhythm in humans. This graph shows the different amounts of two hormones, growth hormone and prolactin, that are secreted at different times of day in a regular, 24 hour cycle. All organisms show these daily cycles in many aspects of their lives, including behavior, body temperature, metabolic rate, and biochemical reactions. The cycles continue even if organisms are kept in permanent darkness so that they are not influenced by normal cycles of light and dark. Little is known about the control or function of these daily cycles.

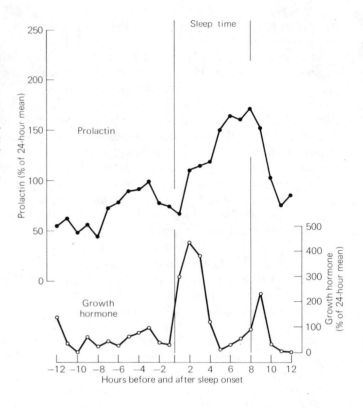

Fig. 23–17

Parts of an electroencephalogram showing electrical activity of the brain of someone awake and in 3 of the different stages of sleep. Each recording covers 20 seconds.

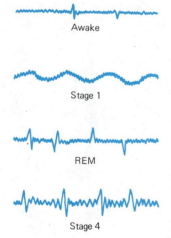

big toes rises by about 5°C; this change has been used in studies as the first reliable sign that a subject was asleep.

The most obvious changes during sleep are seen in the electrical activity of the brain. Electrical "brain waves" can be detected by way of electrodes taped to the head. The brain activity recorded from a sleeping person shows four different stages of sleep (Fig. 23-17). The most striking stage is **REM**, or **paradoxical sleep**, which occurs every 80 to 120 minutes.

REM stands for "rapid eye movements." During REM sleep the brain activity resembles that of someone who is awake, although the sleeper's muscles are more relaxed, and it is more difficult to awaken the sleeper than at any other state. People awakened during REM sleep nearly always recall dreams, although dreaming can also occur during the other stages of sleep. REM sleep seems to be the sleep stage most crucial to our psychological well-being; people deprived of REM sleep become extremely tired and will compensate for loss of REM sleep with longer periods of REM sleep on subsequent nights.

The concentrations of the various transmitters in the brain are different in waking and in different kinds of sleep. For instance, acetylcholine is present in higher concentrations in the brain of someone who is awake, whereas the concentration of the transmitter serotonin rises in the brain of a sleeping subject. Cause and effect are difficult to distinguish in this situation. For instance, it is possible that an increase in the level of serotonin puts one to sleep. On the other hand, the higher levels of serotonin during sleep may occur because neurons that produce serotonin are more active at that time.

It seems likely that sleep permits the brain to restore chemicals depleted by the day's activities and also permits the processing and reorganization of information already present in the nervous system, but how and why these things happen are still unknown.

23-C Muscles as Effectors

Muscles and glands are the body's **effectors**, the organs that "do something." Muscle tissue is like nervous tissue in that its cell membranes can be stimulated to transmit all-or-nothing electrical impulses. In addition, as a result of this membrane excitation, muscle tissue can contract, using specialized proteins to alter its shape.

Vertebrates have three main types of muscle tissue: **cardiac muscle**, which makes up the heart; **smooth muscle**, which lines the walls of many internal organs, including the digestive tract, blood vessels, uterus, vagina, and urinary bladder; and **skeletal** or **striated muscle**, which moves the skeleton and thus can change the position of the body or of its parts (Fig. 23-18). Skeletal muscle is normally the only type of muscle under voluntary control.

STRY-ate-ed

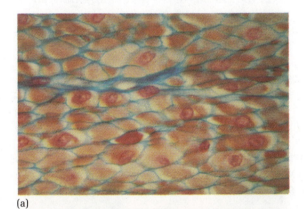

(a)

Fig. 23–18

The three types of muscle in vertebrates. (a) Smooth muscle (from the gut wall). Gaps between the cells are blue and nuclei red. (b) Cardiac (heart) muscle. Each cell is lightly striped; nuclei are black. (c) Striated (striped) muscle (from a leg) is unique in that many cells fuse to form a fiber, each of which is lined by many nuclei (black oblongs.) (Biophoto Associates).

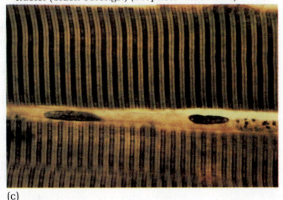

(b)

(c)

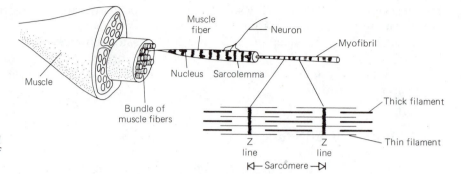

Fig. 23–19

Structure of a vertebrate striated muscle. The muscle consists of bundles of fibers.

Cardiac and smooth muscle are made up of individual muscle cells. Striated muscle is peculiar in that it forms from individual cells which fuse together during embryonic development to produce muscle fibers, each containing nuclei and cytoplasm from many cells within a single membrane.

Muscle contraction. Contraction has been studied most intensively in striated muscle; it is clear, however, that all muscles contract in much the same way. Bundles of **muscle fibers** make up each organ known as a muscle (for example, the biceps of the upper arm or the gluteus maximus of the buttock) (Fig. 23-19). Under the microscope each fiber has a striated (striped) appearance, evidence that structures within the muscle fiber are arranged with some regularity (Fig. 23-18). The organized structures are filaments of contractile proteins and the membranes to which they are attached. The internal structure of the basic unit of a striated muscle fiber, lying between two "Z" lines, is shown in Fig. 23-20. The **thin filaments** consist mainly of the protein **actin**, and the **thick filaments** of the protein **myosin**. When the muscle contracts, "heads" sticking out from the myosin molecules attach to the actin filaments and then swivel so that the filaments move past each other; the movement is much like that of a jack used to raise an automobile when you change a flat tire. As the myosin heads swivel, the actin and myosin move with respect to each other, shortening the distance between adjacent Z lines, and thus shortening the whole muscle (Fig. 23-21).

ACT-in
my-OH-sin

Muscle contraction requires energy in the form of ATP, which is necessary for the swiveling of the myosin heads. One thing that puzzled researchers for a long time is that ATP is needed for muscle relaxation as well as contraction, whereas we might expect one reaction to be active and the other passive. The phenomenon of *rigor mortis* makes it clear that this is not the case. After death, as ATP disappears from the body, muscles lose their ability to relax or contract. In fact they become locked in the position they occupied when the ATP ran out. ATP is needed for muscle relaxation because the myosin heads cannot detach from the actin filaments until new ATP molecules bind to the myosin and release them.

In addition to ATP, muscle contraction requires the presence of calcium attached to the thin filaments. Not surprisingly, calcium is released into the fiber when a muscle is activated and removed when it relaxes.

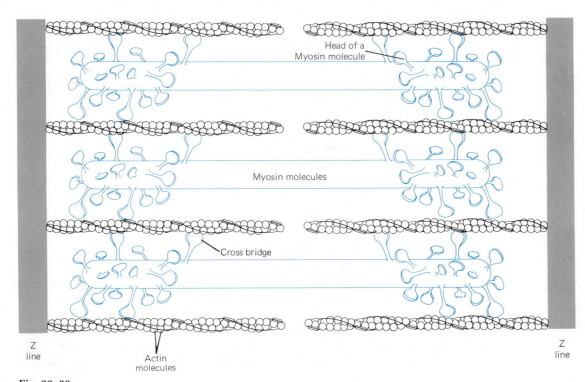

Fig. 23–20

The molecular structure of muscle with thick and thin filaments of contractile proteins between two Z lines.

Control of muscle contraction.   A skeletal muscle fiber normally contracts after a nerve impulse arrives at the neuromuscular junction. The transmitter acetylcholine, released from the end of the neuron's axon, binds to receptors on the muscle membrane and excites the membrane. An electrical impulse starts at the neuromuscular junction and spreads throughout the membrane of the muscle fiber. The impulse causes the membranes of calcium-storing reservoirs in the muscle fiber to "leak" calcium, which then attaches to the thin filaments. Contraction cannot occur until calcium is bound to the thin filaments in this way. If there is enough ATP in the muscle, it will bind to myosin heads, which will attach to the thin filaments and swivel, causing the fiber to contract. Proteins in the membranes of the calcium-storage areas constantly pump calcium back into the reservoirs. Unless subsequent nerve impulses arrive, releasing more calcium, there will soon be too little calcium left on the thin filaments for contraction, and the muscle will relax.

A single nerve impulse arriving at the neuromuscular junction will cause a muscle twitch, a swift contraction followed immediately by relaxation. The smooth, sustained muscle contractions that allow you to carry a cup of coffee across the room without mishap result when the brain sends continuous trains of nerve impulses to the neuromuscular junction. Because each single contraction

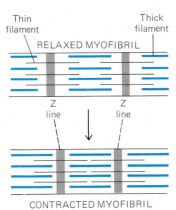

Fig. 23–21

The sliding filament model of muscle contraction. Contraction occurs when the thick and thin filaments slide past each other, reducing the distance between two Z lines. The whole fiber shortens although the filaments themselves do not.

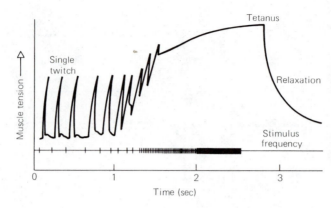

Fig. 23–22

Contraction in a skeletal muscle. The muscle is stimulated by faster and faster stimulation of the nerve supplying it (stimulus frequency). On the left, stimuli are far enough apart so that the muscle contracts and relaxes in response to each stimulus (single twitches). As the stimuli come faster, the muscle does not have time to relax between stimuli and the contractions fuse into a powerful tetanus. After the stimuli cease, the muscle relaxes.

begins before the previous one has reached the relaxation part of the cycle, the muscle remains contracted (Fig. 23-22).

You need more force to carry a whole tray of dishes across the room than to carry just one cup of coffee. If muscle contraction is an all-or-nothing process, how is the strength of muscle contraction controlled? The answer to this lies in the number of muscle fibers called into play at any one time. When little strength of contraction is required, only a few muscle fibers are stimulated by their associated neurons; when more force is needed, more fibers come into action and add the force of their contractions.

Skeletal muscle can be stimulated so fast and so often that it uses up all its ATP and becomes fatigued, a state in which it cannot contract again until it makes more ATP. This is the familiar "weak-kneed" feeling that follows exercising to exhaustion; some of our muscles feel as if they have quit on us. Fatigue occurs only in skeletal muscle; cardiac and smooth muscle, fortunately, do not fatigue.

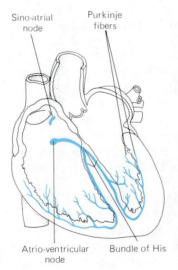

Fig. 23–23

The electrical conduction system of the human heart (color). Electrical activity in the sino-atrial node (the "pacemaker") initiates each heartbeat and spreads throughout the heart via the conduction system so that each part of the heart is stimulated to contract at the appropriate moment.

Cardiac muscle has another peculiar property: it requires no nervous stimulation, but generates its own electrical activity. An isolated heart, or even an isolated cardiac muscle cell, will contract rhythmically all by itself. In the normal heart, the heart rate is set by a pacemaker (Fig. 23-23), the sino-atrial node, which contracts faster than any other part of the heart. The electrical activity started in the pacemaker spreads throughout the heart by way of specialized cardiac muscle cells in such a way that the various parts of the heart contract smoothly, and in the right order. Without a pacemaker in control, the millions of cells in the heart beat to their own individual rhythms, and erratic, ineffective contractions of the whole heart result. In a normal animal, both nerves and hormones affect the heartbeat. However, nerves to the heart are not vital, as has been shown by experiments in which all the nerves to a dog's heart are cut and the dog survives quite normally, able to exercise and otherwise stress the heart.

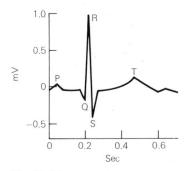

Fig. 23-24

An electrocardiogram (EKG). The waves of electrical activity that spread through the heart and control its beating can be detected by electrodes attached to the skin. The resulting EKG can show whether or not the heart's electrical activity is normal. (P represents electrical activity in the atria, Q, R, S and T, activity in the ventricles).

23-D Glands as Effectors

The secretions of salivary glands, tear glands, mucous glands, and many others travel just a short distance from the gland to their sites of action. By contrast, **hormones**, secreted by ductless **endocrine glands**, travel through the bloodstream and may influence parts of the body far removed from the endocrine gland itself.

Hormones complement the action of the nervous system in controlling processes in the body. On the whole, nerve impulses can initiate more rapid responses than hormones, but a hormone remains longer and affects more organs.

Hormones permit the body to react to changes in the internal and external environments. For an example of a response to an internal change, let us consider some of the hormones (Table 23-1) that help to keep the composition of the body fluids constant. For such homeostatic mechanisms to work properly, the

TABLE 23-1 **SOME VERTEBRATE HORMONES, THEIR SOURCES AND ACTIONS**

Hormone	Where produced*	Stimulates:
Thyroxin	Thyroid	Growth and metabolism
Calcitonin	Thyroid	Decrease of blood calcium by causing its deposition in bone
Parathyroid hormone	Parathyroid	Increase in blood calcium by causing its release from bone
Insulin	Pancreas	Decrease in blood sugar by causing its uptake by cells
Glucagon	Pancreas	Increase in blood sugar by release from liver
Adrenalin	Adrenal medulla	Dilation of some blood vessels; increase in blood pressure; increase in blood sugar

*Positions of some of the glands that produce hormones are shown in Figure 23-25.

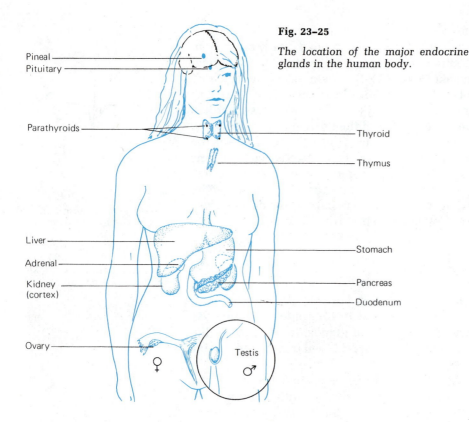

Fig. 23–25

The location of the major endocrine glands in the human body.

Pineal

Pituitary

Parathyroids

Thyroid

Thymus

Liver

Stomach

Adrenal

Pancreas

Kidney
(cortex)

Duodenum

Ovary

Testis

♀ ♂

Fig. 23–26

Feedback control of the level of calcium in the blood.

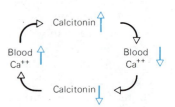

hormones must be secreted into the bloodstream only when they are needed. Secretion of these hormones is under **negative feedback control**, the type of control in which a process automatically limits itself.

In a familiar negative feedback system, a thermostat controls the heating (and cooling) of our homes. The thermostat responds to a drop in temperature by turning on the furnace. When the thermostat detects that the temperature has risen to the set level, it turns the furnace off. Similarly, a rise in the level of calcium in the blood causes secretion of the hormone **calcitonin** by the **thyroid gland**. Calcitonin causes cells in the bones to absorb calcium from the blood, and the blood calcium level decreases; this decrease, in turn, shuts off secretion of calcitonin (Fig. 23-26). This negative feedback loop is one of many control systems by which hormones contribute to homeostasis.

Hormones may also be secreted in response to a change in the external environment. For instance, many animals camouflage themselves by changing color depending on the color of the background. The color change is caused by changes in the size of **chromatophores**, pigment cells of various colors in the skin; this changes the animal's overall color and pattern. In some fish, amphibians, and some reptiles, the pattern of light reaching the retina of the eye controls the release of melanocyte-stimulating hormone (MSH) by the pituitary gland; this hormone in turn changes the size of the chromatophores (Fig. 23-27).

CAL-sih-TONE-in

crow-MAT-oh-forz

mell-AN-oh-site

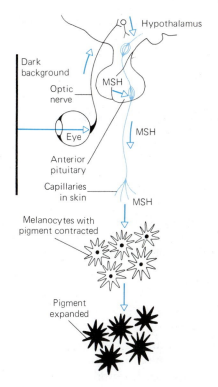

Fig. 23–27

How a frog changes color. The eye detects a dark background. The brain responds to this by causing the pituitary gland to release MSH, which stimulates pigment in cells in the skin (melanocytes) to spread out, darkening the skin. Colored arrows show the path of information flow.

Fig. 23–28

Two photographs of the same chameleon, a reptile famed for its ability to change color. Color change in this animal is controlled by the sympathetic nervous system, not by hormones. The animal does not match its background in either picture; the photographer moved it in each case because it was almost invisible against the background to which it had adapted. (Biophoto Associates)

How are the sense organs that detect changes in the external environment linked to hormone-secreting glands such as the pituitary? As we might expect, the nervous system acts as the go-between, gathering information from the sense organs and sending it to a vitally important area of the brain, the **hypothalamus** (Fig. 23-13). Groups of cells in the hypothalamus are responsible for sensations such as pleasure and pain, and for appetites such as hunger, thirst, and the sex urge. The hypothalamus communicates with other parts of the nervous system and also with the **pituitary**, an important endocrine gland, which secretes many hormones. Each pituitary hormone causes a response of one or more distant organs (Table 23-2).

HI-poe-THAL-uh-mus

TABLE 23-2   **SOME HUMAN HORMONES RELEASED FROM THE PITUITARY GLAND AND THEIR ACTIONS**

Pituitary hormone	Stimulates:
Oxytocin	Uterine contractions and milk production
Vasopressin (= antidiuretic hormone)	Decreased water loss from body in urine
Adrenocorticotropic hormone	Secretion of corticosteroid hormones needed in metabolism by adrenal glands
Thyrotropin	Secretion of hormones for growth, metabolism, blood calcium control by thyroid gland
Follicle-stimulating hormone (FSH)	Production of eggs or sperm
Luteinizing hormone (LH)	Secretion of sex hormones by ovaries or testes; ovulation in females
Prolactin	Mammary gland growth and milk production
Growth hormone	Body growth; increased blood sugar

Reproduction is one area in which the interaction of the nervous system and hormones can be seen most clearly. Consider a toad in early spring (Fig. 23-29). The toad's eyes detect light and pass this information to the brain, which determines that the days are growing longer. The hypothalamus sends the appropriate releasing factors to the pituitary. Here various hormones, including FSH (follicle-stimulating hormone) and LH (luteinizing hormone) are released into the blood. When the testes or ovaries detect these hormones in the blood they respond by growing, producing gametes, and secreting their own hormones—including the sex hormones, testosterone and estrogen. The brain detects the sex hormones and, as a result, sends nerve impulses to the muscles with the result that the animal hops around looking for a breeding site and a mate. So, by the intricate interaction of sense organs, nerves, brain, muscles, and endocrine glands, the animal responds appropriately to the season.

LOO-tee-in-ize-ing

gametes: eggs or sperm.
tess-TOSS-ter-own
ESS-troe-jen

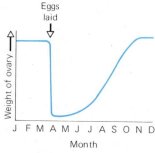

Fig. 23–29

Seasonal changes in the weight of the ovaries of a toad. Ovary growth and egg laying are under hormonal control.

SUMMARY

In order to produce adaptive responses, an animal must be able to detect changes inside its body and in the world around it. Sense organs are collections of receptor cells specialized to react to particular forms of energy by producing electrical impulses in the nervous system.

Information passes along a neuron in the form of an electrical impulse running along the neuron cell membrane. When it reaches a synapse, a nerve impulse is usually passed on in the form of a chemical transmitter that crosses the synapse to the membrane of the next cell, disturbing its electrical balance in turn. The information carried by a neuron is coded largely in terms of the frequency and duration of a train of nerve impulses and the position of the neuron in the wiring of the nervous system.

Most of the movement of an animal's body is due to the action of its muscles. A muscle works by contracting; in the presence of ATP and calcium, filaments of the protein myosin form cross-bridges to filaments of another protein, actin; the cross-bridges then swivel, moving the filaments past each other, thus reducing the length of the muscle. Skeletal muscle fibers contract in response to nerve impulses arriving at the neuromuscular junction. Cardiac muscle cells can generate their own electrical activity and contract without external stimulation, but the rate of the heartbeat is normally speeded or slowed by nerves of the autonomic system. Contraction within the heart is normally coordinated by electrical impulses originating in the specialized cells of the pacemaker.

The interaction between the nervous and hormonal systems depends mainly on the hypothalamus. The hypothalamus receives nerve signals from the sense organs via the rest of the brain, and initiates appropriate responses by way of the nervous system or the pituitary gland, which releases hormones responsible for the maintenance and activity of many of the body's other endocrine glands.

Hormones are involved both in homeostatic mechanisms within the body and in many of an animal's responses to its environment.

SELF-QUIZ

Match the type of receptor on the right with the sense organs listed on the left:

____1. nose
____2. eye
____3. ear
____4. fingertips

    a. chemoreceptors
    b. photoreceptors
    c. electroreceptors
    d. mechanoreceptors
    e. thermoreceptors

5. Draw a simple diagram of the eye, and label the following parts: cornea, pupil, iris, lens, retina.

6. Name the two types of photoreceptors in the normal human eye.

7. Photoreceptors convert light energy into electrical energy by way of:
    a. lenses
    b. enzymes
    c. pigments
    d. photocells
    e. neurons

8. Impulses leave a neuron via the:
    a. dendrites
    b. axon
    c. cell body
    d. nucleus
    e. receptor molecules

9. A neuron codes information about the intensity of a stimulus that it has received by:
    a. changing the size of its impulses
    b. changing the speed at which its impulses travel along its membrane
    c. releasing different types of transmitters
    d. releasing different amounts of transmitters per impulse
    e. releasing more transmitter as a result of an increased number or frequency of nerve impulses

10. Number the following events of information transmission at a synapse in chronological order:
____a. transmitter binds to receptor molecules
____b. enzyme breaks down transmitter
____c. membrane becomes leaky to charged particles (ions)
____d. nerve impulse arrives at synapse
____e. transmitter crosses synaptic gap
____f. transmitter released from vesicles

11. Which of the following is *not* a change seen in the "fight or flight" response?
    a. increased breathing
    b. increased blood flow to brain
    c. increased heartbeat rate
    d. increased blood pressure
    e. decreased blood flow to stomach

12. During muscle contraction:
    a. adjacent **Z** lines move closer together
    b. actin and myosin move past each other
    c. ATP is used
    d. calcium must be present
    e. all of the above
    f. none of the above

13. Number the following events in muscle contraction in chronological order:
    ___ nerve impulse arrives at neuromuscular junction

___ calcium binds with thin filaments
___ myosin filament detaches from thin filaments
___ acetylcholine binds with membrane
___ calcium is released from reservoirs
___ myosin-ATP complex binds to thin filament and swivels

14. Information from internal or external receptors may influence the endocrine glands by passing through the part of the brain known as the _____. This area connects intimately with the _____ gland, which stimulates other glands to secrete hormones by sending out _____ through the _____.

QUESTIONS FOR DISCUSSION

1. During the course of evolution, the cell bodies of the nervous systems of animals have become increasingly concentrated in the front end of the body (the brain). What are some of the advantages and disadvantages of this trend?

2. In what way can muscle activity (cardiac, smooth, and skeletal) be said to contribute to homeostasis?

3. If you are a mystery novel fan, you know that the time it takes for *rigor mortis* to set in after death varies. What are some of the reasons for this variation?

4. Muscles attach to bones via tendons. What are some of the advantages of this arrangement instead of having muscles attach directly to bone?

BEHAVIOR  24

When you have studied this chapter, you should be able to:

1. Explain the difference between innate and learned behavior, and give one example of each.

2. List the selective advantages of innate, learned, and stereotyped behaviors.

3. Give an example of a stimulus and the behavior pattern it evokes; explain why the same stimulus does not always evoke the same behavior in an animal.

4. Describe the characteristics of territorial behavior and give an example.

5. Describe the functions of threat and of appeasement behavior in the maintenance of a dominance hierarchy, and recognize examples of these behavior patterns.

6. Describe conflict behavior and explain why it is thought that threat displays and courtship evolved as conflict behavior.

Biophoto Associates

Fig. 24-1

Vertical bared-teeth display by a chimpanzee. How do we decide what this behavior means? Is the chimp yawning? Is it threatening an enemy? Repeated observation has convinced researchers that this display serves to reassure a smaller or weaker chimpanzee that this chimpanzee will not attack it. (Biophoto Associates)

Fig. 24-2

A frog's huge wraparound eyes and quick reflexes help it to catch a fly in flight, something that we can seldom do. (Biophoto Associates)

AN-throw-puh-MORE-fism

We are inclined to conclude that a dog is "ashamed" when it slinks to the corner with its tail between its legs after a spanking, and "happy" when it wags its tail. This type of description is called **anthropomorphism**—ascribing human emotions to other animals. Another human prejudice about animal behavior shows when we say that a bird sings from instinct, assuming that it is not capable of intelligent behavior. Our view of human behavior as intelligent and that of other animals as instinctive is reflected in our tendency to ascribe actions of which we are ashamed to our "animal instincts." Both of these approaches to animal behavior say more about human prejudices than about why animals behave as they do. Recent research has attempted to study behavior with as little prejudice as is humanly possible. The remarkably complicated behavior of even the simplest animal makes this no easy task.

Behavior is directly and indirectly the result of an animal's genes. If an animal doesn't have wings, it cannot fly, and whether it has wings or not depends largely on its genes. An animal's genes determine the *range* of physical characteristics it can develop, but just which traits develop depends on interactions between genes and environment during the animal's development. The same is true of behavior. However, the development of behavior is more complicated than the development of a physical characteristic such as blood circulation. Physical characteristics are largely complete by the time an animal stops growing, whereas genes and environment may continue to interact and alter an animal's behavior throughout its life.

In this chapter we shall consider some of the ways in which scientists try to disentangle the genetic and environmental influences on an animal's behavior, and the kinds of selective pressures that have produced the varied behavioral repertoires of different animals.

24–A Causes

A frog is sitting in the grass when a fly buzzes past. The frog's tongue flicks out and pulls the fly into the frog's mouth. How and why does the frog behave like this? The question "how" can be answered by a description of the way the frog's eyes, nervous system, and muscles function. A moving fly causes the frog's eye to send nerve signals to the brain, which sends out nerve impulses that direct the tongue muscles used to catch the fly.

The question of "why" the frog catches the fly is different because it can be answered on two levels. The immediate reason "why" the frog catches the fly is that seeing the fly activates a nerve/muscle reflex (see Figure 23-1) that results in the frog's striking at the fly. But there is also an evolutionary answer to the question "why." The frog's fly-catching behavior exists because it has been selected for during the course of evolution.

Three main evolutionary causes, or selective pressures, have brought about the behavior patterns we see today in all animals:

1.   Ultimately, an animal's behavior patterns will be selected for as they contribute to its evolutionary success. This is most obvious in reproductive behavior because an animal that does not reproduce is doomed to evolutionary failure, but it is also true of all other behavior, from feeding to scratching fleas.

2.   On a more short-term basis, behavior patterns must allow an animal to react to and solve immediate problems. A hungry animal must feed, and a hunted animal must escape from predators, if either is to survive and reproduce.

3.   Sights, sounds, and other stimuli continually bombard all animals. Successful behavior patterns permit an animal to react to stimuli that affect survival and reproduction by completing appropriate behaviors. Mechanisms for discriminating between stimuli and for ensuring that an animal completes a behavior pattern are crucial parts of any animal's behavioral makeup.

Fig. 24–3

Reproductive behavior. A blue tit feeding her young. (Biophoto Associates)

Fig. 24–4

Social animals, like these lions in Botswana, learn much of their behavior from other members of the group. (Biophoto Associates, N.H.P.A.)

24–B Instinct and Learning

Instinctive or **innate** behavior is behavior that is genetically programmed into the nervous system and that is next-to-impossible to alter. **Learned** behavior is behavior that is acquired or lost as a result of experience.

The idea of instinctive behavior has caused bitter controversy among biologists. This is partly because it is so hard to test whether a given behavior pattern was instinctive or learned. Instinctive behavior develops without the animal's having to learn it, so the only way you can test for it is to deprive the developing animal of as many environmental stimuli as possible and see if the behavior still appears. Even if the pattern does appear, it may still not be instinctive; the experimenter might merely have failed to remove the stimuli needed to learn the behavior. In practice, it is usually impossible to show whether a particular behavior is instinctive, and there are more illuminating methods of analyzing behavior. However, the idea that a behavior pattern is either instinctive or learned or, more often, a combination of the two is borne out by research into behavior.

It is generally advantageous to an animal to have both innate and learned behavior patterns. Learned behavior has the advantage that it can change with time as the animal's environment changes; the advantage of innate behavior, such as jerking a hand away from a hot stove, is that it occurs rapidly and without error every time it is performed. Innate behavior also does away with the need to spend time and energy to develop a learned behavior pattern (such as skiing) and reduces the chance of fatal mistakes which might occur, say, in learning to move away from, and not toward, a fire.

A species' way of life will often let you predict what mix of learned and innate behaviors its members will develop. Consider a solitary wasp, which hatches from the egg alone and grows up with practically no interaction with members of her species. The behavior by which she finds a male wasp, mates, builds a nest, and lays her eggs must be largely innate in order for her to perform each action perfectly the first, and perhaps the only, time in her life. On the other hand, a social animal, such as a cat, can and does learn much of its behavior from observing other members of its group and practicing (playing) in the security of the family. It would, however, be an enormous oversimplification (and is a common mistake) to say that most of the behavior of the wasp is innate and most of the cat's behavior is learned. In all but the simplest animals both types of behavior are vitally important. Even the solitary wasp learns to search for food, to find her way back to her nest, and many other behavior patterns during her short life. Similarly, all mammals have many innate behavior patterns.

In general, it seems that learning plays a large part in the development of behavior patterns evoked by stimuli which are local or changeable. Every animal with an individual home, for instance, must learn to find that home. Innate behavior, on the other hand, is appropriate where the stimulus that provokes the behavior is always the same, where speed of reaction is important, and where the cost of an initial mistake is high.

24-C Stereotyped Behaviors

A striking feature of the behavior of any animal is its repertoire of **stereotyped behaviors**—acts that are always performed in an essentially identical manner. Reflexes (Fig. 23-1) are the most familiar stereotyped actions. A **reflex** is an action whose performance requires few nerves and muscles and no action by the higher ("conscious") centers of the brain. Stereotyped behaviors which involve more numerous nerve/muscle interactions are just as common as reflexes. For instance, a cockroach or cricket will leap forward in a standard escape reaction when receptors on its abdomen are stimulated by a puff of air. This is an example of an innate stereotyped behavior; others are learned. When rats learn to press a lever for food, each presses the lever with a characteristic gesture that is the same every time. One uses a fist, another its middle finger. Similarly, how you hold your pen, walk, or ride a bicycle is a stereotyped behavior that you have learned; it is very conservative and unique to you.

A stereotyped behavior often goes to completion even if some parts of the behavior are inappropriate at the time. A dog hiding a bone under the living room rug goes through the motions of covering it up with nonexistent earth. Similarly, most dogs turn around several times before they lie down on the floor, although there is no grass to trample down.

Animals with reasonably simple nervous and muscular systems have been studied intensively in attempts to discover the basis of stereotyped behavior. Among the animals studied have been sea slugs, leeches, crayfish, and lobsters (and some very fine dinners were based on the remains of those experiments!). These studies suggest that stereotyped behavior differs from other behaviors in two main ways: a stereotyped behavior is controlled by one or a few "trigger"

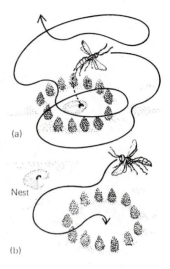

(a)

Nest

(b)

Fig. 24–5

A digger wasp finds her nest by learning to recognize visual landmarks. (a) She makes an orientation flight over the nest entrance that an investigator has ringed with pine cones. (b) The pine cones have been moved in the wasp's absence and she returns to the center of the ring of cones where she "expects" to find her nest.

abdomen: the rear part of an insect's body, behind the thorax, which is the part where the legs attach.

Fig. 24–6

Stereotyped defensive behavior. The sea anemone on the right has drawn its tentacles into the body in response to a poke from the photographer. (Biophoto Associates)

Fig. 24–7

A female three-spined stickleback. (Biophoto Associates)

cells in the nervous system, and it can occur without feedback from the sense organs. When you play a piece that you know well on the piano, for instance, you don't wait to hear what each note sounds like before hitting the next key; however, when you are learning the piece—before the behavior has become stereotyped—you do listen to the notes.

Stereotyped behavior, whether innate or learned, permits an action which is performed often to be completed perfectly without wasting energy to work the action out from scratch each time. Innate behavior has the same advantage, but learned stereotyped behavior has an additional virtue—it can be programmed into the nervous system at any time during life, and not just during development.

24–D Stimuli

It is obvious that we do not respond to every one of the sights, sounds, and smells that continually bombard us. The nervous system and the sense organs filter stimuli, in ways that we do not fully understand, so that only certain stimuli evoke responses.

In the 1940s, Niko Tinbergen, who shared a Nobel Prize for his work on animal behavior, studied a small fish, the three-spined stickleback. The male develops a bright red belly when it is in breeding condition. Tinbergen noted that when a red truck drove past a nearby window, all the male sticklebacks tried frantically to swim through the glass of their tanks, as if they would attack the truck. Because a male stickleback in breeding condition will attack other males, Tinbergen wondered if the red color was the stimulus that provoked attack. He showed various models to the sticklebacks and found that they attacked a crude

Fig. 24–8

Models used to determine the stimulus for attack by a male stickleback. Tinbergen found that males in breeding condition ignore the lifelike stickleback but attack either of the crude models with red undersurfaces. (The eye is also necessary to provoke attack. The presence of an eye is often necessary if an animal is to identify an object as another animal.)

model of a fish with a red belly but not a lifelike model without a red belly (Fig. 24-8). The color red is the stimulus for attack.

This experiment suggests another aspect of stimuli that provoke behaviors: they work only when the animal is in a particular state. In this case the males have to be in reproductive condition (a particular hormonal state) before they will react to red. Similarly, food will normally stimulate feeding in a hungry mouse. If the mouse sees a cat about to pounce at the same time that it sees the food, however, the food stimulus will not provoke feeding; instead, the mouse will avoid the cat. If an animal is really starving, however, it may feed while ignoring stimuli that would normally send it scuttling for safety.

Behaviorists sometimes express the fact that an animal will respond to a stimulus at one time but not at another by saying that an animal's **drive** or **motivation** changes. A hungry animal has a strong drive to feed, but a well-fed animal does not. Since different behaviors are appropriate at different times even when the stimulus is the same, variations in motivation help to ensure that an animal's behavior changes to fulfill its short-term needs.

24–E Learning

Learning is the process that produces adaptive changes in an individual's behavior as a result of its experiences. It occurs in many different ways and there is no evidence that those discussed below bear any relationship to the physiological basis of learning, which is poorly understood.

Habituation is the loss of old responses. Animals learn not to respond to stimuli which occur frequently but are not important to them. Young animals often show alarm behavior at a variety of stimuli, most of which they rapidly learn to ignore. Habituation is advantageous in that it increases an animal's reaction to new stimuli, which stand out against the background of stimuli to which the animal has become habituated. Not reacting to unimportant stimuli also saves energy which would otherwise be wasted on useless behavior.

Conditioned reflexes are behavior patterns evoked by stimuli which an animal has learned to associate with the stimulus that normally elicits the reflex. The Russian physiologist Ivan Pavlov showed that there is a reflex which causes hungry dogs to salivate when they see food. Pavlov rang a bell when he fed his dogs, and after several trials the dogs would salivate when the bell was rung, even when they were not shown food. The dogs had learned to respond to a new stimulus, the bell, to which they had not previously responded. A conditioned response to a negative stimulus (punishment) can be formed in the same way.

Fig. 24–9

Habituation. This African black-footed wild cat kitten reacts with alarm (bared teeth, flattened ears) to its first sight of a human being. When the photographer reappeared every day, the kittens became habituated and stopped producing the alarm reaction. (Biophoto Associates, N.H.P.A.)

Fig. 24–10

Learning. A blue tit spikes the foil top of a milk bottle and drinks the milk. The first tit to perform this behavior presumably learned it by trial and error. Birds now learn this behavior by observing each other, and the habit of opening milk-bottles has spread rapidly through the population. (Biophoto Associates)

Fig. 24–11

Kohler demonstrated insight learning in chimpanzees. Presented with a number of boxes and a bunch of bananas too high to reach, the chimpanzees piled boxes into a stand from which they could reach the bananas.

Trial-and-error learning is what its name implies. An animal's spontaneous movements may produce, by chance, some sort of "reward"; the animal then learns, by trial and error, to repeat that behavior pattern. The reward may merely be the pleasure of performing an action more accurately. Young mammals and birds perfect their prey-catching skills, and people learn to play the piano or ride bicycles, by a trial-and-error form of practice.

All these types of learning are **associative learning**, in which reinforcement (reward or punishment) is important. Another characteristic of associative learning is that it improves with practice.

Latent learning occurs without any obvious reward or punishment. In fact, there is not even any obvious behavior at the time it occurs. For instance, a recently fed animal may give no sign that it has just noticed a new source of food until later, when it is hungry and returns to feed there.

We often think of **insight learning** as the highest form of learning. This is usually what we mean by the ill-defined term "intelligence." Insight learning is drawing on past experience to solve a new problem. Reasoning of this sort has been shown in many mammals and in some birds, although it is often difficult to distinguish from other forms of learning.

Imprinting describes a situation in which an animal learns something, which it is predisposed to learn, only at a particular stage in development. Goslings (young geese) and ducklings learn to follow their parents, and to respond to their parents' signals, during a critical period after they hatch. Konrad Lorenz, another Nobel Prize–winning behaviorist, found that young birds would follow him as if he were their mother if they saw him, rather than the mother, during the critical period when they were susceptible to this imprinting.

Many animals learn what their future mates will look like by sexual imprinting. Lorenz had a tame jackdaw (a bird similar to a crow) which unfortunately became sexually imprinted on him before he understood how the process worked. It caused Lorenz some inconvenience by stuffing regurgitated worms into his ear during its "courtship feeding."

It is characteristic of the development of behavior that animals learn some things at certain times more readily than at others. There is good evidence, for instance, that humans learn languages much more rapidly up to the age of about ten than they do later. Similarly, many birds must hear the song of their own species at certain critical stages of development if they are to sing the song correctly when they become adults.

In addition, animals are genetically predisposed to learn the behavior patterns of their own species. If a juvenile white-crowned sparrow hears only the song of a song sparrow, it will not sing the song of a white-crowned sparrow when it becomes an adult, because it has never heard that song. On the other hand, it will not sing like a song sparrow: it has no genetic predisposition to learn the song sparrow's song (Fig. 24-13).

24–F Territorial Behavior and Courtship

In Chapter 8, we considered the contribution of territory and courtship to reproductive success and the evolution of particular mating systems. Here we use them to illustrate that behavior patterns which evolve in one context may recombine in different ways that give them new meanings during the course of evolution.

Many animals defend territories—areas where they raise their young or where they have a monopoly on the food resources. A territory-holder attacks and drives away other members of the same species. If it does not, the invader can take over the territory and its resources. Aggression in territorial behavior is

Fig. 24–12 (below left)

A jackdaw. (Biophoto Associates, N.H.P.A.)

Figure 24–13 (below)

A white crowned sparrow. (Cornell Laboratory of Ornithology)

precisely controlled. It is to an animal's advantage to defend a territory with as little attack behavior as possible, since every attack carries with it the risk that the attacker will be injured. Various mechanisms that minimize physical injury during territorial encounters have evolved. For instance, actual fighting is rare because there are "rules" about who wins encounters between two individuals.

Consider a male thrush defending a territory before the females arrive in the spring. The male is most aggressive near the center of his territory. As he moves toward the edge, his attacks on a trespassing neighbor become less violent until he reaches a point at which he is as likely to flee as to attack when he sees another male thrush. This point marks the boundary of his territory.

When two neighbors meet at the boundary of their territories, they act as if they have conflicting drives to escape and to attack. These may show themselves as **conflict behavior**, in which tendencies to attack and to escape are visible, or as **displacement** behavior, which always looks rather peculiar because it seems completely unrelated to the situation at hand. For instance, a gull involved in a territorial clash may peck grass violently out of the ground—a completely unproductive behavior. We are all familiar with human displacement behavior. Faced with a difficult examination or an awkward social situation we chew our fingernails or pencils, finger a lock of hair or consume food and drink we don't want—all inappropriate, displacement activities.

Many species have evolved conflict behaviors which are ritualized **threat displays** directed against intruders. Threat is obviously an improvement over actual fighting in that it injures neither party. An experienced observer can usually tell which of two antagonists will win a threat display (which is really a ritualized fight) between two animals by deciding which animal incorporates more attack movements in its display. The loser will eventually move away from the winner. Threat displays sometimes work between species. A growling dog may threaten a human being on its territory. If the human visitor withdraws, the

Fig. 24–14

A threat display by an eagle owl. (Biophoto Associates, N.H.P.A.)

Fig. 24–15

Male red deer fighting. This is not ritualized but a real battle, possibly to the death. The males are disputing leadership of a harem of females. The winner may leave dozens of offspring, but the loser will probably not reproduce at all. (Biophoto Associates, N.H.P.A.)

Fig. 24–16

Many social animals, like these monkeys in India, have dominance hierarchies such that dominant individuals have first choice of mates, food, etc. An individual's position in the hierarchy may change many times in a year. (Biophoto Associates)

Fig. 24–18

Courtship: a male eider duck displaying to the female. (Biophoto Associates, N.H.P.A.)

Fig. 24–17

Individual distance. These cormorants and gulls maintain a minimum distance from each other as they perch on a crag.

dog's threat may become intensified, whereas most dogs stop threatening if one approaches and ignores them.

Threat displays occur in many situations other than territorial disputes. In some social animals, for instance, there is a **dominance hierarchy** or "pecking order," which ensures that dominant individuals have first choice of limited commodities such as food, shelter, and mates. Dominance hierarchies are maintained by threats from the dominant individual and **appeasement**, or peacemaking, gestures by the subordinate individual. Appeasement behavior inhibits a more dominant animal from attacking. Human behaviors such as smiling and shaking hands sometimes act as appeasement gestures and inhibit aggression by the person at whom they are directed.

Courtship behavior permits an animal to identify a member of the opposite sex (and of the same species) with certainty so that sperm and egg are not wasted. Courtship behavior usually shows conflict between tendencies to approach and to retreat, and this appears to be how courtship has evolved. Most animals, even those that live in social groups, maintain a minimum individual distance from each other; swallows sitting on a telephone wire or gulls on a roof are always a certain minimum distance apart. Many of us have been made uncomfortable by another person who comes too close for comfort during conversation, invading the individual distance that we usually maintain. Generally, the invasion of individual distance is a threat calling for attack or retreat from the invading animal. For animals to approach close enough to mate, however, individual distance must be invaded, and so it is not surprising that conflicting tendencies to approach, to attack, and to flee are parts of most courtship rituals.

24–G Migration and Homing

Many animals migrate over hundreds of miles of land and sea. This behavior is peculiarly fascinating because migrating animals can do many things that we cannot do ourselves. A Manx shearwater, which had never been more than 10 miles from home, was removed from her nest on an island off the coast of Wales, flown to Boston, and released. She was back on her nest before the letter announcing her release reached observers in Wales. To perform an equivalent feat, such as sailing from Boston to Wales, a person would have to spend hours learning to use a compass (to determine direction) and a sextant (which measures the altitude of the sun or stars and can be used to calculate one's position on the earth's surface if one knows the time accurately). After crossing the ocean, our sailor would need a map to find the nest on the other side. Birds, monarch butterflies, fish, and salamanders all perform equivalent journeys with no mechanical aids and with little or no learning. Some means of animal navigation are now understood, but others remain a complete mystery.

Many animals, when in familiar country, orient themselves by landmarks which they learn and recognize by sight or by smell. A dramatic case is that of salmon, which hatch in freshwater streams and mature hundreds of miles away in the ocean. Seven years later, when the time comes to spawn, each salmon finds its way back to the very stream in which it hatched. We now know that a salmon can perform this feat only if its chemical receptors are undamaged; it smells out its home stream. Even so, it is hard to imagine that each tiny stream gives off a chemical "fingerprint" sufficiently strong and unique to guide the fish into the mouth of the main river and then past every misleading fork where a tributary joins the river.

Fig. 24–19

Manx shearwaters. (Biophoto Associates, N.H.P.A.)

Fig. 24–21

The homing pigeon that won the 1979 France to England race, with its owner. (Biophoto Associates)

Fig. 24–20

Several species of terns migrating together off the coast of South Africa. (Biophoto Associates, N.H.P.A.)

Many animals can move in a specific compass direction. We can do this ourselves if we remember that the sun rises in the east, sets in the west, and is due south of us at midday. This sort of compass navigation requires the ability to see the sun, and a "clock" to tell the time of day. All animals and plants have the physiological basis for an internal clock in the daily rhythm of their chemical reactions, although human beings are not very good at sensing and interpreting this information. Many other animals accurately sense their internal clocks and can tell where the sun is even on overcast days by detecting the pattern of polarized light (which we can detect only with polarizing lenses) in the sky. Similarly, humans and other animals can find compass directions very accurately at night by the positions of the stars.

The ability to fly on a constant compass course using the stars or the sun, however, does not explain the ability of a Manx shearwater to cross the Atlantic or of a homing pigeon to return to its loft when released in unknown country. To get from A to B using a compass, you have to know whether B is north, south, east, or west of A, and how far away it is. This is called a **map sense**, because it means you must know the relative positions of A and B on a hypothetical map. Many animals obviously have a map sense, but we haven't the slightest idea what it may be. (The human "bump of direction" is not a map sense; it usually means that one has kept subconscious track of turns since leaving a path that pointed in a particular compass direction and therefore knows in which compass direction one is facing.)

When pigeons are prevented from using a sun compass (by fitting them with opaque contact lenses and resetting their internal clocks), they can find their way home using magnetic cues; presumably they can detect the earth's magnetic field. Pigeons do not need their internal clocks for magnetic navigation, but this navigation can be upset by attaching small magnets to the pigeons. Several kinds of insects can also detect the earth's magnetic field, although we have no idea how they do it. The earth's magnetic field can give a crude compass direction, as it does with our own magnetic compasses. We do not know if it can supply an animal with the mysterious map sense.

Another interesting set of animal navigational aids are various types of "echo-sounders," in which animals detect sound or electric currents bouncing off nearby objects and thereby determine where the objects are and, sometimes, what they are made of. Cetaceans (whales and porpoises) and bats have the best-known sonar (echo-sounding) systems. Members of both groups make sounds and then detect their echoes. The best-known electric "echo-sounders" are those of certain fish in muddy tropical rivers, where the visibility is very poor. These fish use their electric organs as their main system for finding food and avoiding obstacles. They create electric fields around themselves and then detect any distortions in the field caused by an object in the water.

Although much has been learned about how animals find their way around, we are still far from understanding this remarkable collection of behavior patterns.

24–H Social Behavior (Sociobiology)

Some animals have very little contact with members of their own kind, but in many species, some degree of cooperation with others is apparent. This may be of limited duration, as in flocks of chickadees and red-wing blackbirds that forage together in winter. If "social" is defined as involving cooperation between

Fig. 24–22

A bat in flight avoids obstacles by a remarkably efficient echo-location sense. (Biophoto Associates)

Fig. 24–23

A gray heron at its nest. (Biophoto Associates)

members of the same species, it is often difficult to tell whether or not such flocks are really social. Studies of gray herons, for instance, have shown that the birds are found in groups because they all fly to where food is available but there is little interaction between individuals (Fig. 24-23). At the other extreme are animals such as honeybees, humans, and wolves, which form cooperative, long-lived societies upon which the individual's very life depends.

To live in a society, an animal must be able to communicate with other members of its species; those animals with more complex societies tend to have more elaborate methods for communicating. Every communication involves action by a communicator and reception by another individual. Human beings, for instance, use sound and hearing (when we speak, clap, or laugh), visual stimuli, and vision (consider advertising posters, dressing up, shaking a fist) among our means of communication. Birds, like humans, have highly developed vision. It is not surprising that they communicate largely by movement and color (also by sound and hearing). Our sense of smell is poor, so we pay little attention to the chemical communication that is so common in other animals. Many mammals, like dogs, mark their territories, determine another animal's mood, find their mates and food and, for all we know, communicate in many other ways, by scent. (We do not even have a common word, equivalent to "blind" or "deaf," for lacking the sense of smell.) A **pheromone** is a chemical whose function is communication between members of the same species. (We have to judge its function by noting whether or not the chemical influences the behavior of another individual since we can never discover the "intent" of an animal of another species.) Female moths emit pheromones that attract males; ants foraging for food mark their paths with trail pheromones, which other ants follow to find the food source.

FEAR-uh-moan

Communication by sound can be highly elaborate even if it does not involve language, which is the most important aspect of our own sound communication. Crickets, frogs, and mosquitoes produce sounds that have two effects—they tell a listening individual whether or not the sound-maker is a member of the same or a different species; and they permit the sexes to find each other in the mating season.

Alarm calls, pheromones, speech, and courtship displays have presumably evolved primarily for their value in communicating with other individuals. Other signals which animals produce, like the electric currents emitted by electric fish and the sonar signals of cetaceans, are also used for communication; however, they probably originally evolved because they conferred other advantages—in these cases for detecting objects around the animal.

Honeybee societies.   Many insects are more or less social, but honeybee (and ant) societies are the most elaborate and widely studied of insect societies. The unit of social organization is a family of related individuals. A honeybee society typically consists of a reproductive female (the queen) and her daughters (and sometimes her sons). The society controls development of the queen's offspring and so determines the "caste" to which each will belong. A colony of honeybees may contain 80,000 individuals, all organized by pheromones. The queen mates once, stores the sperm, and uses it to fertilize the hundreds of fertilized eggs that she lays during her life of seven years or more. Her eggs hatch into larvae cared for by the workers. When the colony is overcrowded, the queen also lays unfertilized eggs which develop into drones (fertile males), which will be available to furnish sperm for the queen of a new hive. The diet fed to the larvae determines whether a fertilized egg develops into a queen or a worker. A new worker usually first serves as a nurse, preparing cells for the larvae and feeding them. After about two weeks she becomes a house bee, cleaning, secreting wax for the honeycomb, and guarding the hive. After this she forages outside the hive for the remaining five or six weeks of her life.

As we would expect from such a complex society, honeybees communicate extensively. Karl von Frisch, who shared a Nobel Prize with Tinbergen and Lorenz, found that foraging bees returning from a successful trip "dance" on the honeycomb, recruiting other bees to harvest a good food source and telling them where to find it. Pheromones permit bees to identify their own hive and serve as trail markers and alarm signals. A pheromone produced by the queen prevents the workers from producing more queens and ensures that all the female larvae are fed so that they develop as workers. This continues until the hive is overcrowded, when the queen stops producing that particular pheromone, and the workers start to raise new queens. Eventually the queen may leave, secreting a swarming pheromone, which attracts many of the workers and keeps them with her. The swarm lands somewhere and may remain several days while scouts search for a new nest site. The scouts return to "dance" a description of the location of a possible new nest. The intensity of her dance conveys the scout's impression of the merits of the site. Other workers go to inspect the sites, and finally a consensus emerges when all the scouts are dancing for one site. The swarm then flies to the new site. This democratic method of making a decision impresses us by its resemblance to some human actions.

Vertebrate societies.   Most vertebrate societies, like those of bees and ants, consist of genetically related individuals; their organization, however, is differ-

Fig. 24-24

The life and times of honey bees. (a) Workers in a hive clustered around their queen (center). (Carolina Biological Supply Company). (b) A worker "fanning." She stands on her forelegs and uses her wings to disperse pheromones secreted by the scent gland on her abdomen. (Biophoto Associates, N.H.P.A.) (c) A worker foraging for nectar. (Carolina Biological Supply Company) (d) a swarm of honey bees on a tree trunk. (Biophoto Associates, N.H.P.A.)

(a)

(b)

(c)

(d)

ent. Typically, such a society consists of a leader (usually a male), his mates, and their descendants. Members of vertebrate societies can identify each member of the group individually, whereas insects probably cannot.

Individual ability has more effect on an individual's role in a vertebrate than in an insect society. There is usually a dominance hierarchy and individuals may fall in rank as a result of age or disability. In one baboon troop, for instance, the top male changed five times in two years. Position in the hierarchy is not determined solely by an individual's size or fighting ability. In many species (probably most primates), having a mother of high social status gives one an initial boost up the social ladder.

The evolutionary advantage of a social hierarchy is probably that it reduces the deleterious effects of competition between related individuals (such as fighting) and ensures that at times when food (or any other resource) is short, some individuals will get all the food they need to survive instead of the whole group becoming half-starved and likely to die, as happens with honeybees. Where members of a group are related, such apparent altruism will be selected for because individuals carry many of the same genes, and an individual that starves while another lives is actually contributing to the survival of his or her own genes in future generations. Males are less necessary than females to evolutionary success (since females raise the offspring) and low-status males will usually starve or be pushed out of the protection of the society first in times of food shortage.

Animal societies, like anything else biological, have been selected for as they enhance reproductive success. A wolf or an elephant seal, for instance, can raise more offspring as a member of a society than as an individual. Once a simple society existed, there was selection for cooperation in hunting, defense, and rearing the young and for the communication that makes such cooperation possible.

SUMMARY

An animal's genes determine the range of behavior patterns that it can develop. Most behavior patterns, innate or learned, will not develop normally if the animal is not exposed to the appropriate environmental conditions during the right period of its development. Thus, gene and environment must interact for normal behavior patterns to develop.

The immediate reason that an animal behaves in a particular way is that it has been exposed to stimuli which induce the behavior pattern. Evolutionarily, behavior patterns that enhance survival tend to become innate, or written into the nervous system (or the tendency to learn such behavior patterns becomes so written). Learning requires time and energy and is reserved for behavior that must be flexible in meeting local or changing conditions. Many behavior patterns, both innate and learned, become programmed into the nervous system as fixed action patterns, triggered by stimuli and controlled by a small number of nerve cells with little feedback from the sense organs.

Animals are always exposed to a variety of stimuli, which may or may not evoke a response. Action or inaction is determined by factors such as the animal's physiological state and its conscious or unconscious ranking of stimuli. Conflict behavior, frequently visible in courtship and territorial displays, is one possible outcome of motivation toward more than one, mutually exclusive, behavior pattern.

Most animals seldom or never cooperate with other members of their own species, but true societies have evolved in a few cases—particularly among insects and vertebrates. A society consists of genetically related individuals; communication between individuals is most highly developed in social animals. Honeybees communicate mainly by pheromones and "dancing," which permit a bee to identify her hive and her sisters, provide information about food sources, and determine the caste in which a female is raised. A queen lays all the eggs in the hive and her pheromones regulate the behavior of her daughters, the workers.

Vertebrate societies are characterized by hierarchies which determine an individual's access to limited resources. The society provides an individual with protection, and its members cooperate in finding food, raising the young, defending the group and its territory (if it has one), and fighting off predators.

SELF-QUIZ

1. True or False? If a behavior develops only after exposure to a certain stimulus, it is a learned behavior.

2. The advantage of learned behavior is that it:
   a. occurs quickly
   b. is performed the same way every time
   c. adjusts response to prevailing conditions
   d. occurs perfectly the first time it is done
   e. doesn't take up space on the animal's genes

3. Which of the following is *not* a characteristic of territorial behavior?
   a. it tends to confine a territory owner to a particular area
   b. it prevents some individuals from using resources in a particular area
   c. it is usually a response to other individuals of the same species
   d. it usually prevents bloodshed among members of one species
   e. none of the above

4. True or False? The establishment of a dominance hierarchy through threat and appeasement behaviors mainly serves to allocate resources in a social group without bloodshed.

5. Courtship behavior is said to show conflict because:
   a. the two mates fight a lot
   b. the mates cannot always agree on a nest site
   c. the mates are sexually attracted to each other but do not normally permit another animal to approach as close as copulation demands
   d. each animal must choose the mate that represents the best compromise from among all the available members of the opposite sex
   e. the mates are in competition for food in a territory of limited size

QUESTIONS FOR DISCUSSION

1. In what ways are dominance hierarchies and territorial behavior similar? What are the advantages of living in a social group with a dominance hierarchy, compared with those of living in a pair (with offspring) in a territory?

2. How far do you recognize human society in the description in this chapter of vertebrate societies? What signs of a dominance hierarchy do you see in the society in which you live?

SUGGESTED READINGS

Alcock, J. *Animal Behavior*. Stamford, CT: Sinauer Associates, 1979. Excellent behavior text.

Asimov, I. *The Bloodstream: River of Life*. New York: Collier Books, 1961. An entertaining introduction to our vital fluid.

Boston Women's Health Book Collective. *Our Bodies, Our Selves*. N.Y.: Simon and Schuster, 1971. A book about the female reproductive system and sexuality.

Bullock, T. H. *Introduction to Nervous Systems*. San Francisco: W. H. Freeman and Co., 1977. An excellent text on how the nervous system works.

Chapman, C. B., and J. H. Mitchell. "The physiology of exercise." *Scientific American*, May 1965.

Harlan, J. R. "The plants and animals that nourish man." *Scientific American*, September 1976. An interesting discussion of human food plants and animals; where they originated and how they have become tamed or cultivated.

Hasler, A. D., and J. A. Larsen. "The homing salmon." *Scientific American*, August 1955.

Katchadourian, H. A., and D. T. Lunde. *Fundamentals of Human Sexuality*. N.Y.: Holt, Rinehart and Winston, 1975. Covers all aspects of human sexuality in readable fashion.

Lorenz, K. Z. *King Solomon's Ring*. London: Methuen, 1942. A delightfully written autobiographical account of life with animals.

Scrimshaw, N. S., and V. R. Young. "The requirements of human nutrition." *Scientific American*, September 1976. A list of human nutritional requirements and a discussion of factors that affect them; interesting and readable.

Watt, B. K., and A. L. Merrill. *Handbook of the Nutritional Contents of Foods*. New York: Dover Publications, 1975. The United States Department of Agriculture sponsored this collection of data, giving approximate composition and content of vitamins and minerals for raw, processed, and prepared foods.

White, T. H. *The Book of Merlin*. New York: Berkeley Publishing Corp., 1977. A literary figure's view of animal and human societies.

Wilson, D. *Body and Antibody: A Report on the New Immunology*. New York: Alfred A. Knopf, 1972. Extremely readable, including anecdotes, history of immunology, and informal comments by immunologists; written for the lay person.

ANSWERS TO SELF-QUIZZES

Chapter 2:
DIVERSITY OF LIFE

1. a. cilia
2. a. iv. mollusc
   b. ii. cnidarian
   c. iii. echinoderm
   d. i. arthropod
   e. v. segmented worm
3. 1) air-borne pollen
   2) seeds containing multicellular embryo and food supply, with protective covering
4. a. robin— v. bird
   b. shark— i. cartilaginous fish
   c. codfish— ii. bony fish
   d. snake— iv. reptile
   e. toad— iii. amphibian

5.

	Amphibians	*Reptiles*
Body structure:	no claws	claws on toes
	no scales	scales on skin
	legs out to sides	legs under body
	thin bones	thick, heavy bones
Reproduction:	eggs lack shells	eggs have waterproof shells
	eggs laid in water	eggs laid on land

6. 1. food for people and animals
   2. extracts used in processed foods (ice cream, pudding)
   3. extract (agar) used as lab growth medium
7. b. marshmallows

Chapter 3:
NATURAL SELECTION AND EVOLUTION

1. c. (only populations, or their characteristics, can evolve; individuals can't)
2. d.
3. e. (only reproduction counts)
4. b.
5. d.

Chapter 4:
THE DISTRIBUTION OF ORGANISMS

1. a.
2. b.
3. (1) short fugitive plants
   (2) taller, perennial plants
   (3) shrubs and pioneer trees
   (4) climax trees
4. they are separated by great distances; they have similar climates, which exert similar selective pressures (in this case, nothing much taller than grasses can grow with so little rainfall).
5. c. cacti and euphorbs are the only examples given of similar organisms in similar but widely separated habitats

Chapter 5:
ECOSYSTEMS

1. release of nutrients, from dead bodies and wastes, in a form plants can use
2. (see diagram at right)
3. a. (trophic level measures number of feeding steps to solar energy)
4. b.
5. c.
6. c.
7. Check your answer by comparing it with Figure 5-7; be sure you included carbon dioxide in the atmosphere, producers, decomposers, and consumers.
8. sunlight
9. a.
10. c.

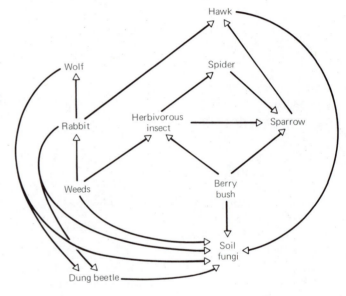

Chapter 6:
POPULATIONS

1. b.
2. d.
3. c.
4. (see diagram at right)
5. b.
6. False
7. c.

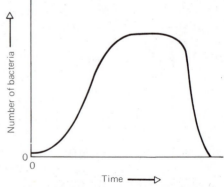

The population grows exponentially, levels off, and then declines rapidly to extinction in this "closed" environment.

Chapter 7:
HUMAN EVOLUTION AND ECOLOGY

1. b.
2. d.
3. True

4. c.
5. d.
6. b.

Chapter 8:
EVOLUTION AND SEX

1. b.
2. favorable; unfavorable
3. c.

4. d.
5. a.
6. d.

Chapter 9:
THE CHEMISTRY OF LIFE

1. A sodium ion has one less electron than a sodium atom has. A sodium atom bears no electric charge; a sodium ion does (its charge is +1).
2. a.
3. covalent
4. covalent
5. a.
6. a. H_2O
   b. NaCl
   c. CO_2
   d. O_2
7. b.
8. e.
9. a. Monosaccharides

   b. 1. Energy supply
      2. Structural support in plants
   c. Lipids
   d. Fatty acids, glycerol
   e. Proteins
   f. 1. Enzyme catalysts
      2. Structural elements, etc. (see Table 9–2)
   g. Nucleic acids
   h. Nucleotides
   i. 1. Carries hereditary information (DNA)
      2. Participates in protein synthesis (RNA)
      3. Energy source for chemical reactions
10. c.

Chapter 10:
THE LIFE OF A CELL

1. k.
2. d.
3. b.
4. g.
5. c.
6. l.
7. h.
8. a.

9. i.
10. j.
11. All
12. Prokaryote
13. Plant and Prokaryote
14. All
15. Animal and Plant
16. c.

17. a. Water left the frog's cells and moved into the more concentrated external solution applied by the students. The cells died, and therefore the frog died too.
    b. Osmosis
    c. No. There is no known carrier for water; it passes freely across cell membranes.
18. a.
19. c., a., b., e., d., f.

Chapter 11:
PHOTOSYNTHESIS

1. a. iii.
   b. i.
   c. iii.
   d. iv.

2. a.
3. b.
4. d.
5. b.

Chapter 12:
CELLULAR RESPIRATION

1. b.
2. a.
3. d.
4. a.
5. c.

6. 2; $^{34}/_{36} \times 100\% = 94\%$
7. False. Carbohydrates not needed immediately to supply energy are converted to glycogen or fat and stored until their energy is needed.

Chapter 13:
ORIGIN OF LIFE

1. b.
2. oxygen; hydrogen
3. 1. organic monomers; 2. proteinoids; 3. fermentation; 4. photosynthesis; 5. aerobic respiration
4. c.
5. Opportunity: ability to expand greatly because organisms could make their own food rather than being limited to food produced by non-biological means in the environment. Threat: destruction by oxygen.
6. Nutrient depletion created strong selective pressures favoring organisms with any kind of autotrophic ability.

Chapter 14:
DNA

1. a.
2. c.
3. d.

4. 1. thymine (T)
   2. guanine (G)
   3. adenine (A)
5. d.

Chapter 15:
PROTEIN SYNTHESIS

1. d.
2. a. mRNA: A-U-G-U-U-C-A-U-G-A-A-C-A-A-A-G-A-A
   Met-Phe-Met-Asn-Lys-Glu

   b. mRNA: A-U-G-C-A-A-C-G-A-C-G-G-A-C-G-G-C-C
   Met-Gln-Arg-Arg-Thr-Ala
3. c.

4. 1. mRNA attaches to a ribosome.
   2. tRNA brings the first amino acid and binds to the first codon.
   3. another tRNA brings the second amino acid and binds to the second codon.
   4. the first amino acid is joined to the second amino acid, on the second tRNA.
   5. the empty first tRNA leaves and the mRNA and second tRNA move on the ribosome, bringing the third codon into place on the ribosome.
   6. the tRNA bearing the amino acid specified by the third codon binds to the third codon.
   7. the growing chain of amino acids is attached to the newly-arrived amino acid, still attached to its tRNA.
   8. steps 5 through 7 repeat with each amino acid in turn until a "Stop" codon on the mRNA is reached.

5. a

Chapter 16:
MENDELIAN GENETICS

1. a. ³/₄ tall: ¹/₄ short (= 3 tall: 1 short = .75 tall: .25 short, etc.)
   b. all tall
   c. ¹/₂ tall: ¹/₂ short
2. a. The gene for normal wings is dominant to the recessive dumpy-wing gene.
   b. Both were heterozygous, with one normal-wing and one dumpy-wing gene.
3. ¹/₂ × 80 = 40 (¹/₂ of the offspring, or 40 out of 80, would receive the male's normal-wing gene.)
4. a. ¹/₂ × 500 = 250
   b. ¹/₄ × 500 = 125
5. a. Sniffles: probably homozygous for colored coat (which must be dominant)
      Whiskers: certainly heterozygous, as he sired both colored and albino young.
      Esmeralda: homozygous albino (recessive), because she shows this trait in her phenotype.
   b. ³/₄ colored: ¹/₄ albino
   c. ¹/₂ colored: ¹/₂ albino
6. He should mate his dog to bitches known to carry the retinal atrophy gene. If any of the resulting pups is blind, he should not use the dog for stud, because blind pups would indicate that the dog carries a retinal atrophy gene. If no blind pups appear, it is possible that they were all lucky enough not to inherit the retinal atrophy gene, but the larger the number of normal pups, the more confident the owner can be that his dog does not carry the gene.

7. a.

	DH	Dh	dH	dh
Dh	DDHh	DDhh	DdHh	Ddhh
dh	DdHh	Ddhh	ddHh	ddhh

b.

	DH	Dh
Dh	DDHh	DDhh
dh	DdHh	Ddhh

c.

	DH	Dh	dH	dh
dh	DdHh	Ddhh	ddHh	ddhh

   d. ¹/₂
8. a. ³/₁₆
   b. ³/₁₆
   c. ⁹/₁₆
9. a. 480
   b. 160
   c. 40

10. a. $\frac{1}{8}$
    b. $\frac{1}{8}$
    c. $\frac{3}{8}$
11. let T^A = crosswise stripes
    T^L = lengthwise stripes

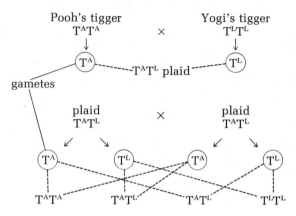

Pooh's tigger T^AT^A × Yogi's tigger T^LT^L

gametes

T^AT^L plaid

plaid T^AT^L × plaid T^AT^L

T^AT^A $T^AT^{L'}$ $T^AT^{L'}$ T^LT^L

1 crosswise : 2 plaid : 1 lengthwise

12. a. clover patch: $\frac{1}{2}$ roan, $\frac{1}{2}$ white
    alfalfa field: $\frac{1}{2}$ red, $\frac{1}{2}$ roan
    cornfield: $\frac{1}{4}$ red, $\frac{1}{2}$ roan, $\frac{1}{4}$ white
    b. it doesn't matter; $\frac{1}{2}$ the calves will be roan in any case
13. a. $\frac{1}{4}$
    b. $\frac{1}{2}$
    c. $\frac{1}{4}$
    d. $\frac{3}{16}$
    e. $\frac{1}{8}$
14. The genes are linked: in the female parent, the genes for sable body and normal wing are on one chromosome, and the genes for normal body and miniature wing are on its homologue. This arrangement is indicated by the large numbers of offspring with the combinations sable body + normal wing and normal body + miniature wings.
15. The genes are probably unlinked; this is shown by the almost-equal numbers of offspring in each phenotype category.

Chapter 17:
INHERITANCE PATTERNS AND GENE EXPRESSION

1. a. $\frac{1}{4}$
   b. 50% normal: 50% brachyphalangic
2. a. Yellow mice are heterozygous.
   b. The yellow gene is lethal in the homozygous condition.
   c. Homozygous yellow mice die during gestation and are resorbed into the mother's body.
   d. Mate yellow mice and perform surgery on the females early in pregnancy to demonstrate existence of dead embryos.
3. (Hint: start at the bottom and work backward.)
   a. $I^A i$
   b. $I^B i$
   c. $I^A i$ or $I^A I^B$
   d. $I^B i$
   e. $I^A I^A$, or $I^A i$, or $I^A I^B$
   f. $I^A i$
   g. $I^A i$, $I^B i$, or ii
   h. $I^A i$
   i. $I^B i$
4. Mr. Rae could have been the father; the jury must consider other evidence to decide whether he is the father.
5. $\frac{1}{4}$ normal daughters: $\frac{1}{4}$ hemophilia-carrier daughters: $\frac{1}{4}$ normal sons: $\frac{1}{4}$ hemophiliac sons.
6. If the mother is a hemophilia carrier or a hemophiliac.
7. a. Mother: X chromosomes homozygous for color blindness; Father: normal color vision gene on X chromosome plus one Y chromosome.
   b. $\frac{1}{2}$
   c. $\frac{1}{2}$
8. 2 females: 1 male
9. No. Baldness is not sex-linked, and so can be inherited via either parent.

Chapter 18:
PLANTS: THE INSIDE STORY

1. a. b.
2. a. stomata
    b. xylem
    c. phloem
    d. guard cells
    e. cuticle
3. a.
4. a. hold water; supply mineral nutrients

    b. needed for roots' respiration
    c. release mineral nutrients from organic matter
    d. supply nutrients; condition soil
5. e.
6. phloem
7. e.

Chapter 19:
GROWTH AND REPRODUCTION OF FLOWERING PLANTS

1. a-d-c-b
2. b.
3. a.
4. a. both
    b. both
    c. dicotyledons
    d. dicotyledons
    e. monocotyledons
5. c.
6. stamen
    stigma (of the pistil)

pollen tube
ovule
ovary
fertilization
embryo
seed coat
ovule
fruit
ovary
7. d.
8. b.

Chapter 20:
HUMAN REPRODUCTION AND DEVELOPMENT

1. i.
2. h.
3. d.
4. b.
5. a.
6. e.
7. g.
8. e-i-f-h-a-g-b
9. a.
10. f.
11. d.
12. b.

13. a.
14. c.
15. a.
16. nuclear
    cytoplasmic
    eggs
    sperm
17. C
18. N
19. G
20. d.

Chapter 21:
NUTRITION AND DIGESTION

1. f.
2. a.
3. a., b., c.
4. c.
5. d.
6. b.

7. f.
8. c.
9. e.
10. c.
11. a.

Chapter 22:
REGULATION OF BODY FLUIDS

1. a. excretory
   b. lymphatic
   c. respiratory
   d. circulatory
2. b.
3. d.
4. a.

5. a.
6. d.
7. a.
8. Carbon dioxide: a.
   Water: a., b., c.
   Salts: b., c.

Sugars: c.
Hormones: c.
Spices: a., c.
9. b.
10. c.
11. a.

Chapter 23:
ANIMAL COORDINATING MECHANISMS

1. a.
2. b.
3. d.
4. d., e.
5. Check your answer against Figure 23-3.
6. rods and cones
7. c.
8. b.
9. e.
10. a. 4
    b. 6
    c. 5
    d. 1
    e. 3
    f. 2

11. b.
12. e.
13. a. 1
    b. 4
    c. 6
    d. 2
    e. 3
    f. 5
14. hypothalamus
    pituitary
    hormones
    bloodstream

Chapter 24:
BEHAVIOR

1. False
2. c.
3. e.

4. True
5. c.

INDEX

Numbers in *italics* refer to illustrations; *t* indicates tables.